IN VIVO

The Cultural Mediations of Biomedical Science

PHILLIP THURTLE and ROBERT MITCHELL, Series Editors

IN VIVO: THE CULTURAL MEDIATIONS OF BIOMEDICAL SCIENCE is dedicated to the interdisciplinary study of the medical and life sciences, with a focus on the scientific and cultural practices used to process data, model knowledge, and communicate about biomedical science. Through historical, artistic, media, social, and literary analysis, books in the series seek to understand and explain the key conceptual issues that animate and inform biomedical developments.

The Transparent Body: A Cultural Analysis of Medical Imaging
by José Van Dijck

Generating Bodies and Gendered Selves:
The Rhetoric of Reproduction in Early Modern England
by Eve Keller

The Emergence of Genetic Rationality: Space, Time,
and Information in American Biological Science, 1870–1920
by Phillip Thurtle

Bits of Life: Feminist Studies of Media, Biocultures, and Technoscience
edited by Anneke Smelik and Nina Lykke

Life as Surplus: Biotechnology and Capitalism in the Neoliberal Era
by Melinda Cooper

HIV Interventions: Biomedicine and the Traffic between Information and Flesh
by Marsha Rosengarten

Bioart and the Vitality of Media
by Robert Mitchell

Affect and Artificial Intelligence
by Elizabeth A. Wilson

Darwin's Pharmacy: Sex, Plants, and the Evolution of the Noösphere
by Richard Doyle

Darwin's
PHARMACY

SEX, PLANTS, AND THE EVOLUTION OF THE NOÖSPHERE

Richard Doyle

UNIVERSITY OF
WASHINGTON PRESS
Seattle and London

Design by Thomas Eykemans
16 15 14 13 12 11   5 4 3 2 1

UNIVERSITY OF WASHINGTON PRESS
PO Box 50096, Seattle, WA 98145, USA
www.washington.edu/uwpress

LIBRARY OF CONGRESS CATALOGING-IN-PUBLICATION DATA
Doyle, Richard, 1963–
Darwin's pharmacy : sex, plants, and the evolution of the noosphere /
Richard M. Doyle.
     p. cm.
Includes bibliographical references and index.
ISBN 978-0-295-99094-1 (hardback) —
ISBN 978-0-295-99095-8 (paperback)
1. Biology—Philosophy.
2. Gaia hypothesis.
3. Biosphere.
4. Hallucinogenic plants—Psychic aspects.
5. Hallucinogenic drugs—Psychological aspects.
6. Consciousness.
7. Sexual selection in animals.
8. Rhetoric—Philosophy.
I. Title.
QH331.D688 2011   570.1—dc22   2011001201

FRONTISPIECE
Illustration by Richard Doyle (1824–1883), *In Fairyland*, 1870.

For as the earth comes from the waters,
plants from the earth
and man from plants,
so man is speech
and speech is OM.
    —Chandogya Upanishad

The introduction of open-minded, multiple-level, continuously
developing, on-line, operational, dynamic, economical, expanding,
structural-functional, field-jumping, field-ignoring theory is needed.
    —John C. Lilly,
    Programming and Metaprogramming
    in the Human Biocomputer

.

Contents

Acknowledgments

THIS BOOK WAS NOT REALLY MY IDEA, BUT EVOLVED OUT OF AN ELAB-
orate encounter with a sentient mesh. I have always been grateful for
the plant ecosystem from whence we have manifested in order to move plant
genes around and explore the space of all possible experiences, but I now
realize that I *am* that mesh, and for that unmistakable teaching I give thanks
to the late Norma Panduro. From the very beginnings of the project all doubts
were brought low before my admiration for the psychonautic researchers I
was fortunate enough to meet and read. Thanks MAPS! Gracias Erowid!
Sasha and Ann Shulgin welcomed me to their kitchen table many years ago
as well as visited my seminar at UC Berkeley, encouraging me beyond all
words with their joyful spirit and immense intelligence. I knew that following these Bodhisatvas in their research could not be the wrong path. Dennis
McKenna exemplifies the fierce inquiry of open minded science and coura-
geous investigation of self—someday there will be a Nobel for Psychonautics,
may it be named for the Shulgins and given to the McKennas. . . .

My family, with whom I am ever more intensely and joyfully enmeshed,
taught and loved me so well that I survived and flourished through this
extraordinary healing ordeal. Violet, Jackson, and Amy: Groove Hug!
Thanks to my dad, who taught me to follow the data, my late mother, who
counseled me even in the rain forest, my sister, the Indomitable, and my late
brother, who set me on this path twenty-three years ago. I'm getting there,
brother!

Kathleen Pike Jones, Paul Bowersox, and Jacquie Ettinger were of
immense help in putting this logos flow between covers.

Cue the Spiral of Friends & Allies:

Brian Rothman Talbot David Michael Fortun Elizabeth Wilson Jeff Nealon Philip Thurtle Rob Mitchell Stanley Salthe Paul Gerome Lao Tzu Trey Conner Lindsey Talbot David Michael Fortun Elizabeth Wilson Jeff Nealon Philip Thurtle Rob Mitchell Stanley Salthe Paul Gerome Wrye Sententia James Palazzolo Andrew Schelling Light Nykorchuk Francesca LaBate Robert Caserio Cohen Tsong Khapa Steve Beyer Stefan Helmreich Pam Jackson Sunil Aggarwal Diana Slattery Big Class St. John Drew Halley Jacobson Jeff McGonigal Freeze Sarah Schwartz Leisha Jones Raul Guerra Andrew Schelling The Burroughs Guru Wittje Peter John Rieger Pilsch Jane Yarber Robert Yogananda Drewchin Evgenia Fotiou David Bob The Bike Gavin Shoemaker Alan Fox Keller Robert Shahi Awais Nicholas Tara Michelle Akritas Proctor Erik Davis Hoy Evelyn Royster Glen Boire Socrates Richard Stan Shostak Pruchnic

DARWIN'S PHARMACY

Sex, Plants, and the Evolution of the Noösphere

Twenty minutes in, like clockwork, the visions begin. They are strong, an admixture of force and light, a flickering current of rhythm that passes through me like electricity. I was expecting them this time and found myself less shocked than prepared for the plant gnosis no doubt in my near future as I let go . . .

Norma, the vegetalista who so astonished me with her care, skill, and knowledge during my first ceremony two nights before, had packed a big bowl with a knot of the local Nicotina rustica and blown curling, whistling smoke over a plastic liter bottle filled with an opaque orangish liquid I knew to be ayahuasca, the potent brew of tryptamines and MAO inhibitors that has been prepared in the upper Amazon for perhaps sixteen thousand years. I knew it to be ayahuasca since I had, after all, helped mix it the day before, pounding a kilo of the soft but woody vine of fresh B. caapi and tossing about fifty green glossy leaves of P. viridis, a DMT containing a relative of coffee, into the black cauldron simmering over a wood fire on the shores of the Yanayacu River, one of the eleven hundred tributaries of the Amazon. Back home this could be a felony. Here, I now understood, it is a medicine.

The smoke whistle is a trope, a refrain that often begins or ends an icaro, the beautiful songs sung and whistled during the entire four-hour ceremony. Smoke and its whistling inflection act as protocols to open up a spirit portal while keeping unwelcome entities—what I think of as affect—at bay. After my first session, I had also learned that the songs serve to orient the ayahuasca drinker. The songs seem to mime and sample the birdsong of the region, an ecosystem with over two thousand species of birds and polyrhythms of insect chatter. I held on to, and was held by, the icaros, giving intense thanks for the whistled orientation.

I took the coffee mug and fearfully eyed its contents. My first contact with ayahuasca had been perhaps the most difficult experience of my life that didn't involve someone (else) dying. For I had indeed palpably, and unmistakably, died during that experience. To remix Mark Twain: the accounts of ego death were not at all greatly exaggerated.

Nonetheless, here I was, two days later, again looking into the flickering, refracted, and reversed image of myself I think I spied in the mug, lit only by candlelight in the Amazonian night. The liquid was dark and iridescent, but I now knew that tales of its horrid flavor were something like an urban legend from the rain forest. My first gulp of ayahuasca tasted like nothing so much as my first pint of draft stout slurped in Ireland at the age of seventeen with my now departed brother. And it was equally shamanic.

Still, I was fearful and full of respect for this plant intelligence with which I had seemingly interacted. The mug appeared nearly two-thirds full, easily as large a dose as the first, most difficult, night. I had secretly hoped for a tinier tourist dose, but now I really had no choice but to drink down the cup I was offered.

As a result of my extensive research into the ceremony (as well my inquiries into scientist John C. Lilly's "science of belief"), I carefully addressed the ayahuasca to orient my journey. Having played with the I Ching as a writing tool, I was comfortable posing questions to nonhuman entities as a rhetorical experiment, a practice of rhetorical invention that seeks interaction with other forms of order and its disruption. Among other things, I asked how I could possibly integrate the knowledge from my first journey into my life back in North America. Then I threw it back like a fat shot of tequila, opening my throat to the entirety of the viscous flow.

Like I said, twenty minutes of meditation later and the visions began; the same as the first night. A pixelated doorway appeared in my closed-eye visuals and I went through it. "Here goes," I thought to myself, "What have I done?"

INTRODUCTION

Glimpsing the Peacock Angel

If we recognize the plant as an autonomous power which enters in order to put roots and flowers in us, then we distance ourselves by several degrees from the skewed perspective which imagines that spirit (Geist) is the monopoly of human beings and doesn't exist outside of them. A new world-picture has to follow the planetary leveling; that is the task which the next century will take up.
—Ernst Jünger, "The Plant as Autonomous Power"

CRAWLING WITH TRANSACTIONS, THE CONTEMPORARY EARTH WHIRLS and whorls,[1] uncannily bereft of human agency. The global ecosystem, undeniably in crisis due to the presence and activities of humans and their fossil-fuel familiars, maintains itself far from equilibrium, surfing diverse gradients through raised ocean levels and proliferating vectors of disease; malarial mosquitoes have followed thickening sea levels and achieved new highs, more than doubling the altitude at which they can survive and reproduce. Global warming is no longer debated, but instead yields a muted and contentless call for "adaptation" while tens of thousands die of heat stroke in "old Europe."[2] A 2003 Pentagon report identifies sudden climatic change as a plausible "challenge to U.S. security in ways that should be considered immediately" (Schwartz and Randall), and an enormous hurricane provoked what the BBC called the biggest failure of the U.S. government since the Great Depression, itself echoed by a financial crisis linked to our use of fossil fuels that flow into the Gulf of Mexico even as I write. While trillions of dollars are spent in the pursuit of "security," a ubiquitous superpower—Gaia—launches global-defense operations against Homo sapiens of every demographic. Has yet another security briefing gone unheeded?[3]

The apparent inability of humans to perceive the densely interconnected nature of their habitat threatens not only said ecosystem but the very self-definition of humanity itself as *homo faber*, an organism *actively creating*, rather than created by, her environment. Faced with overwhelming evidence of climatic change, one would expect an outburst of human agency, an ordering of the world according to the specifications of Homo sapiens—the species who, after all, knows what it is doing. And yet humans—or at least, the only ones deserving of the slur—call for a strange acquiescence to the agency of the Earth:

> The United States is a world leader in addressing and adapting to a variety of national and global scientific problems that could be exacerbated by climate change, including malaria, hunger, malnourishment, property losses due to extreme weather events, and habitat loss and other threats to biological diversity. (U.S. Global Change Research Information Office)

This is a book that is, in part, about rhetoric, so let's zoom in on the paradox: Watch as the alleged lone superpower "leads" not through "resolve" or "will," but by speaking to, "addressing," even "rehearsing," "adaptation."[4] Even as Paul Wolfowitz dreamed of a global empire ordered according to the "interests" of the United States, the Bush regime beat a hasty retreat before the very real activities of bioterror their own report suggests could plausibly unfold. Instead of repeating the usual algorithm of empire—"make it so!"—the bloated über power called for nothing but an address or a "rehearsal," a simulacrum of "adaptation," yet another retraining that teaches humans how to respond to their devastated environment. It is in this rather absurd context that a discussion of plant agency or "power" alluded to by the German writer and botanist Ernst Jünger must take place. Ethnobotany has long devoted itself to the relations between humans and plants, as has the shamanic medicine that has served the vast majority of Homo sapiens in history and the present. This book will suggest that indeed in responding to global climatic change we must less adapt than evolve, and this evolution begins with the recognition of plants, and the Earth itself, as a power, perhaps a superpower worthy of the name.

Though this phrasing may sound a bit odd to some, this claim is unlikely to remain controversial for long, as the massive effects of climatic change become slowly and unmistakably visible. Part of the adaptation called for by the Bush administration would entail a submission to a world-governing body—the world's body—whose weaponry is temperature change, rising ocean levels, and emergent and proliferating diseases rather than shock and awe.

But if it is easy enough to say we must "recognize the plant as an autonomous power," even a superpower, we must somehow do the more difficult work necessary to inhabit this space where plants present a paradoxical and uncanny "autonomy" when we are most directly and unmistakably linked to them. The future of Gaian biodiversity and a modicum of global stability depends precisely on a thoroughgoing and practiced re-articulation of human autonomy in the experience of imbrication with global ecosystems, including capital and information flows as well as the carbon cycle. In short, in order to alter what we do, we must re-engineer and re-imagine who we are. Across the life and climate sciences, the news is this: You are deeply implicated in the global ecosystem in ways scientific and technical practices are only beginning to comprehend and model. If the breakthroughs in medical and global imaging systems have provided us with revelations, they reveal that our separateness from ecosystems is itself an illusion, and that we are membranes inseparable from a global ecology. In 1943, Soviet geoscientist V. I. Vernadsky offered a new "continuously connected" model of human biology:

> Man is elementally indivisible from the biosphere. And this inseparability is only now beginning to become precisely clear to us. In reality, no living organism exists in a free state on Earth. All of these organisms are inseparably and continuously connected—first and foremost by feeding and breathing—with their material-energetic environment. (2005)

The news of this imbrication can, of course, be communicated in a cognitive fashion, but its persuasiveness—as measured by the emergence of a vision response-able to biodiverse futures—seems to hinge on an *experience* of this interconnection as well as an *understanding* of it. If the Upanishads instruct that *"Tat Tvam Asi,"* "You are that," and they do, "that" is an ecosystem subject to sudden volatility and massive extinctions even as it is increasingly interconnected with an otherwise dynamic, even lively, cosmos. It is therefore a rhetorical challenge to make this perception available to those humans who so violently cling to visions of autonomy even as they are forced to adapt. Rhetoric is the practice of learning and teaching eloquence, persuasion, and information architecture by revealing the choices of expression or interpretation open to any given rhetor, viewer, listener, or reader. Robert Anton Wilson offers a definition of rhetoric by example when he focuses on the word "reality" in his book *Cosmic Trigger:*

> "Reality" is a word in the English language which happens to be (a) a noun and (b) singular. Thinking in the English language (and in cognate Indo-European languages) therefore subliminally programs us to conceptualize "reality" as one block-like entity, sort of like a huge New York skyscraper, in which every part is just another "room" within the same building. This linguistic program is so pervasive that most people cannot "think" outside it at all, and when one tries to offer a different perspective they imagine one is talking gibberish. (iii)

It is unavoidably time to practically and continuously imagine the earth as a mesh of systems with which we are entangled. Rabbi Zalman Schachter-Shalomi describes a different kind of rhetorical framework and its effects:

> It has to do with Gaia. There is a situation where I'm very clear that I'm not acting for myself. And when I realize that, that I am a cell, a living cell of the global organism. . . . If I see myself as a living cell of a living planet, that creates a whole different set of ethical norms than the ones that I find when I'm speaking from the Patriot Act. (Goldman)

As the ill effects of our illusion of separability become clear, can we avoid once again missing the signs of an enormous global security threat before it is too late? What rhetorical choices and means of persuasion do we have to make the perception of the Earth as a political plant planet palpable?

There are no doubt diverse answers to these questions. Biologist Christopher Uhl's extraordinary book, *Developing Ecological Consciousness,* maps out and offers an array of practices that help to cultivate a relation to the plant superpower, a superpower whose main characteristic is dense interconnection. For Uhl and his attentive reader, becoming sensitive to the night sky, pondering the surface area of Earth's plankton and tracking the spoor of a beetle all are recipes for interconnection, a calm and sometimes oceanic apprehension of the *immanence* proper to biological systems. Immanent systems are massively interconnected with themselves, neither subject nor predicate, but web. For Uhl, the development of ecological consciousness is contingent upon *experiencing* the interconnection of earth and cosmos. This fact of interconnection appears to be a feature of life as essential as DNA itself, a densely intertwined bundle of nucleic acids.

Lest we think that such immanent visions depend upon the context of nature, recall that the Apollo space program provoked "cosmic consciousness" or the "overview effect" in those astronauts lucky and attentive enough to experience such interconnection while encapsulated in their military hardware. No doubt the practices of immanence are as diverse as the planet itself; fasting, inhaling carbon dioxide, and even working with latex have all provoked encounters with immanence, suggesting that in some fashion human perception is indeed "wired" for a periodic recognition of the dense imbrication of organism and environment and is highly tuneable by our practices.

One such practice is thought experiment, as in Uhl's treatment of the usually optical and now haptic night sky. Uhl reminds us that while gazing "up" at a night sky, one in fact hangs off the planet and near the edge of a galaxy, vertiginous, suspended over the infinity of space. Uhl quotes cosmologist Brian Swimme:

> As you lie there feeling yourself hovering within this gravitational bond while peering down at the billions of stars drifting in the infinite chasm of space, you will have entered an experience of the universe that is not just human and not just biological. You will have entered a relationship from a galactic perspective, becoming for a moment a part of the Milky Way galaxy, experiencing what it is like to be the Milky Way galaxy. (Uhl, 13)

Note that this perception is often scalar in character—a shift from the metric realm of human perception (about 10^{-3} meters) to the galactic magnitudes of 10^{22} meters. Looking not at the night sky but at the daylight earth from space, astronaut Edgar Mitchell of Apollo 14 inhabited a galactic perspective through an involution of his perspective into interconnection. Apollo 14's mission was more or less complete, and Mitchell had a brief moment to relax, so he looked out the window:

> Then looking beyond the earth itself to the magnificence of the larger scene, there was a startling recognition that the nature of the universe was not as I had been taught. My understanding of the separate distinctness and the relative independence of movement of those cosmic bodies was shattered. There was an upwelling of fresh insight coupled with a feeling [of] ubiquitous harmony—a sense of interconnectedness with the celestial bodies surrounding our spacecraft. Particular scientific facts about stellar evolution took on new significance. (2008, 74)

Fundamental to this insight is a perception of dwelling in an evolutionary space. Mitchell's vision—which he hastens to point out was neither "religious" nor "otherworldly"—momentarily, but irreversibly, rendered the interdependence of the cosmos palpable to him. This "startling recognition" substitutes a sensory and even affective imbrication, "there was an upwelling of fresh insight coupled with a feeling of ubiquitous harmony," for the everyday perception of the [alienated] distinction between subject and object, earth and its others, human and universe. Mitchell's vision offers perhaps an equally startling irony: it was only by taking on a literally extraterrestrial perspective that the moon walker overcame alienated perception.[5] This vision was productive of learning and knowledge as well: Mitchell still grapples with his perception of evolutionary interconnection, and has offered a physical theory of consciousness to account for it. Integral to this theory is John Bell's 1964 non-locality theorem, the apparent fact of imbrication or "entanglement" essential to matter itself and constitutive of that apparent bubble of experience, subjectivity:

> The basis of subjective experience is rooted in the quantum attribute of nature called non-locality. I will use the word "perception" in its most generic sense to denote a basic subjective experience at all levels of complex matter. Thus the non-local quantum correlation between entangled quantum particles is considered the root cause of the phenomenon experienced as perception in more complex matter, but the non-local quantum hologram is the non-local carrier of information for molecular and larger scale matter.

Thus, perception is not an object but rather the label for a nonlinear process involving an object, a percipient and information." (Mitchell n.d.; emphasis mine)

The physics of non-locality are notoriously difficult to comprehend, but specific to the core of Mitchell's claim here is that subjectivity—the human feeling of being an observer on a continuous world of duration—is essentially and paradoxically non-local. Here holographic information emerges from an evolutionary process distributed over the universe and subject to "nonlinear" transformations—such as his own "startling recognition" that the perception of "separateness" is a label or snapshot of an enormously dynamic system. Even as it appears to confer a sense of interior and exterior to human experience that certainly feels awfully local and "distinct," the universe does so in a thoroughly informational and non-local fashion. Much can be "done" with information, but confining it to a single location is seldom a tractable strategy. Like the mind apprehending it, information "wants to be free" if only because it is essentially "not an object," but rather "the label for a nonlinear process involving an object, a percipient and information."[6] It is worth noting that Mitchell's experience induces a desire to comprehend, an impulse that is not only the desire to tell the story of his ecodelic imbrication but a veritable symptom of it.[7]

Hence this insight seems to involve not only an act of perception, but an action whose achievement makes legible the nature of perception itself as a nonlinear and highly distributed system not "ownable" by a self and navigable only through its practiced but always irreducible dissolution, the sometimes shattering detachment from "distinctness" before which a sense of interior and exterior dissolves in awareness and awe. This awareness of interconnection occurs in and with what Vernadsky dubbed the "noösphere"—the aware and conscious layer of the earth's ecosystem and, perhaps, feeds back onto our ecosystems as we become conscious of our interconnections with them.

Yet while billions of dollars of hardware and support provided the context for Mitchell's close encounter of an interconnected kind, millions of humans experience and write incessantly of this sense of subjective, ecological, non-locality after ingesting compounds derived from plants, such as tryptamines and phenethylamines. Consider this recent posting on The Vaults of Erowid, a Web site devoted to ethnobotanical information and harm reduction.

After ingesting ten inches of a legal mescaline cactus species (San Pedro), Smokeloc writes:

> The basic principal of the experience is "EVERYTHING EXISTS WITHIN ITSELF" MEANING: all of reality is so basic yet so infinite. Think about space, the universe . . . the galaxies exist within it and the solar systems exist within that . . . planets exist in that . . . living animals exist within the earth bacteria and cells live within the animals and atoms exist within that . . . EVERYTHING EXISTS IN AN ETERNAL PLANE. I could see everything as eternal. . . . So beautiful and so simple. I have never experienced such a strong sense of peace with living and existing. The next day I went outside and noticed things like birds singing and the wind steadily blowing, everything became so beautiful and I had a change of life views and was no longer jealous about what other people had. I lost my anger that was deep inside of me . . . and it feels GREAT![8]

At a moment when "egoic" consciousness—that form of human experience that insists on the radical distinction between self and cosmos, as the former insists on incessantly consuming and colonizing the latter—seems to have reached a pandemic, a humble cactus enables the news of our fundamentally nested nature. What *are* psychedelics such that they seem to persuade humans of their interconnection with an ecosystem?

Terence McKenna's 1992 book recursively answered this query with a title: *Food of the Gods*. Psychedelics, McKenna argued, were important vectors in the evolution of consciousness and spiritual practice. In his "shaggy primate story," McKenna argued that psilocybin mushrooms were a "genome-shaping power" integral to the evolution of human consciousness. On this account, human consciousness—the only instance we know of where one part of the ecosystem is capable of reflecting on itself as a self and acting on the result—was "bootstrapped" by its encounter with the astonishing visions of high-dose psilocybin, an encounter with the Transcendental Other McKenna dubbed "a glimpse of the peacock angel." Hence for McKenna, psychedelics are both a food fit for the gods and a food that, in scrambling the very distinction between food and drug, man and god, engenders less transcendence than immanence—each is recursively implicated, nested, in the other.

This book samples, remixes, affirms, and periodically denies McKenna's increasingly paradigmatic account as it inquires into the swarm of ontological, epistemological, and ethical questions provoked by psychedelic experience in the context of the global ecological crisis: What *are* these compounds and how do they reliably produce experiences of interconnection? What do the states that they induce suggest about the nature of human minds? How should we respond to the claims of psychonauts that these materials give them nothing less than an encounter with alternate and perhaps divine realities? Do we have a robust notion of how to live in a reality that has itself become plural in the context of an ecosystem increasingly saturated with information amidst dwindling biodiversity? What can contemporary science and technology learn from psychedelics? Can psychedelics help cultivate a new and paradoxical outburst of interconnected and hence transhuman agency on Earth, one that embraces and even enhances ecological imbrication?

This proliferating array of questions is of course dizzying, and they are all ethical questions, and not only because the United States is now in its third decade of a war against plants, by far our longest and most expensive war. Since disgraced President Richard Nixon declared a War on Drugs in 1972, millions of people have been incarcerated and hundreds of billions of dollars have been funneled into self-propagating structures sprouting out of the constantly proliferating prison industrial complex.[9] If one of the great evolutionary achievements of bacteria was the conversion of sunlight into sugars through photosynthesis, then the carceral ecology has achieved a no less adaptive transduction: the conversion of fear into money, drugs into assets. Since 1984, law enforcement in the United States has opened up entirely new domains of capital for a large-scale organization: the seized assets of putative drug offenders.

This "war" has perhaps metastasized, and certainly amplified, what the historian Michel Foucault characterized as "biopower," a complex of forces by which "life" became both an object of technoscientific control and a widespread feature of political governance. The state's increasingly fine grained interest in the "life," "health," and "well-being" of its citizens accompanied and enabled the increased capacities of the nascent life sciences to control and predict living systems. Population policies such as eugenics are one way in which the polity has sought to exert control at the molecular level—which

DNA shall be replicated?—but the drug war extends this "interest" into states of consciousness or what William James has pluralized more helpfully as "states of mind,"[10] Aldous Huxley mapped as "Mind-at-Large," and Vernadsky dubbed "noösphere." This war is, of course, a global one; in order to regulate the health and minds of its own citizens, the United States carries out military operations in Columbia and the DEA opens foreign offices in Rangoon, Merida, Hong Kong, Islamabad, Kabul, and Moscow.

DMT, KITCHEN CHEMISTRY, AND THE PSYCHEDELIC COMMONS

> *A labyrinth is said, etymologically, to be multiple because it contains many folds.*
> —Gilles Deleuze, *The Fold: Leibniz and the Baroque*[11]

The war, too, is both global and intensely local. The legal contexts for the cultivation and medical use of marijuana, for example, vary enormously even throughout the United States. They are intensely local in another fashion, as well: they are practices that demarcate and patrol the very contours and phenotypes of human bodies, transforming the human-plant coevolutionary extended phenotype into the clean, logical, but not biological categories of plant and human. Evolutionarily speaking the emergence of widespread animal life on earth is not separable from a "mutualistic" economy of plants, pollinators, and seed dispersers.

> The basis for the spectacular radiations of animals on earth today is clearly the resources provided by plants. They are the major primary producers, autotrophically energizing planet Earth . . . the new ecological relationships of flowering plants resulted in colonizing species with population structures conducive to rapid evolutionary change. (Price, 4)

And if mammalian and primate evolution is enmeshed in a systemic way with angiosperms (flowering plants), so too have humans and other primates been constantly constituted by interaction with plants.

Despite the fact that states and nations and global organizations share an interest in the enormous global traffic in inebriants, they have acted in concert to prohibit even the desire to study many of these compounds and plants while subsidizing research on others.[12] Indeed, if we are to make any sense at all of a prohibition affecting an enormous sector of the economy, we must look at contemporary biopower as a business model in competition with other molecular and ecological practices.

Bioprospecting, the attempts of pharmaceutical giants to exercise ownership and control over the incalculable array of compounds produced in the global botanical commons, is better mapped not as innovation but extraction—an extraction of value from the commons. Jünger noted that it was precisely extraction that was at issue in industrial modernity's relation to plants:

> The whole nineteenth century is interspersed with this precipitation and concentration of active principles from organic substances. It began with the extraction of morphine from the juice of the poppy by the twenty-year-old SERTURNER, who thereby developed (*entwikelte*) or rather unwrapped (*auswickelte*) the first alkaloid. (Jünger, 35)

This practice of intensification, of course, is part of the transformation of the pharmacopeia into a commodity form, a rendering of the meshed relations of ecology into a thing available for ownership and control. Yet Jünger rhetorically concentrates his account of this process by intensifying our understanding of extraction itself. This concentration and amplification is, for Jünger, best understood as an "unwrapping," a peeling away of layers. This peeling away is not simply a reification or revelation, a cutting off of the alkaloid from its context, but an unwrapping that also envelops the one who unwraps: concentrated morphine now much more effectively entangles the addict. Navigating our implication with both plants and their precipitates might begin, then, with the startling recognition of plants as an *imbricated* power, a nontrivial vector in the evolution of Homo sapiens, a power against which we have waged war. "Life is a rhizome," wrote Carl Jung, our encrypted ecological "shadow" upon which we manifest as Homo sapiens, whose individuation is an interior folding or "involution" that increases, rather than decreases, our entanglement with any given ecosystem.

This unfolding that enables an enfolding is exemplified in the contemporary psychonautic practice of an alkaloid-base extraction of DMT from plants, the only practicable way most psychonauts can obtain this Schedule One compound endogenously produced by the human brain. Separated from a plant source such as *Mimosa hostilis* rootbark, smokeable DMT envelops the kitchen-chemist psychonaut in an ecology of amateur science where over the counter non-polar solvents, lye, common phalaris grasses, and online anonymous messages enable a grassroots global production of relatively pure DMT (Noman). Experimenting with the intensities of the DMT experience demands an investigation of all these nonhuman actants and their capacities to transform, a faculty described by French ecologist and psychoanalyst Felix Guattari as chaosmosis—acts of dissensus that put the usual habits of a self into disarray while allowing for the stability and repetitions of "existential refrains"—practices that help the diverse bodies of which we are made up dwell together and repeat even as they undergo massive transformation. Trip reports after smoking DMT often describe becoming "part of the intergalactic information network," but even more concentrate on the network of online denizens, hardware stores, and turkey basters needed to render any of the alchemical "elf spice" from the numerous and abundant plants where it can be found.[13]

Hence this precipitation and concentration is fundamentally a conjunctive practice—a deterritorialization that is also a re-territorialization—separating an alkaloid from its plant substrate, one becomes mycelially connected to a community. Concentrating and intensifying cocaine, crack kitchen chemistry spreads the network of addiction ever further.[14]

Perhaps community itself is the most important "lab apparatus" of contemporary psychedelic practice, a large-scale, dissipative structure for the ongoing investigation of the mind. Law professor and Creative Commons founder Lawrence Lessig has argued persuasively that the open source ethos and relation to intellectual property was essential to the emergence of the Internet and its innovation, a practice of tinkering that suspends protocols of autonomous authorship in order to enable a "collective enunciation" of code, the digital commons. As "free" software—one may use it without charge, and tinker with it at will—GNU/Linux exemplifies the emerging strength of business models that rely less on extraction than on distribution and the emergence of collective intelligence.

Clearly, it is not only a desire to be done with ownership that drives the success of this "open source" model—fundamental to the digital commons is a strong belief in the right to tinker. Indeed, the right to alter one's tools is the very *telos* of open source, the main rationale for the creation of a code commons such as Linux (a free and open-source operating system competitive with Windows both on the desktop and as a server) and Apache (the free open-source software that serves over 60 percent of the Internet.) Chemists Alexander Shulgin and David Nichols have tinkered with phenethylamines and tryptamines and yielded over two hundred new psychedelic compounds since the psychedelic prohibition began in 1966.[15] This commons, too, emerges from an extraordinary desire to alter and test diverse cognitive tools—molecular as well as rhetorical practices linked to the experience of diverse modalities of consciousness, a biodiversity of mind. The tools themselves—including online forums, trip reports, as well as the patent free content of Shulgin's compounds—operate through an ethos of mutual aid, one in fact contra (often necessarily) to some of the secret histories of mystic practice. Harvard Professor Timothy Leary's group, for example, shared practices of more covert traditions when offering their trip manual, *Psychedelic Experience:*

> In publishing this practical interpretation for use in the psychedelic drug
> session, we are in a sense breaking with the tradition of secrecy and thus
> contravening the teachings of the lama-gurus. (Leary et al., 13)

So too did Leary's early investigations into the management of psychedelic states—stratagems crucial to the scientific investigations done by Leary and others—draw from the "archaic" traditions of ayahuasca curanderos.[16] By bringing this knowledge into an economy of exchange, Leary's efforts were nearly indiscernible from the spread of global capital and its demands of exchange and extraction. Yet insofar as psychedelic culture seemed to induce gatherings and group ownership experiments of all kinds, it also summoned a commons, an ecology whose links to the digital commons itself is nontrivial.

Leary's group did some of its most important and fundamental work in psychedelic science as a collective at Millbrook. The commons has thus hardly ceased its activity. But certainly, in some sense, the prohibition and control tactic has "worked." It has created massive economic growth in sectors of society and the planet that are most at home in this habitat of devastation and denial: global crime syndicates and terror networks find perfect ecosystemic niches in the incessantly escalating violence and volatility of global black markets in inebriants. In an attempt to disrupt an always mobile and highly distributed and evolutionary network of distribution and research, the War on Drugs has grown a dissipative structure of crime, suffering, and addiction. A self-sustaining, autocatalytic network of prisons, police apparatus, and about-to-be-criminal subjects proliferate in this ecology of attempted centralized control of increasingly mobile "substances."[17]

So too has the prohibition and widespread dismissal of extraordinary states of mind done much to help cultivate the present system of preemptive war and accelerated corporate accumulation: the criminalization of psychedelic states is a criminalization of the political thought and practices of the 1960s themselves, precisely insofar as the alleged pathology of psychedelic states serve as a synecdoche not only for the resistance but the ethos and practices of *experimentation* at the heart of the cultural and political revolutions of the 1960s.

Of course the upheavals of the 1960s were about justice, but the paths to justice, like any path in a complex ecology, are essentially and unavoidably devoted to experiment. The criminalization of psychedelic states has worked to erode any historical right to experiment at the core of human evolutionary experience. Indeed, we must engage in *gendankenexperiment* and rethink our memory of the 1960s as a politics not primarily of resistance, but of experiments with the sacred: If psychedelics left any consistent trace on the literature of trip reports and the investigation of psychedelic states, it is that "resistance" is unlikely to be a useful tactic and that experiment is unavoidable. Leary, whose own "setting" was consistently clustered around practices of the sacred, offered this most compressed algorithm for the manipulation ("programming") of psychedelic experience, a script asking us to experimentally give ourselves over to the turbulence: "Whenever in doubt, turn off your mind, relax, float downstream."[18] Such an experiment begins, but is not completed, by a serene letting go of the self under the pull of a transhuman and improbable itinerary. This letting go, of course, can be among the greatest of human achievements, the very goal of human life: Meister Eckhart, the fourteenth-century German heretic, reminds us that this *gelassenheit* is very old and not easily accomplished.

I say moreover: If the soul is to know

God it must forget itself and lose itself, for as long as it

contemplates self, it cannot contemplate God. (Eckhart, 22)

FLASH FORWARD:
OUR PSYCHEDELIC FUTURE,
OR "THINK PEACOCK"

The excitement here arises from our present position: we are probably
on the threshold of a physical basis of consciousness. Perhaps our
times are analogous to those at the beginning of the century, which
culminated in today's clear concept of the physical basis of heredity.
   —Lynn Sagan, "Communications: An Open
   Letter to Mr. Joe K. Adams"[19]

But if we look at the putative goals of the criminalization of psychedelic states in particular, the policies of incarceration and prohibition have indeed not worked at all. They have served only as intensifications of psychedelic culture even as they have crowded out more sustainable practices of education and justice. Today, there are well over two hundred different reliable psychedelic agents produced by an open-source psychedelic movement whose innovation on some measures far exceeds that of the multibillion dollar official pharmaceutical industry.[20] Cannabis is the largest cash crop in the United States, and addicts crowd methadone and other "treatment" programs to capacity. The annual economic cost of "the drug problem" exceeds five hundred billion dollars a year. The alleged remedy, as many have pointed out to no avail, is indeed worse than the disease.

   And the question of how to respond to the testimony of psychedelic experience is ethical not only because of these all too literal police actions on consciousness, wars seeking to eradicate a practice deeply integrated into human experience.[21] It is ethical precisely because our responses to psychedelic experience represent a case study in our response to a transhuman, transpersonal, or post-human condition. In the research for this book I have reviewed thousands of reports about psychedelic experience, from the Rigveda to Havelock Ellis to DJ Short, and suggest that a signature of these varied and incessantly ineffable experiences has been what I will call the "ecodelic" insight: the sudden and absolute conviction that the psychonaut is involved in a densely interconnected ecoystem for which contemporary tactics of human identity are insufficient. As with Mitchell and Smokeloc's accounts above, this ecodelic symptom entails a sudden apprehension of immanence, a connectivity that exceeds the rhetorical capacities of an ego and simultaneously summons transpersonal characters who, at the very least, function as rhetorical tactics for managing the strangeness of ecodelic experience. Machine elves compete with aliens to prankishly rule the interdimensions revealed to smokers of DMT, and visions of evolution itself become a continual trope or scene with which to frame and articulate psychedelic experience. All these touchstones of contemporary psychedelic testimony make up a transpersonal, even transhuman interpellation—a palpable encounter with Darwin's "tangled bank" of evolution. Even Richard Dawkins, a popular evolutionary theorist whose work has perhaps most thoroughly embraced the "informatic vision" of Neo-Darwinian reductionism, seems unable to avoid describing evolution through recourse to the psychedelic. In his last trope of a six-hundred-fourteen-page "pilgrimage" (a sacred trip), Dawkins compares his amazement at evolution to the, for Dawkins, tamer hallucinogens:

> This pilgrimage has been a trip, not just in the literal sense but in the countercultural sense I met when a young man in California in the 1960s. The most potent hallucinogen on sale in Haight or Ashbury or Telegraph Avenue would be tame by comparison. If it's amazement you want, the real world has it all. Not to stray outside the covers of this book, think of Venus's girdle, migrating jellyfish and tiny harpoons; think of the platypus's radar and the electric fish; of the horsefly larva with the apparent foresight to pre-empt cracks in the mud; think redwood; think peacock. (613)

Yet Dawkins seems to forget, "the potent hallucinogen on sale in Haight or Ashbury" is indeed part of the real world and its amazing evolutionary splendor. "Think Five Grams of Dried Psilocybin Cubensis!" In other words, *psychedelics are (a suppressed) part of evolution.* As Italian ethnobotanist Giorgio Samorini put it "the *drug phenomenon* is a natural phenomenon, while the *drug problem* is a cultural problem" (87). Indeed, as Dawkins's recourse to the "trip" trope makes clear, evolution is itself both literally and figuratively psychedelic. According to its coiner (scientist Humphry Osmond) "psychedelic" comes from the Greek for "manifesting mind," and evolution's dizzying and diverse array of life-forms emerging over time on the surface of one planet indeed "manifests" minds sometimes capable of "thinking" about evolution: "think of Venus's girdle . . . think of the platypus's radar and the electric fish . . . think redwood; think peacock." And the visual riot of possibility and differentiation represented by even a short list of life-forms practically requires the comparison to the absurdity, awe, and splendor attributed to psychedelic experience.[22]

And in awe we forget ourselves, becoming aware of our context at much larger—and qualitatively distinct—scales of space and time. And over and over again we can read in ecodelic testimony that these encounters with immanence render the ego into a non sequitur, the self becoming tangibly a gift manifested by a much larger dissipative structure—the planet, the galaxy, the cosmos. For many, this "grokking of Gaia" enables a joyful survival beyond the fragility of a human, all too human, identity and calls for their commitment to something other than themselves. Ecodelics can introduce unprecedented states of consciousness inscribed into texts and images in modes not subject to the usual protocols of either identity or objectivity. Subjectivity, as a paradoxically transhuman phenomenon of awareness rendered only in ecologies, is rendered into inscriptions and images even as no self is adequate to the report.

Faced with a technology whose effect was to amplify inhuman or transhuman presences in our environment and consciousness, a technology whose effects included the attempt to make consciousness itself—and, hence, science itself—an item of scientific investigation, the United States sought to regulate and indeed prohibit the most proximate agent of that transhuman experience: psilocybin, LSD, and mescaline as well as more recently scheduled research compounds such as 5-MEO-AMT. Indeed, even DMT, an endogenous and very real product of the human brain, has been "scheduled" by the federal government. DMT would be precisely, by most first person accounts, "the most potent hallucinogen on sale in Haight or Ashbury or Telegraph Avenue" and is a very real attribute of our brains as well as plant ecology. We are all "holding" a Schedule One psychedelic—our own brains, wired for ecodelia, are quite literally against the law.

Has this tactic worked? Is regulation and criminalization the "best practice" for a global culture's reception of a novel technology whose very existence puts ordinary understandings of human ontology and experience into momentary but unmistakable disarray? This book will argue that indeed the United States's reaction to psychedelics, while perfectly understandable, is anything but a best practice. It is analogous to the initial panic familiar to the psychonaut, a response to be worked through, shrugged off, or ignored if one is to learn something from the experience. It is time to learn from our ecodelic experiences and work toward harm reduction in the context of transhuman technologies.

The first principle of harm reduction with psychedelics is therefore this: one must pay attention to set and setting, the organisms for whom and context in which the psychedelic experience unfolds. For even as the (re)discovery of psychedelics by twentieth-century technoscience suggested to many that consciousness was finally understandable via a molecular biology of the brain, this apex of reductionism also fostered the recognition that the effects of psychedelics depend on much more than neurochemistry.[23] If ecodelics can undoubtedly provoke the onset of an extra-ordinary state of mind, they do so only on the condition of an excessive response-ability, a responsiveness to rhetorical conditions—the sensory and symbolic framework in which they are assayed. Psychologists Ralph Metzner and Timothy Leary made this point most explicitly in their discussion of session "programming," the sequencing of text, sound, and sensation that seemed to guide, but not determine the content of psychedelic experiences:

> It is by now a well-known fact that psychedelic drugs may produce religious, aesthetic, therapeutic or other kinds of experiences depending on the set and setting. . . . Using programming we try to control the content of a psychedelic experience in specific desired directions. (5; reversed order)

Leary, Metzner, and many others have provided much shared code for such programming, but all of these recipes are bundled with an unavoidable but difficult to remember premise: an *extraordinary sensitivity to initial rhetorical conditions characterizes psychedelic "drug action."*

Leary and Metzner establish the initial rhetorical conditions of their famous trip manual, *The Psychedelic Experience*, modeled after the "Bardo Thodol," with a dedication, introduction, and a series of tributes. The dedication is to Aldous Huxley, and comes in the form of "profound admiration and gratitude" to the then recently departed Huxley and a rather lengthy sample from *The Doors of Perception*, the title of which is itself a sample from the work of William Blake. Beginning with a discussion of beginnings, the passage focuses on the need to discipline attention if the experience of ego death, like the experience of death, is to be endured:

> "If you started in the wrong way," I said in answer to the investigator's questions, "everything that happened would be a proof of the conspiracy against you. It would all be self-validating. You couldn't draw a breath without knowing it was part of the plot." (Leary et al., 5)

Note that the nature of the psychedelic experience is contingent upon its rhetorical framing—what Leary, Metzner, and Richard Alpert characterized in *The Psychedelic Experience* as "the all-determining character of thought" in psychedelic experience. The force of rhetorical conditions here is immense—for Huxley it is the force linking premise to conclusion:

> "No I couldn't control it. If one began with fear and hate as the major premise, one would have to go on the conclusion." (Ibid.)

Rhetorical technologies structure and enable fundamentally different *kinds* of ecodelic experiences. If the psychonaut "began" with different premises, different experiences would ensue. Still sampling from the sample, Laura Huxley prescribes a disciplining of attention in response to the powers of initial rhetorical conditions:

> "Would you be able," my wife asked, "to fix your attention on what *The Tibetan Book of the Dead* calls the Clear Light?"
>
> I was doubtful.
>
> "Would it keep the evil away, if you could hold it? Or would you not be able to hold it?"
>
> I considered the question for some time. "Perhaps," I answered at last, "perhaps I could—but only if there were somebody there to tell me about

the Clear Light. One couldn't do it by oneself. That's the point, I suppose, of the Tibetan ritual—somebody sitting there all the time and telling you what's what." (Ibid.)

The navigation of ecodelic experience, then, is not only the work of a self—it is instead available only to a being willing to submit to the perspective of another, a transpersonal, exterior agency that helps "hold" the attention of the psychonaut and programs or orients the turmoil of ego death. This "holding" can be more than figurative:

> The tender, gentle, supportive huddling together of participants is a natural development during the second phase. Do not try to rationalize this contact. Human beings and, for that matter, most all mobile terrestrial creatures have been huddling together during long, dark confused nights for several hundred thousand years. Breathe in and breathe out with your companions. We are all one! That's what your breath is telling you. (Ibid., 48)

Such programming by the commons emerges essentially from interaction—meditation, yoga, reading, or creative practice before and after the experience and interaction during it—and instills a new capacity: the capacity to focus on and pass through altered states of mind, forming a collective body and distributed mind of breath or *prana*. "Somebody sitting there all the time and telling you what's what" was not only comforting, but pedagogical:

> The experienced voyager, trained to recognize non-symbolic patterns, is better able to recognize and direct his awareness and better able to deal with the flow of stimuli, whether planned or accidental. (Metzner and Leary, 6)

Metzner and Leary's notion of "non-symbolic patterns" is instructive, as it reminds us that the able navigation of psychedelic experience hinges on an ability to let go of interpretation and control, to think of patterns not as bundles of "symbolic" meaning but as glyphs that "stand for" nothing but the experience itself—the patterns are a veritable map for and of the experience, refrains that are themselves often sufficiently complex to organize the attention of an otherwise vertiginous subject navigating the observation of their observation of their transhuman essence: *Tat Tvam Asi*. Listening to icaros, the whistled songs that accompany the upper Amazonian ayahuasca practice, the drinker is guided by rhythm, cadence, and pitch as the disciplined attention to the constantly shifting icaro provides a sonic handhold for the drinker, a place to dwell within the multiplicity presented by ayahuasca and competing for focus.[24] Along with ethnomusicologist Fred Katz, the anthropologist Marlene Dobkin De Rios compares the function of music in entheogenic shamanic practice to a "jungle gym."

> Once the biochemical effects of the hallucinogenic drug alter the user's perception, the music operates as a "jungle gym" for the consciousness during the drug state. It provides a series of pathways and banisters through which the drug user negotiates his way. (Dobkin De Rios 1996, 208)

Dobkin De Rios and Katz refine the analogy—in contrast to the freedom of the jungle gym, where the itinerary of the gymnast is flexible and open-ended, the shamanic programming of music increases constraint and enables a new threshold of freedom. The itinerative awareness enmeshed with music in the psychedelic experience literally does not have a moment for interpretation or choice in any usual (egocentric) sense, as it necessarily follows a dynamic flow and cannot comfortably take leave of it.[25] In such a context, one searches, perhaps in futility, for handholds in the reports of others and borrows from them, forming a commons, until finally letting go. As with Huxley before him, ethnobotanist Jonathan Ott samples from William Blake when summoning this encounter with a dynamic infinite: "I saw no God, nor heard any, in a finite organical perception; but my senses discover'd the infinite in everything."[26]

Given the importance of such endlessly tuneable rhetorical conditions (and traditions) to psychedelic practice, it is perhaps not surprising that even the name for these compounds becomes important and oft contested. Osmond coined the term "psychedelic" in 1956 in order to amplify mescaline and LSD's capacities to actualize or "manifest" "mind" (*psyche*), offering it in correspondence with Aldous Huxley as at once a corrective to the older non sequitur "psychotomimetic" and a counter point to Huxley's putatively "too beautiful" offering, "phanerothyme." More recently (1979), Gordon Wasson, Jonathan Ott, Carl Ruck, and other neologists offered the term "entheogen," suggesting that "psychedelic" had become too "invested with connotations of sixties culture" to provide a term general enough to encompass both contemporary and ancient shamanic practice. Ott writes, "the neologism derives from an obsolete Greek word meaning "realizing the divine within" (19). Given the non-local feeling of the experiences I have narrated thus far, this topological inflection of "entheogen"—it's reliance on a "within" and "without"—is less than ideal. And the reference to the "divine" will, for many readers, simply substitute one mystery for another. But it is worth noting that according to Ott, entheogen was the "term used by the ancient Greeks to describe states of poetic or prophetic inspiration" (19). Hence psychedelic/entheogens/hallucinogens/ecodelics may not only challenge rhetorical faculties—what Havelock Ellis rendered as mescaline's "indescribeableness"—but are essentially productive of them—Havelock Ellis's "indescribeableness."

Clearly a mixture of terms is called for and into the mix I want to whorl "ecodelic," a name that both samples from tradition and highlights an important, but less discussed, effect of the compounds for inducing sudden bouts of interconnection, the perception of being enmeshed by the terrestrial and extraterrestrial ecology, a tuning to the eco-logos. This is an insight at least as old as the Upanishads—"Thou art that"—with a tiny difference: ecodelics recall ongoing participation with a living ecoystem and raise the old logo of ecology in the ongoing fossil-fuel wars. Ecologies are essentially open source, so if "ecodelics" becomes a brand name, let us hope it becomes a brand whose stake holders live in our complicated and thoroughly participatory future. With and after Leary, I am a hope fiend, and I hope that this book and its ecodelic offering can be part of a distributed and ubiquitous movement toward harm reduction and creative experimentation in the context of our ecoystems. As a rhetorical practice, it has within it rhythms, chants toward efforts at successfully shaping the set and setting of the future by focusing our collective attention on the "problem" of our interconnection: How can we grow sustainable subjectivities, human or otherwise, for a transforming planet? As part of a toolkit for sustainability, appropriately deployed psychedelics can do much more than avoid harm—when used with the care, craft, and respect they require, ecodelics are tools for creatively thriving in a future modeled on ecosystemic rather than egoic practices.

THOU ART YOUR BRAIN, THOU ART YOUR BRAIN
ON TROPES: MERLIN DONALD'S GENOMIC SORCERY

> *Eyes close . . . back to dancing workshop . . . joy . . . incredible beauty . . . the*
> *wonder, wonder, wonder . . . thanks . . . thanks for the chance to see the dance . . .*
> *all hooked together . . . everything fits into the moist, pulsating pattern.*
>   —Timothy Leary, "Programmed Communication
>   During Experiences with DMT"[27]

Already, our brains—and not just Richard Dawkins's—are on tropes. The extended and distributed phenotype of the mind that will emerge and converge here is offered as an antidote to the more isolated perception of the "brain in a box," analyzed most authoritatively by evolutionary psychologist Merlin Donald.

> The central assumption of cognitive solipsism is that the mind may be,
> indeed must be, conceptualized as a system that is contained entirely inside
> a box. In the case of vertebrates like ourselves, that box happens to be the
> brain. (Donald 2000, 22)

Donald, a long time solipsist and practicioner of the "myth of the isolated mind," now describes a tangled bramble of interconnections between brain and world, a tangle of "wisteria vine" (or, perhaps, *B. caapi*) whose complexity renders heuristic any treatment of the mind as encapsulated from culture. While such a webby map of mind might appear driven by the *parataxis* of digital information technologies and their drift, Donald introduces this sense of exteriority through the media technology that put Plato or, at least, the character of Socrates, into a panic: writing.[28] In contact with practices of literacy, Donald argues, the brain undergoes a transformation of its *functional anatomy*—the scribal mind now emerges from a distributed network. Donald treats the evolution of "external symbolic storage" as the brain's new memory ally, a transformative one that enabled new kinds of prediction, social organization, and identity. For Donald, human capacities for prediction and anticipation were drastically amplified by the invention and deployment of written records, a developmental transformation of the human ecology of mind as substantial as shifts in the more properly corporeal anatomy of the brain such as the neo-cortex. For the "problem" with brains is that they only work when in contact with other symbol-using systems: "the problem is that our brains can never produce truly symbolic acts unless they are imposed from the outside" (ibid., 32).

Primate brains only become capable of great feats of open-ended learning when they become intensely response-able. Fundamentally mimetic, the enormous plasticity of the human brain emerges from its sensitivity to outside interactive "programming."

> Our recent work verifies the profound importance of social interaction
> on human language learning. In a recent study, we exposed 9-month-old
> American infants to a foreign language (Mandarin Chinese) spoken by
> native speakers who read books and played with toys. The infants were later
> tested on phonetic units contained in Mandarin but not in English. Infants
> learned phonetic units from the foreign language with <5 h of experience,
> but, interestingly, they failed to learn when the same speakers using the
> same materials were prerecorded and presented from audiovisual or audio-
> only DVDs. (Kuhl)

In this aspect of their nature, humans copy the African Gray Parrot: According to researcher Irene Pepperberg, parrots learn only from live, rather than recorded instruction. Language acquisition begins with a perceptual mimesis that is only later cognitive. This perceptual capacity is also a capacity for replication:

> Replicative mechanisms are central to evolutionary theory because natural
> selection acts on the entire process of replication, including its nongenetic
> components. The replicative mechanism of the human mind is, by defini-
> tion, responsible for transmitting our cognitive architecture across genera-
> tions. . . . The prehistory of the human mind, even at the earliest stages of
> hominid emergence, must acknowledge the evolution and role of symbolic
> culture as an integral part of cognitive evolution. (Donald 2000, 22)

Clearly, Donald's view articulates a Copernican shift in our technoscientific map of cognition and its evolution. The "cognitive architecture" of humans is distributed in both space and time, making our rhetorical practices uncannily efficacious—rather than representational—agencies of a distributed mind whose vectors are multiple and replicating. In this context, the symptoms of plant "intoxication" are also states of mind replicated over time and emerging from a plant-human interaction. The states likely introduced by interactions with plants can themselves be mimed and altered, as the "programming" of psychedelic experience through visuals (e.g., the light shows of what Gene Youngblood called "an expanded cinema") *modeled* on states of mind makes clear. Yet at times Donald's mimetic understanding of the brain seems strangely closed to influences other than human ones. For some birds, for example, the replication of song need not emerge from contact with the same species:

> In young zebra finches, visual interaction with a tutor bird is typically
> required to learn; in fact, the impact of social interaction is so potent that
> young zebra finches will learn an alien song from a Bengalese finch foster
> father who feeds them. White-crown sparrows will also learn an alien song
> from a live tutor, even though they reject those songs when presented on
> audio tape. In barn owls and in white-crowned sparrows, the duration of
> the sensitive period for learning is altered by the richness of their social
> environments. (Kuhl, 9645)

Hence what seems crucial to the generative plasticity of some bird and human brains is participatory interaction itself, a dynamic entanglement with the "richness" of an environment. Such interaction cannot, it appears, transform completely into symbol—a recording will not enable birdsong acquisition, enormous complexity haunts even the simplest of computer algorithms—but must instead be enacted. Rather than simply rendered or explicated by symbolic culture, then, we must entertain the notion that the human brain is constituted by what Aristotle described as "all available means of persuasion"—an alien song, not navigable by symbols, to which our ears were, are, tuned.

> Ever since it began, our community of increasingly conscious brains has
> coexisted with an exploding process of enculturation. All else followed from
> this elemental brain-culture chemistry. (Donald 2000, 35)

Donald's latter phrasing suggests something like a portmanteau description of the bramble of connections that make up an actually functioning brain in Donald's model. "Brain" and "culture" are both distinct and connected—a mixture whose rhetoric, at least, is linked to "chemistry." But given the importance of "chemistry" itself to contemporary understandings of consciousness, it seems odd that its dynamics are otherwise absent in Donald's treatment of the evolution of cognition. This book will seek to build on Donald's exocentric model of mind by linking his vision of a replicative model of culture to the use and replication of molecules, coevolutionary transformations of brains and ecosystems analogous to those observed between lactose-tolerant humans and cattle milk protein genes.[29]

For chemistry—the access to and use of concentrated and mixed compounds from labs as well as rain forests—is clearly an element of human culture, and an ancient one at that. If natural selection consists in the sorting of different replicative capacities—DNA, as well as the extended rhetorical structures of the literacy brain—then so too must we consider the actual chemistry of this interconnected brain and the extended pharmacopiea of culture.

Recent studies of video gaming, for example, are increasingly focused on the effect of video gaming on neurotransmitters, their agonists, and their reuptake, in dynamic relation with gaming. In this sense, human consciousness can be said to be continually "playing" its brain chemistry. What I will suggest here is that the brain, having been let out of its box, is now unmistakably and repetitively on plants as well as tropes. For if writing, as Donald would have it, is a devoted ally of human memory, what are cultures memorializing? A pharmacopiea:

> One culture invents sailing, another throat singing. One culture insists on a capacious oral memory to remember entire pharmacopieas, whereas another demands great dexterity with external symbols to manage an electronic universe. We are asked to handle now broadswords, now biplanes, now remote microsurgical hands. (2000, 36)

We are essentially flexible organisms, due in part to the enormous dexterity of the mind-hand nexus invoked here by Donald in his rhythmic *repetitio* of "now."[30] So too is mind essentially flexible and subject to persuasion and alteration by the entire bramble of "brain-culture chemistry." As Donald's exuberant recourse to the rhetorical triplet "now . . . now . . . now" might remind us, even such technical communication involves the transformation of consciousness through chanting rhythms as well as more properly semantic information.

Why all this chanting repetition: "Think Venus Girdle . . . Think Peacock . . . Think Now . . . Now . . . Now . . . "? Even "ordinary" consciousness is essentially *alterable* consciousness, and when asking after the evolution of our capacities it is worth recalling which altered consciousness is repeated (in this instance memories, memories troped and recalled through repetition). And, from the evolutionary perspective adopted by Donald: Which ones allow humans to survive and leave more progeny? Psychologist Roland Fischer indulged in what he calls the "temptation" to understand such memories in evolutionary terms:

> One would be tempted to theorize that myth, fairy tales and narratives
> which contribute to the survival of the species, survive, i.e., are selected as
> behavioral templates to be re-written and re-composed time and again, to
> assist in the survival of the next generation. (1978, 66)

So too would those narratives and other memory devices that allow for the recall of efficacious pharmacopiea be selected for under such a rubric. Many such pharmacopeia even contain putative memory adjuncts, further encouraging their recall and replication. Yet if replicative culture is crucial to Donald's account in the evolution of cognition, might we be justified in asking where replication, rather than language, comes from? While the contemporary imagination is well prepared to envision evolution as a great competition machine, it comes up short when imagining how different replicative practices are sorted out. In short, how can one fairy tale possibly enable more progeny than another? Has Donald, in his turn toward networks of replicative rather than autonomous mind, substituted one mystery for another?

A return to our example of bird brains is instructive. If, with Charles Darwin, we remember that zebra finches and humans both likely replicate song and language in response to the pressures of sexual selection, we are then in a position to understand the strange persistence of ecodelics in the human toolkit. For practices of sexual selection—think the rhythm of grasshopper stridulation, think the flutter of a peacock feather, think the practiced warble of birdsong, the bugle and musk of deer—cultivate replicable traits in the evolution of reproduction by selecting not only for survival, but for *seduction*. Chapter 3 will feature an investigation into current theories of Darwinian sexual selection and its mechanisms, but for now I want to note that these practices often induce the breakdown between self and other associated with the ecodelic insight, and that these outbursts of symmetry breaking—as in the tragic abuse of date-rape drugs such as Rufinol and alcohol, as well as the more ecstatic domains of rave culture—are unmistakably both culturally induced and biological in both mechanism and outcome. The biology of ecstatic (or even spiritual) experience focuses on the tryptamines and the phenethylamines, so these compounds and the plants that synthesize them become proximate causes for what Terence McKenna called a "glimpse of the peacock angel" and Neal Cassady described as a peacock feather eye massage. "I've done the magic mushroom in Oaxaca, you understand, and felt the rainbow peacock tail brush my eyeballs" (Leary 1983, 52). That peacocks are poster creatures for sexual selection is perhaps no accident; their plumage is a highly replicable and very effective rhetorical technique in sexual selection—the sorting of peacock and peahen mates over time—and it is not surprising and perhaps quite instructive that they have troped psychedelic experience. As Lynn Margulis and Dorion Sagan write:

> Drugs, pure chemicals with highly specific and predictable effects, have profoundly influenced sexuality since Archean times when bacteria ruled Earth. The antibiotic tetracycline, for example, when placed in a bath of bacteria, increases the rate of bacterial sex a thousand times. (1990, 176)

The usual thought experiment summoned when imagining the evolutionary emergence of traits is danger—a tiger is often invoked as the ultimate challenge to the homind brain. No doubt such danger was an intensive vector—the "killer frisbee" and "Swiss army knife" of early humans, the handaxe, was clearly a hunting adjunct, carcass cleaning tool, and personal defense technology. Yet seduction, and its continual use of drugs, also informed and entrained the human brain into its contemporary capacities. Note that the cultural invention most associated with the brain in Donald's treatment is memory, and for what is the memory used? The recall of "entire pharmacopieas" is perhaps not merely one memory among others, but a memory whose (rhetorical) replication is deeply implicated in the differential survival and evolution of human beings. Along with many other organisms, we have evolved a medicine whose pharmacopiea includes aphrodisia, an extended phenotype of the brain where compounds both endogamous and exogamous contributed to the evolutionary fitness of humans though amplifications of desire and fertility. While Dobkin De Rios and many others in the anthropological and ethnobotanical community note that "hallucinogenics" could have provided benefits as a hunting ally due to increased visual accuity, she also notes that such a molecular ecology would "endanger the lives and the reproductive success of any animal so involved." More recent work on the abundance of aphrodisia in nature reminds us that such adjuncts might also, like so many unwieldy peacock feathers, *increase* sexually selected fitness.

RELAX, IT'S ONLY A SHIFT IN THE SENSORY-MOTOR RATIO: TUNING THE PSYCHEDELIC MIRROR

> *We can only say that these limits can only be transcended*
> *by the ongoing evolution of the species.*
> —Roland Fischer, "On Separateness and Oneness"[31]

Has this coevolution of rhetorical practices and humans ceased? This book will argue that psychedelic compounds have already been vectors of technoscientific change, and that they have been effective precisely because they are deeply implicated in the history of human problem solving. Our brains, against the law with their endogenous production of DMT, regularly go ecodelic and perceive dense interconnectivity. The human experience of radical interconnection with an ecosystem becomes a most useful snapshot of the systemic breakdowns between "autonomous" organisms necessary to sexual reproduction, and, not incidentally, they render heuristic information about the ecosystem as an ecosystem, amplifying human perception of the connections in their environment and allowing those connections to be mimed and investigated. This increased interconnection can be spurred simply by providing a different vision of the environment. Psychologist Roland Fischer noted that some aspects of visual acuity were heightened under the influence of psilocybin, and his more general theory of perception suggests that this acuity emerges out of a shift in sensory-motor ratios.

For Fischer the very distinction between "hallucination" and "perception" resides in the ratio between sensory data and motor control. Hallucination, for Fischer, is that which cannot be verified in three-dimensional Euclidean space. Hence Fischer differentiates hallucination from perception based not on truth or falsehood, but on a capacity to interact: if a subject can interact with a sensation, and at least work toward verifying it in their lived experience, navigating the shift in sensory-motor ratios, then the subject has experienced something on the order of perception. Such perception is easily fooled and is often false, but it appears to be sufficiently connective to our ecosystems to allow for human survival and sufficiently excitable for sexually selected fitness. If a human subject cannot interact with a sensation, Fischer applies the label "hallucination" for the purpose of creating a "cartography of ecstatic states."

Given the testimony of psychonauts about their sense of interconnection, Fischer's model suggests that ecodelic experience tunes perception through a shift of sensory-motor ratios toward an apprehension of, and facility for, interconnection: the econaut becomes a continuum between inside and outside. Correlating molecular mechanisms of psychedelic compounds with ecodelic testimony and the fine-grained description of the "non Euclidean space" of the hallucination-perception continuum could provide us with a veritable imaging system of the imagination, one which suggested to early 1960s scientists such as Lynn Sagan that indeed consciousness, like life, was on the brink of being a technoscientific object and thus subject to radical transformation as mind achieves a putatively new threshold and connectivity—its own physical manipulation and evolution through feedback.

This new connectivity—mind pondering, designing, and actualizing its own physical manipulation and transformation—felt revolutionary to many in the 1950s and early 1960s, and the wide-spread evidence for this "recursivity"—the ability of evolution to alter itself through consciousness, and for consciousness to alter itself through evolution—suggests a radical need for increased sensitivity to the interconnections of our decisions and inventions.

Julian Huxley, Aldous's biologist brother, coined the term "transhuman" in English to describe this emerging awareness of evolutionary processes:

> It is as if man had been suddenly appointed managing director of the biggest business of all, the business of evolution—appointed without being asked if he wanted it, and without proper warning and preparation. What is more, he can't refuse the job. Whether he wants to or not, whether he is conscious of what he is doing or not, he is in point of fact determining the future direction of evolution on this earth. That is his inescapable destiny, and the sooner he realizes it and starts believing in it, the better for all concerned. (J. Huxley, 13; cf. also Teilhard De Chardin 1959)

Despite the banality and dated gendering of Huxley's otherwise grand analogy, this sudden realization of the coevolutionary future of human consciousness seems to have no counter argument in the early twenty-first century. While Huxley looks for "belief" in this destiny, our present suggests that we need less to "start believing" than to "stop the denial" of our response-ability toward living systems. This includes, for Huxley, a response-ability to the biodiversity of human experience:

> We need to explore and map the whole realm of human possibility, as the realm of physical geography has been explored and mapped. How to create new possibilities for ordinary living? What can be done to bring out the latent capacities of the ordinary man and woman for understanding and enjoyment; to teach people the techniques of achieving spiritual experience (after all, one can acquire the technique of dancing or tennis, so why not of mystical ecstasy or spiritual peace?); to develop native talent and intelligence in the growing child, instead of frustrating or distorting them? (J. Huxley, 15)

In other words, if humans are to evolve, they must learn to be ecstatic. Yet as Huxley's exploration analogy unwittingly suggests, such exploration is not without its own tendencies toward colonization and replication; rather than "transcending himself," humans have explored and mapped the planet while unconscious of their interconnection with it. Researcher Myron Stolaroff—a Stanford engineer whose research on psychedelics is astonishing for its breadth and care—suggests that the therapeutic use of ecodelics can be effective tools in the Jungian quest to "make the unconscious conscious," allowing us to reflect on and work through "shadow" material and tune ourselves toward more affirmative, and perhaps sustainable, lives. Reagents of interconnectivity, ecodelics appear to amplify the human perception of immanence.

Yet this insight does not yield the usually perceived increase of control associated with technoscientific advances. Along with this ecodelic perception is a radical displacement of the human into the context of the "tangled bank" of evolution—psychedelics may be, as I will continue to argue, agents of sexual selection, but they most certainly and repetitiously stage evolution as a motif. In this sense, psychedelics pose an unavoidable encounter with the question of human *being* and its nature: who we have been and what, in short, we are capable of becoming.

TRIP REPORT, OR THE PROGRAM

> *The program is a voyage chart, a series of signals, which, like the pilot's radio,*
> *provides the basic orienting information required for the "trip." The program*
> *need not be followed exactly, but it can be of great help in orienting oneself*
> *in unfamiliar territory or finding one's way if lost in rough weather.*
> —Ralph Metzner and Timothy Leary,
> "On Programming Psychedelic Experiences"[32]

The problem of the trip report joyfully haunts every page of this book. If a legion of psychonauts have not succeeded in producing anything like a consensus description of the psychedelic experience because of an alleged deficit in representation—not even the *name* of these compounds achieves consensus—how can I hope to offer even a simulacrum of a scientific argument concerning these plants and compounds and their putative role in human evolution? In chapter 1, I assay the idea that despite their ineffability, trip reports persistently present rhetorical software or *programs* with which to replicate the psychedelic experience.[33] Hence while there is little confidence in language's capacity to represent the effects of psilocybin, mescaline, or ayahuasca, trip reports, anthropological testimony, and oral traditions are all nonetheless implicitly oriented to the linguistic *management* of psyche-delic states. This management extends to the use of trip reports themselves to orient the psychedelic experience, to act as the recursive "set and setting" of psychonautic practice. Indeed, contemporary psychonautic experimenta-tion—a network of researchers that are the veritable Linux of consciousness science—suggests that speech itself might plausibly emerge as nothing other than a symptom and practice of early hominid use of ecodelics. Aldous Hux-ley's famous description of his mescaline experience in *The Doors of Perception* becomes then less a revelatory text providing the keys to the infinite and more a text for programming and tuning the reader's attention toward our evolutionary situation: imbrication with a whole, an interconnection whose hallmark is often phenomenologically infinite ecstasies.

The induction and management of ecstatic states through rhythm, song, gesture, and feather was precisely the interest of Charles Darwin in *The Descent of Man,* and in chapters 2 and 3 I become fascinated by the evidence that psychedelics have been used in a context selected as much by sexual selection—and its practices of excessive interconnection and the fluttering breakdown of usually well-defined borders—than our ordinary understand-ings of natural selection as the struggle for autonomous survival. Recent work on sexual selection by biologist Geoffrey Miller and others echoes Darwin and suggests that the evolution of the human brain is entangled not only with survival but with practices of courtship and seduction. Other recent work also notes that evolution excels in the production of aphrodi-sia—from the eloquence of birdsong to the frenzy of catnip—and it is in this context that, I will argue, we must begin to comprehend the human use of psychedelics. It may be that evolution, as biologist J. B. S. Haldane famously put it, has "an inordinate fondness for beetles," but it's rather fond of diverse sexual practices as well. Perhaps it is obvious that the molecules and plants flowing through the "Summer of Love" ought to be treated as aphrodisia, but if so they work not only to excite desire but to redistribute and blend it into new forms. Competition for attention, sexual selection is a fundamen-tally rhetorical domain, a multimedia display of genetic information, and in altering this display, new domains and systems of traits, their sorting and combination are opened up to (sometimes "runaway") evolutionary change.

This "tuning" of attention again reminds us that like the peacock feather, psychedelics can act as *attention attractors*. Chapter 4 looks carefully at the ways in which LSD caught the attention of Swiss chemist Albert Hofmann in 1943 and the rhetorical challenge he faced sharing his research with colleagues. The challenges continued, and LSD research became increasingly rendered in the language of nascent molecular biology, and before long the initials DNA became a veritable password for putting psychedelic experience into language. As the newfound "secret of life," DNA thus became a key to articulating the mysteries of ecodelic practice. This use of DNA as a frequent rhetorical software for framing ecodelic experience is echoed by Nobel Laureate Kary Mullis, who has linked his creation of Polymerase Chain Reaction to the action of LSD. Contextualizing Mullis's claim, the chapter looks to the history of science archives and learns that before their prohibition, mescaline, LSD, cannabis, and other compounds and plants were championed as effective agents for "breaking creative log jams." The chapter will look to some recent claims by scientists that psychedelics were an important part of their lab apparatus and contextualize them with the extensive pre-prohibition literature on creative problem solving and psychedelics. Finally, the reverberation of LSD with DNA continues as a journalist comes forward to testify that indeed DNA was itself envisioned by Francis Crick under the influence of LSD. As Alice might put it, the tale of DNA and LSD gets curiouser and curiouser. Could the problem set presented by globalization and climatic change be fruitfully researched with these tools for "toying with concepts" and solving engineering problems? *More research is needed.*

Ayahuasca, the plant complex brewed by ayahuasceros of the upper Amazon and the sacrament of several syncretic churches in Brazil, Europe, and North America, is essentially a mixture, and is sometimes mixed with the vine of the strangle fig to act, together with the sonic programming of an icaro, as a kind of love potion, binding the lovers together until one dies. While this conjunction of constraint and love might appear somewhat too *film noir* to mesh with contemporary ideals of romance, it nonetheless demonstrates the implication of this shamanic medicine with technologies of love and fertility management—ayahuasca is quite plausibly implicated in the distribution of human DNA in the upper Amazon. Such fertility management is, of course, the domain of sexual selection, and chapter 5 looks to the rhetorical practices of ayahuasca—both those that seek to manage the eocdelic state and those that seek to describe it—and finds there a pedagogical practice oriented toward divining and learning to love infinity—a plant love that does not leave human dreams of autonomy intact and binds us to the planet unto its, our, death. An attention vortex, love is unavoidably participatory and hence requires a difficult but transformative "tuning" of attention toward the hyperbolic, beyond the pre-given categories of "true" and "false"—a capacity to navigate and joyfully endure the sheer difference of another, even, especially, a plant other.

Cannabis cultivation and genetics ought to be the envy of more mainstream biotechnology, and chapter 6 links the remarkable success of cannabis cultivators to their capacity to blur the borders between plant and human agency, a blurring that aptly visualizes the emerging relationship between humans and biotechnological transformation in general. The new practices of DNA distribution that light up transgenic zebra fish in bioluminescence are novel forms of sex whose selection matrices are difficult to predict and are likely driven by runaway processes, a blur of pleasure, image, and money. Dancing with cannabis, contemporary breeders create unprecedented lifeforms, green familiars who sometimes persuade us that we are persuaded by them to cultivate them.

Chapter 7 continues this emphasis on a pedagogy of plants and looks to the success of Ibogaine polydrug addiction therapy and asks what a west African root might teach us about contemporary habits of the pharmaceutical industry. According to a now extensive literature, Ibogaine, a west African psychedelic used by mandrills and gorillas as well as humans, blocks withdrawal symptoms for heroin addicts and provokes a "cognitive review" conducive to behavioral change: something like one's life passing before one's eyes. Given the success of Ibogaine and other open-source psychedelics, perhaps the human experience with Iboga should also provoke a cognitive review of contemporary technoscience and its attachment to restrictive practices of intellectual property. The chemical and botanical commons has vastly out innovated more mainstream pharmacological research, and this innovation points the way toward a more ecodelic and sustainable model of property and accumulation in the context of an insurmountable need for the commons and tools for fostering ecodelic practices through our ecosystemic distress and an infoquake. Such cognitive review, though, requires a narrative logic with which to stitch together the enormous amount of information presented when caught up in the epic states of psychedelics and social change, and the use of Philip K. Dick's text *VALIS* by participants in the attempt to mainstream the use of Ibogaine becomes a case study in the importance of information compression techniques in navigating the domains of interconnection seemingly amplified by the careful and practiced use of ecodelics.

This recursive blur—plants growing humans growing plants convincing humans to grow more, and more diverse, plants—becomes fiction as the epilogue ponders an ecodelic future where a clone of Charles's grandfather, Erasmus Darwin, encounters the very telos of the cosmos. In dialogue with Alfred Wallace, Darwin stumbles across the odd effect of psychedelics to foster new forms of social organization, perhaps even an organization capable of sustaining the planet.

As "Darwin's Dreams," the epilogue suggests that we should respond to the ontological suggestions of ayahuasca and other ecodelics with more care than our previous responses to indigenous peoples and the ecosystems with which we are enmeshed. As data received first and foremost as subjective experience, dreams are, like trip reports themselves, neither true nor false, but instead solicit our curiosity and our open-minded inquiry. This inquiry itself hinges on a respect for the subjective aspect of human experience. Subjective experience in no way *lacks* objectivity, but must instead be approached as the excluded "shadow" material of "western" science, a shadow with which science is again becoming fascinated. And the enormity of this exclusion is hardly measurable—technoscientific practice is enmeshed with ways of being and becoming that are also part of the 13.7 billion year legacy of this cosmos. We can't respond to "others"—the others of the upper Amazon, as well as the others of the future—without responding to "their" knowledge and its effects, and this book seeks to connect with that knowledge while neither romanticizing nor trivializing it.

I may or may not succeed in this attempt, but I offer my exploration of an ecodelic hypothesis as a record of my own inquiry, to be shared, compared, and remixed. For the astonishing and joyful exposure to these "archaic" technologies and life-forms, I experience enormous gratitude; a gratitude accessible only with a thoroughgoing dose of humility. Neurophysiologist John C. Lilly, whose work informs the entirety of this composition, wrote a concise algorithm of humility: "Humility starts within one's own structure. One is not one's whole structure. One is only an inhabitant of that structure" (1975, 7).

Lilly's practice is a rhetorical one, a reflection on a favorite trope of technoscience, and indeed, consciousness itself: the substitution of part for whole, the rhetorical move or "trope" called synecdoche. By experimenting with synecdoche and its familiar idea that humans are, indeed, the whole structure, some humans conclude that they *mind* the world rather than are minded beings. "We"—the hyper-elite governing swarm—have likely made a massive error in our judgment of the technologies of ancient humans and our connections to the cosmos, as in Mitchell's "startling recognition" above. We are suckers for synechdoche. This book offers, I hope, a holographic response: an induction to recall the Whole even in the parts.

Reflections on being troped seem to be a crucial feature of ecodelic experience, a sudden but often fleeting experience of interconnection with language, as if one were living one's life according to a certain "script" that indeed needed revision, a task for which another is often no luxury. Crucial to the "archaic" technologies of ecodelia is their reliance upon something like a human relay, a curandera, brujo, shaman, psychologist, sophist, prankster, trip meister, DJ. As numerous North American ayahuascos have learned through their own innovations and practices with the brew, ayahuasca rewards those who rhetorically manage the altered states of mind through practices of rhythm, music, and eloquence.

I am astonished by the sudden depth and tactility of vision. The pixelated doorway is less an image than an active talisman: I more or less "double click" on it and am suddenly enfolded by a labyrinthine space through which I, a pattern of forces, pass, unfolding.

Perspective in the usual sense is a non sequitur. I do not regard a space separable from me, but rather become physically aware of an n-dimensional ecology in which I live. I have not "gone" anywhere but am instead navigating an all encompassing complexity not usually available to my perception.

Given my first night's experience, the ayahuasca drinker braced himself for an onslaught of entities, but found no self there. Perhaps he had remembered the tutorial he had received from the ayahuasca on how to empty his mind and create "Diamond Mind," an emptiness that is the cosmos. There had truly been multitudes—the rather Socratic voice of the ayahuasca itself, the dead, a plant intelligence, a host of light beings, and even some goggle- (or is it Google?) eyed aliens who told me, rather prankishly, not to believe them. They'd even told me they'd enjoyed my last book, especially the last chapter, about them. I did my best not to believe them. I had little choice with the others.

Yes, perhaps I had learned to focus my attention a bit or asked better questions. Or the ayahuasca had learned my constraints and context. Or Norma's songs had a different effect. This time a bird deity matter-of-factly presented itself to me, told me it has an ancient Indian technology to heal me, and can it please get to work now? It begins making passes over my prone body as it unwinds my DNA (RNA?) and buffs it. Phlegm and mucus head for the exits—I let go of some vomit and an enormous amount of phlegm from my lungs. Sinuses flow into the bucket, suffering. I learn, again, how to breathe from the iridescent bird teacher who had somehow wielded a vibrating wand and made passes over a skein of nucleic acids he seemed to unwind from me. I did what I was told and breathed deeply.

Garuda: No, not from there.

Flash to the interior of my wet lungs. Stretching, heaving and labyrinthine, they open up as I pull air from my diaphragm. I am told that this is not a cure but a treatment, and that I must continue to treat myself by learning how to inhale and exhale properly and clear my passages. The bird prompts me to breathe in and out, until I am breathing with the entirety of the universe. Never have I had such a breath. Indeed, "I" didn't—the universe in its unfolding, did. I am being "played" by the universe, the icaros, Norma, like a piano: an almost unbearable fractal montage ensues as I am being spoken, chattering beings telling me to get ready, because here it comes:

I am.

As I exhale, the scene changes.

1

THE FLOWERS OF PERCEPTION

Trip Reports, Stigmergy, and the Nth Person Plural

*This plant, although itself hardly mobile, casts a spell over what is wakened,
for instance, by tea, tobacco, opium, often just by the mere scent of flowers."*
—Ernst Jünger, "The Plant as Autonomous Power"

*Even though, as fungi, mushrooms do not blossom, the Aztecs referred
to them as "flowers," and the Indians who still use them in religious
rituals have endearing terms for them, such as "little flowers."*
—Richard Evans Schultes, Albert Hofmann,
and Christian Rätsch, *Plants of the Gods*

IF WE ARE TO RESPOND ETHICALLY (NOT TO MENTION SCIENTIFICALLY) to the presence of psychedelic technologies in our cultures, then it would seem a good idea to evaluate what they are. This ontological question— "What *are* psychedelics?"—would seem to come before any reasoned or practical response to the systematic alteration of consciousness in interaction with plant compounds or their simulacra. We must at least pose, if not answer, questions about the kinds of practices they are if we are to respond with care to these ancient but nonetheless transhuman technologies.

As compounds, we could quite usefully group psychedelics under the categories phenethylamines and tryptamines. Alexander and Ann Shulgin used these categories as a way of organizing both their chemical research and their (fictional) writings about said investigations.[1] Even the chemistry of these "families" of compounds itself provides a kind of contact high, as the differentiation of various isomers or substitutions renders visions of a tinker-toy universe through which one can move imaginatively and, if one is a chemist of the caliber of Alexander Shulgin or various denizens of www.hive.org, endlessly toy with. Prospecting for novel psychedelics and the pathways for their synthesis becomes an endlessly differentiating glass-bead game, the leading edge of a molecular evolution whose bounds are uncertain but ongoing.

The pathways to synthesize these compounds are remarkable works of creativity in their own right, and carefully following the cascades of reactions that lead one from aluminum foil to MDMA involves a cognitive and perceptual gymnastics no less intensive than listening to a symphony. Consider, for example, the group tinker of the "Brazil Caper," wherein Shura (Alexander Shulgin's fictional alter ego) presided over the preparation of a batch of MDMA for a barely equipped clinic in Rio:

> There was an old rotary evaporator which wobbled alarmingly; Shura patched it with duct tape. There was a superb Büchner Funnel among the treasures, but no sign of filter paper of the right size, so our boys cut out a substitute from a piece of paper towel. . . . Hooking up, pulling down, fitting on, screwing in, both of them were delighted with the challenge one moment, despairing the next, hoping all over again, and mildly hysterical most of the time. (Shulgin and Shulgin 1997, 92)

Yet however engaging the synthetic pathway of any given compound is—here we are watching MDMA *become*—the scientific literature of psychedelics is clear on this point: the compounds are interesting precisely and perhaps only because of their effects on a human consciousness. Alice—Ann's fictional alter ego—shares in "Shura's dance" even as she narrates it: "Hooking up, pulling down, fitting on, screwing in," even the fabrication of MDMA seems to induce contagious and connective rhythms.

Hence while remarkable, the tinker toy, molecular definition of psychedelics is insufficient for any real apprehension of what they are, or why MDMA does or does not belong to the category "psychedelic." This universe of play is quite helpful in the replication and differentiation of psychedelic compounds, yet even participating in this play tells us (almost) nothing about why anyone would replicate psychedelic experience in the first place. It teaches us the joys of molecular evolution and the *sensus communis* often at play in the psychedelic community, but it tells us little about anything specific to the molecular evolution of psychedelics, whether that evolution takes place in a plant ecoystem or a lab. Creativity, theologian and perennial philosopher Matthew Fox tells us, is "Divine." But are psychedelics themselves involved in the divine in any fashion distinct from, say, the same sort of tinkering with a computer, a bicycle, a house, a greenhouse? This is the ontological question raised by "entheogen."

For evidence concerning the nature of psychedelics, then, we must turn to the only kind of data we have: first person testimonies about experiences with these materials. In this context, trip reports are the "readout" produced by a human consciousness and a particular compound. This readout is itself usable in multiple ways, and interpretable only in the context of the possible differentiations of set and setting embedding any particular compound or plant. Thus when the Shulgins "wrote up" the diverse tryptamines and phenethylamines, it was not only the itinerary from one compound to another—the steps of synthesis—that demanded articulation; they also needed to provide character development and community analysis if they were to situate the effects of the compounds on any given psychonaut such that they could be usefully repeated. It is in this latter sense that the voices of *Tihkal* and *Pihkal* are as vital to psychedelic science as their famous chemical recipes: the multiply voiced, famously and obviously pseudonymic characters of Shura and Alice provide fluctuating but actual rhetorical frameworks for the investigation of psychedelic experience, softwares for further research.

And yet how are we to read this data? Testimony concerning psychedelic experience faces several rhetorical challenges. In a later chapter we will watch as Albert Hofmann struggles even to speak, let alone speak well and persuasively about his discovery, LSD-25. And language itself appears to be insufficient to the synesthesias and beyond of psychedelic research. Few elements in a trip report are more predictable than a version of "words fail me." To make matters worse, psychonauts find themselves rhetorically challenged on yet another front: trip reports are written by users of psychedelic drugs. Why would we listen to such babble?

And yet properly contextualized, all of these features of psychedelic testimony can be viewed as aspects of ecodelic experience. The continual disavowal of language *in language*, for example, becomes a site for analysis—a rhetorical move made by psychonauts in response to psychedelic experience perhaps no less predictable than a reindeer's head twitch under the influence of *Amanita muscaria* mushroom. So too does the complicity of psychonauts with their own reports become a strength, as we avoid an all too common weakness in discourse about psychedelic "drugs"—researchers writing about compounds and plants they have themselves never assayed.

PARSING THE NTH PERSON PLURAL: HOW TO "READ" A TRIP REPORT

If our inquiry into psychedelics leaves us to read page after page of graphomaniacal claims to ineffability, an enormous question is nonetheless begged, not to mention repeated: *How* are we to read these texts? For besides the enormity of our task—the online repository The Vaults of Erowid alone offers over twenty thousand reports on over four hundred compounds and plants—there are also some very basic rhetorical questions to be asked about the trip report: to whom are they addressed? What, if anything, do they mean? What can they teach us about the nature of psychedelics and their frequent capacity to induce ecodelia?

Consider, for example, The Vaults of Erowid devoted to a phenethylamine, 2C-I. Invented by Alexander Shulgin in his Lafayette, California, lab, 2C-I is now being researched by an amateur network of worldwide psychonauts who synthesize the compound with the help of clandestine, online chemistry sites and Internet providers of research chemicals. A global market has emerged for this rare phenethylamine, with the bulk of production reputedly taking place in China. A sometime "club drug" in the United Kingdom and Europe, 2C-I has yielded, among other things, four hundred and thirteen experience reports posted to Erowid as of June 2010. Clicking from one report to another, I immediately come across a trope recognizable from other psychonauts: the mistake. The hilariously handled Nanobrain writes of an ill-disciplined tongue indeed:

> T 0:00 The crystalline powder was diluted to a 1mg/g concentration in an aqueous vodka solution, which required vigorous shaking to dissolve completely. I drink 12 g of this solution, containing 12 mg active. Then, I make the mistake. I lick the inside of the paper envelope formerly containing the 2C-I, as well as swallowing a couple very small flecks that were noted near the scale.[2]

The mistake is legion in trip reports—hence Nanobrain's metadiscourse on the mistake—"Then, I made *the* mistake," rather than "a mistake" reads like a refrain because it is one. Psychonautic research, like other techno-scientific investigations, often involves making mistakes faster than usual, such that they might not be repeated. This is not to say that psychonauts are a particularly careless lot—since most research chemicals such as 2C-I are usually sold not as individual dosages but in gram and half gram quantities of loose (and, in the case of 2C-I, often sticky) powders and crystals, many have high-accuracy scales and lab equipment with which to pursue their research and ensure the purity of unknown and little studied compounds. Rather, when dealing with a mixture of psychedelic states and high potency compounds—an average dose of 2C-I is a mere 15 milligrams—the mistake amplifies the contingency of the psychonautic experience: even extreme care is insufficient. A two-hundred-dollar scale is only accurate (even when properly used) within two milligrams, while even a friendly compound like 2C-I has a sufficiently steep response curve to make two milligrams significant indeed. Although this mistake would not be lethal, it could radically alter the qualitative experience, reminding us that tuning psychedelics is an exceptionally delicate affair—they are extraordinarily sensitive to initial rhetorical conditions, including that favorite rhetorical practice of Protagoras, the "measure." As we will see with the example of Albert Hofmann's accidental ingestion of LSD-25, the mistake can be seen to be foundational to modern psychedelic discourse itself, and is perhaps linked to the ego death that is frequently discussed in it.[3]

This atmosphere of extreme care and consistent error reminds us of the audience for Erowid trip reports and orients our (rhetorical) questions. While trip reports are certainly inflected by the gorgeous confessional style of Thomas De Quincey, whose opiated transmissions from the opera are among the most sublime texts in nineteenth-century British literature, they also carry an enormous burden of practicality and education. Psychonauts browse the psychedelic vaults to learn about the experience of other psychonauts with different compounds, dosages, programming, and intent, and carry out their experiments accordingly. Trip reports are, then, first and foremost protocols, scripts for the better or worse ingestion of psychedelic compounds and plants.

Thus within all the chatter of ineffability, it would be easy enough to miss an essential feature of the trip report: they must be repeatable. Generic to the trip report from Havelock Ellis to Murple is a quantitative description of dosage and delivery, as well as immensely detailed descriptions of the context and mindset of the psychonaut. Both capture repeatable elements of a psychonaut's trip—as do the frequent discussions of musical and artistic investigations under the influence. Perhaps more than anything else in his remarkable writings on psychedelic experience, it is Aldous Huxley's treatment of paint and light in Vermeer that is most persuasive in its rendering of entheogenic-mind states for this researcher, if only because it is a treatment often and easily repeated by psychonauts. Consider, for example, this trip report for another phenethylamine, 2C-E:

> (with 20 mg) The view out of the window was unreal. The garden was painted on the window, and every petal of flower and tuft of grass and leaf of tree was carefully sculptured in fine strokes of oil paint on the surface of the glass. It was not out there; it was right here in front of me. The woman who was watering the plants was completely frozen, immobilized by Vermeer. And when I looked again, she was in a different place, but again frozen. I was destined to become the eternal museum viewer. (Shulgin and Shulgin 1991, 518)

What is the function of a Vermeer—or, a frozen menagerie of woman, plants, and water—for this researcher? Vermeer works as an avatar, a visualization technology for rendering the psychedelic experience. Although psychedelic experience presents itself with great immediacy—"when I looked again"—it is nonetheless actualized on an itinerary—"she was in a different place." Here Vermeer—as an algorithm for making the dynamism of life still—helps render a vision by spatializing and visualizing an irreducibly dynamic sequence. That Vermeer located a "style" in which to do so was an enormous rhetorical achievement—to arrest life into a moment while attending to its luminosity is no minor innovation in the picture plane. Hence we can understand Vermeer's usefulness as a sample almost as much as the more ubiquitous treatment of the temporal unfoldings of psychedelic experience into spatial terms: the trope of a psychedelic experience as a "trip."

What does the "trip" trope do? The ineffability of ecodelic states seems to be linked to their capacity to smear our usual rhetorical categories: inside/outside, self/other, before/after. By contrast, an itinerary contains indubitable locales—first here, then there. By mapping the whorl of space-time characteristic of psychedelic experience, "trip" recuperates a psychonaut's capacity to articulate by compressing a thoroughly distributed experience into a serial one. Writer Philip K. Dick treated the psychedelic experience in more temporal terms, but the rhetorical effect is quite similar: a compression of an ineffable experience into a repeatable algorithm.

> To put it in Zen terms, under LSD you experience eternity for only a short period. . . . (If you'd prefer to undergo the experience of LSD without taking it, imagine sitting through *Ben Hur* twenty times without the midpoint intermission. Got it? Keep it.) (Dick 1996b, 177)

The effect of this rhetorical practice of overwhelming visual repetition is to summon a reader willing to enact it. As an engagement with what contemporary psychonaut Jim DeKorne calls the "imaginal realm," even the simulation of such a cinematic marathon erodes any ordinary serial experience of time—the beginning, middle, and end of the movie—and instead solicits waves of attention and confusion: *Didn't I just see that scene with the Timex anachronism a minute ago?* While Dick's imaginative rendering of an LSD experience is remarkably effective, it is by no means easy, as his query and challenge "Got it? Keep it." would seem to imply. Like any good kōan, Dick's thought experiment of sending the reader to a repetitious theater resists interpretation even as it summons our attention. Only by both carrying out the rhetorical recipe, "I am seeing the chariot again," and letting go of it, "Where was I?," can one satisfactorily render the magnitude of ecodelic experience.

Perhaps this is why psychologist Roland Fischer suggests that Gödelian (self-referential) paradoxes are unresolvable only in non-ecstatic states; artifacts of our own (illusory) distinctions between self and world, such paradoxes are for Fischer less qualities of the universe than feedback on our premises, such as our inclination toward success or failure.

> It is not generally recognized that Gödel's theorem refers only to the consistency of the normal state of daily routine associated with low levels of ergotropic arousal, or, in other words, to physical space-time with its constancies . . . and "Aristotelian" (yes-no, true-false) logic and language. The exalted states associated with the highest levels of arousal, as well as

the meditative states at the lowest arousal levels, are the vantage points of a different dimension, with its symbolic (multivalued) logic, from which the subject literally "looks down" upon the "object-ive" normal state. (Fischer and Landon, 1972, 166–67)[4]

When even the premise of a self/non-self distinction is relegated to a different level of analysis—one where the self persists, but as an immanent rather than transcendental feature of a system, the next step in a recursively nested holoarchy—Fischer suggests that the paradoxes disappear as they become resolved in a framework of a higher logical type (Bateson). Hence for Fischer the egoic self, as a way of rendering the world, also entails very particular limitations, such as the problem of self reference:

Our point is to emphasize that the limitations established by Gödel, Turing, Church, and Tarski are inherent in man's self referential nature and may refer only to the internal consistency of a particular state of consciousness. (Ibid., 167)

Fischer here raises the interesting possibility that different states of consciousness may present different sorts of paradoxes and, if so, the careful "tuning" of a mathematician's consciousness may harbor tremendous potential, a technological enhancement perhaps analogous to the effect of physicist Stephen Wolfram's "Mathematica" software on the practice of mathematics.[5]

Yet even if we agree with Fischer that the isolated and alienated self is only a particular (quotidian) form of consciousness with its own capacities and incapacities, it is nonetheless to quotidian consciousness that trip reports, such as Dick's, are addressed. In this sense they face a situation similar to the rendering algorithms of computer graphics, where the actualization of a more abstract "wire frame" schema of an image requires computationally and rhetorically intense practices that would fill out the abstraction, actualize the virtuality into an image. Trip reports face a no less difficult task of compression—transforming the immanent experience and abstracting it into the alienated context of the separated experience of self, separable from an ecosystemic enmeshment. *Trip reports are fundamentally rendering algorithms, clusters of recipes to be tried out, sampled, and remixed by psychonauts, a rhetorical treatment of distributed consciousness in serial terms.* Hence to read a trip report is both to bear witness to a particular episode and provide context for further iterations of research. Indeed, in this context even the active avoidance of trip reports becomes an interesting practice in "Diamond Mind." As such, trip reports such as those posted on The Vaults of Erowid compose a remarkable database for the programming of psychedelic research. They provide both pointers and contexts for the investigation of psychedelic plants and compounds while soliciting experiences capable of navigating the insights suggested by the prolix and often eloquent testimony. These experiences would include the imaginative attempt to "grok" immanence even from the perspective of an ego as well as various self experiments necessary to inhabiting this immanence, such as meditation, dancing, dieting, and the ingestion of ecodelics.

It may seem that the visions—as opposed to the description of set and setting or even affect and body load—described in the psychonautic tradition elude this pragmatic dynamic of the trip report. Heinrich Klüver, writing in the 1940s and Benny Shannon, writing in the early twenty-first century, both suggest that the forms of psychedelic vision (for mescaline and ayahuasca respectively) are orderly and consistent even while they are indescribable. Visions, then, would seem to be messages without a code (Barthes) whose very consistency suggested content.

Hence this general consensus on the "indescribableness" (Ellis) of psychedelic experience still yields its share of taxonomies as well as the often remarkable textual treatments of the "retinal circus" that has become emblematic of psychedelic experience. The geometric, fractal, and arabesque visuals of trip reports would seem to be little more than pale snapshots of the much sought after "eye candy" of visual psychedelics such as LSD, DMT, 2C-I, and mescaline. Yet as deeply participatory media technologies, psychedelics involve a learning curve capable of "going with" and accepting a diverse array of phantasms that challenge the beholder and her epistemology, ontology, and identity. Viewed with the requisite detachment, such visions can effect transformation in the observing self, as it finds itself nested within an imbricated hierarchy: egoic self observed by ecstatic Atman which apprehends itself as Brahman reverberating and recoiling back onto ego. Many contemporary investigators of DMT, for example, expect and often encounter what Terence McKenna described as the "machine elves," elfin entities seemingly tinkering with the ontological mechanics of an interdimension, so much so that the absence of such entities is itself now a frequent aspect of trip reportage and skeptics assemble to debunk elfin actuality (Kent 2004). And McKenna's own testimony resonates with Leary's earlier missives from DMT experience:

> Eyes close . . . back to dancing workshop . . . joy . . . incredible beauty . . . the wonder, wonder, wonder . . . thanks . . . thanks for the chance to see the dance . . . all hooked together . . . everything fits into the moist, pulsating pattern . . . a huge grey-white mountain cliff, moving, pocked by little caves and in each cave a band of radar-antennae, elf-like insects merrily working away, each cave the same, the grey-white wall endlessly parading by . . . infinity of life forms . . . merry erotic energy nets. (Leary 1966, 89)

The presence or absence of odd elfin entities is now an organizing practice used by many to orient an often overwhelming experience. Such is the alterity of experiences with a compound like DMT that the arrival of a familiar, no matter how "alien," can become a source of order with which to organize a narrative.

Hence I want to suggest that trip reports are fundamentally scripts, what I have called elsewhere rhetorical software: linguistic, visual, musical, and narrative sequences whose function resides less in their "meaning" than in their capacity to be repeated and *help generate patterns of response*. They are part of the psychonautic apparatus and not a supplement to it. They are compositions that suggest, but do not exhaust what one may very well *become* in contact with entheogens. Unknown and irreducible, the outcome of a psychonautic experiment is written not from the first person, but the Nth person—the paradoxically non-local aspect of subjectivity, the unknown identity and perspective that emerges when, with Smokeloc or Mitchell, the mescaline eater or astronaut experiences *istigkeit*, the fundamentally distributed nature of being: "*Tat Tvam Asi*," Fischer's non-ordinary state of consciousness that observes its entanglement with the system being observed in a more than cognitive way.

The replication of this experience of entanglement depends not only on plausibility but on effectiveness—what ways of describing psychedelic experience render it persuasive and endurable? What texts, images, and sound make readers want to carefully and respectfully experiment with the plant or compound? Perhaps most important, which ones make it possible to experiment with them again? The perspective of molecular evolution poses a question to which these scripts are, in part, an answer: Which compounds shall be synthesized by a global network of plants, humans, and labs, which shall remain unknown or even unmade? It is this capacity to repeat research that is most salient. Indeed, in this context, any communication "about" ecodelic experience is likely to induce the very distinction between self and other, one experience and another, which these experiences suggest is illusory and difficult to (intentionally) avoid. Continuing with the psychonautic tradition around mescaline, I want to map more specifically how mescaline's ecodelic effects might be made contagious even while they refuse representation.

MAKING SENSE OF MESCALINITY

The contrast between the writings of Henri Michaux and Aldous Huxley on mescaline is immense—*Miserable Miracle* and *The Doors of Perception* describe a diabolical and sacramental experience respectively; Huxley encounters the Dharma body of immanence and Michaux its hideous robotic simulacrum. And yet both implicitly employ their writings as algorithms, less explanations of the ineffable experience of mescaline than protocols for its dissemination. Because Huxley's work played such a large role in the dissemination of knowledge about psychedelic states in scientific, literary, and lay contexts of the twentieth century, I will focus on his famous trip reports as a way of understanding what is distinctive to this quasi-scientific genre. But it is nonetheless Michaux, in fact, who best teaches us how to read Huxley against himself—against a disappointment with language's capacity to report on psychedelic experience and toward language as a recursive symptom of psychedelic experience, an opening to the psychedelic through language:

> As if there were an opening, an opening which would be an assembling,
> which would be a world, which would be that something might happen,
> that many things might happen, that there is a crowd, a swarming of what is
> possible, that all the possibilities are seized with pricklings, that the person I

vaguely hear walking outside might ring the bell, might enter, might set the
place on fire, might climb up to the roof, might throw himself howling onto
the pavement of the courtyard. Might everything, anything, without choosing, without any one of these actions having precedence over another. I am
not particularly disturbed by it either. "Might" is what counts, this prodi-
gious urgency of possibilities, which have become incalculable and continue
to multiply. (Michaux 2001, 9–10)

Michaux writes in and about a style that would above all transmit psychedelic experience rather than *report* upon it. With other psychonautic
practices, this style is dependent on an experimental risk to the psyche such
that it might be manifested and transformed. Michaux's text resounds with
repetition. In his transmission, Michaux connects drawing and writing in
terms of their common debt to vibration, thousands of tiny tropisms such
as the incessant and rhythmic "might" which resounds "without choosing"
and yet in pure freedom: anything "might" happen with the essentially programmable (Leary) ecodelics. This vibration with the future and its sudden
variation registers itself into the very itinerary of Michaux's transcribing
hand and, indeed, his life:

> Those who can read handwriting will learn more than from any descrip-
> tion. As for the drawings . . . they were done with a vibratory motion that
> continues in you for days and days though automatic and blind, reproduces
> exactly the visions to which you have been subjected, passes through them
> again. (Ibid., 5)

Michaux's writings host us; they suggest we become affected by the very
rhythms and cadences of his writing and drawing, mad zig-zagging lines and
refrains acting in desperate counterpoint to and contagion with the mescaline emergency. "More tangible than legible," Michaux suggests that it is
as veritable seismographs, transducers of his miserable mescaline miracle
that we are to read his texts and drawings. Hence when looking closely at
Huxley's *The Doors of Perception*, I want to suggest that to read trip reports for
what they can teach us about psychedelic experience; we must read them as
if they are less failed signs of the ineffable than symptoms of, and subsequent
frames for, psychedelic states.

Michaux's Seismograph

Huxley, no doubt, abounds in symptoms. Along with earlier writers on mescaline such as Havelock Ellis, S. Weir Mitchell, and Louis Lewin, Huxley faced the uneasy task of rendering iridescence into ink, disciplining visuals as dynamic as they were beautiful and horrifying into a sequence of type. After a decoction of three peyote buttons, Ellis wrote:

> The visions never resembled familiar objects; they were extremely definite, but yet always novel; they were constantly approaching, and yet constantly eluding, the semblance of known things. I would see thick, glorious fields of jewels, solitary or clustered, sometimes brilliant and sparkling, sometimes with a dull rich glow. Then they would spring up into flower-like shapes beneath my gaze, and then seem to turn into gorgeous butterfly forms or endless folds of glistening, iridescent, fibrous wings of wonderful insects, while sometimes I seemed to be gazing into a vast hollow revolving vessel, on whose polished concave mother-of-pearl surface the hues were swiftly

changing. I was surprised, not only by the enormous profusion of the imagery presented to my gaze, but still more by its variety. (132)

Ellis's vision of a mineral, transhuman landscape of becoming—"always novel; they were constantly approaching, and yet constantly eluding, the semblance of known things"—echoes Philadelphia poet and physician S. Weir Mitchell's earlier observations and anticipates Huxley's strategy for orienting the to and fro of a mescaline trip into rhythms capable of holding the attention of a reader and, perhaps, a psychonaut, facing Michaux's mescaline multitudes of "might": jewels, folds, and flowers all serve to both capture and transmit the iridescent fluctuation reported as well as induced by Michaux—they are rhetorical attractor states for orienting readerly attention. Even Klüver, searching for the stability of taxonomy with which to articulate mescaline's symptoms, confirms this fluctuation:

The author's analysis of the hallucinatory phenomena appearing chiefly during the first stages of mescaline intoxication yielded the following form constants (a) grating, lattice, fretwork filigree, honeycomb or chessboard; (b) cobweb; (c) tunnel, funnel alley, cone, or vessel; (d) spiral. *Many phenomena are, on close examination, nothing but modifications and transformations of these basic forms.* (Klüver 1966, 42–43)

While taxonomy—division according to kinds—is usually associated with categorical reasoning,[6] Klüver's division of mescaline forms is itself dizzying, resembling nothing so much as Borges's sampling of a reference to a (possibly apocryphal) Chinese encyclopedia, a sheaf of categories whose borders are at best ad hoc:

On those remote pages it is written that animals are divided into: (a) those that belong to the Emperor, (b) embalmed ones, (c) those that are trained, (d) suckling pigs, (e) mermaids, (f) fabulous ones, (g) stray dogs, (h) those that are included in this classification, (i) those that tremble as if they were mad, (j) innumerable ones, (k) those drawn with a very fine camel's-hair brush, (l) others, (m) those that have just broken a flower vase, (n) those that resemble flies from a distance. (103)

Indeed, Klüver seeks something like a series of categories out of Michaux's "miracle," however miserable, of mescaline. But the distinction between these dynamic forms is so thin that one of Klüver's informants might be describing its composition:

> At times the boundaries are represented by lines so thin that it may be impossible to say whether they are black or white. Many observers have stressed the fineness of the lines. . . . As Moller has pointed out, the "absolute one dimensional" appears to have become a reality. (43)

If we are reminded of Edward Abbot's eloquent *Flatland* when the "absolute one dimensional has become a reality," we should note that this reality seems to have induced a rhetorical crisis in researchers—"it may be impossible to *say* whether they are black or white" (emphasis mine). In search of a grid with which to organize the highly participatory content of mescaline experience, Klüver and other mescaline researchers find it impossible even to confirm the allegedly more fundamental category of black or white. As Roland Fischer has suggested, perhaps binary languages simply do not map onto the perceptual frameworks enabled by ecodelics, to say nothing of their capacities for articulating "ordinary" states of mind.

So perhaps we ought not be surprised that it took a literary author to organize mescaline visions not in terms of logical categories but with biological and literary forms, ordering images with sufficient consistency to provide a kind of screen or filter for mescaline experience while remaining seductive enough to hold the reader's, the psychonaut's, attention. For this is no doubt part of the problem with the more taxonomic refrains of Klüver—they do not hold the attention even of the researcher, so much so that the formula seems to be insufficient to communicating even the visual data of psychedelic experience:

> The visual effects, as well as the conditions under which they have been attained, are often so incompletely described that even available data cannot be interpreted. (Ibid., 45)

And the closer researchers looked, the more they found that the visual domain was itself insufficient to any transcript of the psychedelic experience:

Any general theory, however, will have to go beyond a consideration of visual mechanisms per se. The mescaline-produced phenomena demonstrate this point in a striking manner. Mescaline induces changes not only in the visual field but also in other sensory spheres, particularly in the somatosensory sphere. "Haptic hallucinations" and other somatosensory phenomena may dominate the symptomology to the exclusion of phenomena in the visual sphere. (Ibid., 46)

In this context, Huxley's rhetorical symptoms—what I will map out as his recourse to the gem, the fold, and above all the flower—help make *sensory* the articulation of ecodelic experience, since they all trope and provoke an essential *imbrication* at play in the literature on mescaline, and they help manage the immanence of the mescaline experience characterized by early researchers as an "osmosis of reality and the imagination,"[7] an overlay notoriously difficult to differentiate and full of supple and subtle implications. With Huxley himself, we will begin with flowers. What is it about flowers that allow them to so thoroughly and, even as we shall see, *too easily*, render imbrication, even in texts?

Even within botany itself, the category of flowering plants (angiosperms) could be viewed as a foothold with which to orient the researcher attempting to navigate the sheer multiplicity and plurality of life-forms in "the plant world."

The plant world is so enormous that not even all its members have been classified, with estimates ranging up to 800,000 for the total number of species in the floras of the two hemispheres. (Furst, 33)

Anthropologist Peter Furst goes on to note that the angiosperm is a crucial taxonomical distinction in the classification of psychedelics:

The hallucinogens [*sic*] among them are concentrated mainly in two families: (1) The fungi—from which the Clavicepts, the ergot parasite of rye and other Old World grasses, to the sacred mushrooms of Mesoamerican Indians and the spectactularly beautiful Amanita Muscaria, or fly-agaric, of Eurasian shamanism. (2) The angiosperms, that vast family of plants whose seeds are enclosed in an ovary. In contrast, the gymnosperms . . . ferns;

lichens; algae; bacteria; and bryophytes . . . all seem to be lacking in psyche-delically active members. (Ibid., 33)

Hardly devoid of form, angiosperms could be said to visualize taxonomy itself. Flowers, of course, exist fundamentally as rhetorical devices; they capture the attention of a pollinator and enable reproduction, transpecies aphrodisia of scent, vision, and even sound that enable the fertilization of flowering plants. Practices of aphrodisia abound in nature, as in this exam-ple from Olivia Judson's *Dr. Tatiana's Sex Advice to All Creation*, a book that synthesizes recent research in evolutionary biology by remixing it into the genre of the sex advice column:

> Take red deer. During the breeding season, stags spend most of their time roaring: A stag's roar is a long, low rumble; each roar is a single exhalation of breath. . . . It's worth it, though. For females, roars are an aphrodisiac: females exposed to vigorous roaring come into heat sooner than females who are not. (135)

Flowers have themselves long been a seductive site for the description of rhetoric itself, a way of capturing the myriad forms of rhetorical practice and organizing them into a veritable botany. Forms of rhetoric worthy of culti-vation and productive of eloquence have been known as "flowers" at least since the thirteenth century, and it is easy to see the force of the compari-son. Linguist Jack Goody, in *The Culture of Flowers*, notes that floral patterns offered the possibility of a language in which one could communicate loves not representable in speech or writing:

> The Language of Flowers from the beginning was "secret" or else a "discov-ery"; it was a means of communicating clandestinely or at least semi-openly between lovers. (247)

Here Goody writes about the practice of sending and arranging flowers for the purpose of relaying an unspeakable and unwritable message, itself something of an aphrodisiac, as it both offers and withholds communication. What's striking in Huxley's discussion of mescaline experience in *The Doors of Perception* is his dependence on the figure of the flower as both the physical and rhetorical means of organizing his difficult-to-describe trip and its well-distributed report, a trope and management technique that would find itself into many a trip manual. A vase with three flowers becomes the very background against which the difference of mescaline is observed by Huxley:

> I took my pill at eleven. An hour and a half later, I was sitting in my study, looking intently at a small glass vase. The vase contained only three flowers—a full-blown Belie of Portugal rose, shell pink with a hint at every petal's base of a hotter, flamier hue; a large magenta and cream-colored carnation; and, pale purple at the end of its broken stalk, the bold heraldic blossom of an iris. Fortuitous and provisional, the little nosegay broke all the rules of traditional good taste. At breakfast that morning I had been struck by the lively dissonance of its colors. But that was no longer the point. I was not looking now at an unusual flower arrangement. I was seeing what Adam had seen on the morning of his creation—the miracle, moment by moment, of naked existence. (A. Huxley 2004, 16)

Note that the flowers serve as Huxley's visualization technology for mescaline—its difference is the sign of the difference mescaline has made to his perception. What had appeared compelling in the service of a vital discord had now become the sign of an overwhelming order: "the miracle, moment by moment, of naked existence."

How is the miracle seen and endured? Taken in a little at a time, moment to moment, the miraculous unfolding is nonetheless discrete, as one moment becomes another. Hence Huxley cannot be said to be regarding a form per se, static and devoid of content. Its content is hence not simply existence but life itself, a visualization of the dynamic and not still life of rose, carnation, and iris:

> I continued to look at the flowers, and in their living light I seemed to detect the qualitative equivalent of breathing but of a breathing without returns to a starting point, with no recurrent ebbs but only a repeated flow from beauty to heightened beauty, from deeper to ever deeper meaning. (Ibid., 18)

This continuation, then, is a repetition, a steadfast observation through the moment to moment transition of a truly inspired plant. Yet this moment to moment transformation, while not devoid of rhythm, cannot be said to obey a narrative logic; this "breathing" of the plant and its qualities obeys a topology not of beginning and end, inside and outside, but one of imbrication, with beauty layered upon beauty without beginning or end, *ad infinitum*, a layering that transforms "naked existence," and perhaps naked life, into a non sequitur. Yet just as the human perceptual matrix ably transforms any three sufficiently proximate points into a face, so do we (egoically) greet the arrival of multiply layered imbrication with the quest for causality and origin.[8] Perhaps it is not surprising, then, that Huxley seems to understand this breathing without beginning or end in a religious fashion:

> Words like "grace" and "transfiguration" came to my mind, and this, of
> course, was what, among other things, they stood for. (Ibid.)

Yet note that Huxley at once both ascribes agency to language and undoes its univocality; "coming" to his mind, "grace" and "configuration" are indeed part of what the flowers "stood" for, "among other things." But these words are quickly supplanted not by the agency of the words but by the flowers, which send his eyes on a journey. Having been "struck" by the flowers that morning, the cluster of blossoms now send his eyes on a trip:

> My eyes traveled from the rose to the carnation, and from that feathery
> incandescence to the smooth scrolls of sentient amethyst which were the
> iris. The Beatific Vision, *Sat Chit Ananda*, Being-Awareness-Bliss for the first
> time I understood, not on the verbal level, not by inchoate hints or at a dis-
> tance, but precisely and completely what those prodigious syllables referred
> to. (Ibid.)

No "I" narrates this travel; Huxley's eyes are pulled by the strange attractions of plumed light and petrified texts. A "feathery incandescence" gives way to the smooth, sentient, and textual in form. If this be enlightenment, it is an enlightenment whose vector is synesthesia. The mixture is multiple; tactile light gives way to alliteration, "smooth scrolls of sentient amethyst" whose crystal destination brings together even that classically distinct taxonomy, mineral/vegetable. This synesthetic mixture renders, for Huxley, a knowledge of ontology—an "is-ness" explicitly devoted to becoming:

> *Istigkeit*—wasn't that the word Meister Eckhart liked to use? "Is-ness." The
> Being of Platonic philosophy except that Plato seems to have made the
> enormous, the grotesque mistake of separating Being from becoming and
> identifying it with the mathematical abstraction of the Idea. (Ibid., 17)

Hence the "confusion" of sensation marked by a "feathery incandescence"
is not merely scrambling the taxonomy of perception—the visual, haptic
realm of the Peacock Angel. More than sensory modalities are mixed in this
istigkeit; all that is becomes connected even in its difference, as being and
becoming share the same confused territory of "feathery incandescence";
ontology itself becomes not only dynamic but synesthetic.

In re-mixing the distinction between being and becoming and highlighting their dense and synesthetic interconnection, Huxley must nonetheless
meet the difficult burden of visualizing such a becoming for readers in what
philosopher Martin Heidegger called "The Age of the World Picture"—an
epoch that incessantly and precisely separates being from becoming, often
visually. Subjective reality is evacuated of any value precisely at the moment
of increased human control over the "external" world; the condition of the
radically knowable world in "The Age of the World Picture" is that the self
itself becomes unknown. Nietzsche writes in the *Genealogy of Morals*, "We
knowers are unknown to ourselves. . ." How does Huxley render such an
interconnected ontology when objectivity is the idol, "the world appears as
picture," and the very existence of subjectivity is in doubt? Filled with compassion for that "poor fellow" Plato, Huxley's visual aid is again the flower:

> He [Plato] could never, poor fellow, have seen a bunch of flowers shining
> with their own inner light and all but quivering under the pressure of the
> significance with which they were charged; could never have perceived that
> what rose and iris and carnation so intensely signified was nothing more,
> and nothing less, than what they were—a transience that was yet eternal
> life, a perpetual perishing that was at the same time pure Being, a bundle
> of minute, unique particulars in which, by some unspeakable and yet self-
> evident paradox, was to be seen the divine source of all existence. (Ibid.)

What does Huxley, with the help of the flower, see here in this "is-ness"?
Although by convention such experiences are frequently characterized as

"visions," even gnostic ones, Huxley's regard here is a peculiar admixture of intensity and signification. "Nothing more, and nothing less, than what they were," the flowers less communicate than become in an alliterative, and hence tropic, "perpetual perishing." Less about signification than intensity and repetition (of sounds), Huxley's treatment of the flower does not report but solicits into imbrication, the nested semiotic "bundle" through which the osmosis of being and becoming is shared.

While Huxley's appeal to the paradox of a becoming that is nonetheless "pure being" could be seen as a failure of language—the inability of rhetorical forms to render the seemingly contradictory dynamics of mescalinity—or, worse, a scholastic retreat into semantics, I would instead argue that paradox should be encountered as a pedagogical tool, a practice for engaging the interface between language and reality, a fundamentally non-local, interconnected, and, hence, synesthetic cosmos revealed by quantum physics, twentieth-century biology, and mescaline.[9] As in Zen tradition, where the paradoxical, rhetorical form of the kōan is deployed to enable detachment from the repetitous grasping of a persistent ego on the becoming of samsara, Huxley's synesthesia and paradox demand that the attentive reader—and, perhaps, psychonaut—suspend their usual two valued (binary) logic and engage the actuality of reality, "all that is" including their imbrication with it as an observer.

While synesthesia is classically treated as a transfer or confusion of distinct perceptions, as in the tactile and gustatory conjunction of "sharp cheese," more recent work in neurobiology by V. S. Ramachandran and others suggests that this mixture is fundamental to language itself—the move from the perceptual to the signifying, in this view, is itself essentially synesthetic. Rather than an odd symptom of a sub-population, then, synesthesia becomes fundamental to any act of perception or communication, an attribute of realistic perception rather than a pathological deviation from it.

And yet how is this multiplicity of sensory modalities navigated? Ordinary consciousness, to the extent that it is paying attention at all, confronted with the paradox of a "perpetual perishing," oscillates between a thought of the infinite and the finite, a fluctuation that can find no resolution in either "perpetual" or "perishing," and is linked by the very sonic repetition that stitches this trope together. Huxley, or at least his sensorium, is challenged to somehow navigate this fluctuation through another layer of abstraction that would integrate the paradox, and he perhaps does so by dwelling in a rhetorical space beyond meaning but saturated with attention-orienting rhythm.[10] Huxley's trippy tune now has a refrain—the flower, whose effect is to snare the attention of Huxley, if only for a moment, from himself, orienting him to what he calls his "blessed Not-I," that One that is at once perpetual and perishing, the first without a second:

> Of course the Dharma-Body of the Buddha was the hedge at the bottom of the garden. At the same time, and no less obviously, it was these flowers, it was anything that I or rather the blessed Not-I, released for a moment from my throttling embrace, cared to look at. (A. Huxley 2004, 19)

Such revelations of Brahman can indeed be *anything*, provided that the suffocating "embrace" of the I is relaxed or "released" and care is present. The flowers, as snares of perception, allow access to an imbrication less obvious, but no less portentous, than those other rhetorical devices, Huxley's next example and analogy—books:

> The books, for example, with which my study walls were lined. Like the flowers, they glowed, when I looked at them, with brighter colors, a profounder significance. Red books, like rubies; emerald books; books bound in white jade; books of agate; of aquamarine; of yellow topaz; lapis lazuli books whose color was so intense, so intrinsically meaningful, that they seemed to be on the point of leaving the shelves to thrust themselves more insistently on my attention. (Ibid.)

While the flowers serve as refrains, orienting devices, for these revelations of imbrication and immanence constantly competing for attention, the agencies of the books are no less insistent that we regard their jeweled surface on the brink of incipient action: "they seemed to be on the point of leaving the shelves to thrust themselves more insistently on my attention." The attentive reader will not view this page itself as immune from this demand for attention, recursively implanting the reader in the osmosis of being and becoming, as she reads about a book about a book on the brink of action. . . . Note that the flower—with its multiple and layered invocation, a figure-ground dissolution of inside/outside that works as a snare for insect attraction—the (now self-referential) book, and the jewel share a vocation: organized, and thoroughly seductive, fluctuation.

The to and fro from the flowers elsewhere continues Huxley's itinerary; the refrain of the flower seems to provide Huxley with a paradoxical touchstone for a distributed immanence—self, not-self, everything is connected, especially the flowers, whose attraction collapses the very distance between self and not-self even as it ontologically dissolves into a chair:

> I was back where I had been when I was looking at the flowers—back in a world where everything shone with the Inner Light, and was infinite in its significance. The legs, for example, of that chair how miraculous their tubularity, how supernatural their polished smoothness! I spent several minutes or was it several centuries?—not merely gazing at those bamboo legs, but actually being them—or rather being myself in them; or, to be still more accurate (for "I" was not involved in the case, nor in a certain sense were "they") being my Not-self in the Not-self which was the chair. (Ibid., 22)

For Huxley, interacting with mescaline would seem to suspend observation itself, if observation depends upon a distinction between a subject and the object of fascination. Klüver samples a report from an earlier researcher, Kurt Beringer, whose research subject became enmeshed with a fretwork organizing both the self and the not-self:

> The subject stated that he saw fretwork before his eyes, that his arms, hands, and fingers turned into fretwork and that he became identical with the fretwork. There was no difference between the fretwork and himself, between inside and outside. All objects in the room and the walls changed into fretwork and thus became identical to him. While writing, the words turned into fretwork and there was, therefore an identity of fretwork and handwriting. "The fretwork is I." All ideas turned into a glass fretwork, which he saw, thought and felt. He also felt, saw, tasted and smelled tones that became fretwork. He himself was the tone. On the day following the experiment, there was Nissel (whom he had known in 1914) sitting somewhere in the air, and Nissel was fretwork. "I saw him I felt him Nissel was there. Nissel as I." (Klüver, 32)

In contact with a textual production, mescaline summons an interconnectivity which, in its management, fosters further connection. Words, while not devoid of effect, cannot achieve escape velocity on the (fretful) experience of interconnectivity: "While writing, the words turned into a fretwork . . . an identity of fretwork and writing." The striated space of the fretwork, organized into defined segments, becomes a proliferating web, a dynamic navigation of a distributed network that both confirms and denies identity. Confirmed, the subject became fretwork, became rendered into synesthetic tones, became a friend not seen for years. Denied, identity loses its monopoly on a body and enters into a fluctuation: "Nissel as I." It is perhaps crucial that it is only on the condition of becoming beside—oneself that this interconnection emerges. The synesthetic nexus—"He himself was the tone"—offers a habitat less for a self than for a fretful commons: "All ideas turned into a glass fretwork, which he saw, thought and felt." The orderly, even geometric patterns of the dynamically shifting fretwork are templates for an extra-ordinary multiplicity, as the body less breaks the "law" of non-contradiction—"A" is simultaneously rendered as "A" and "Not A"—than becomes increasingly response-able to a multitude.

"Nissel as I" recapitulates *Tat Tvam Asi*, an ecosystemic reality principle from the Upanishads that renders the separation of world and self into the optical and topological illusion that it is. This apprehension of non-locality is also congruent with the non-visualizable fact of non-locality modeled by numerous researchers since physicist John Bell's 1964 work, and as such cannot truly be described as a hallucination, neither the traditional "perception without an object" nor Fischer's more cybernetic "non verifiable information." On Fischer's hallucination-perception continuum, the inter-activity of the fretwork seemingly induced by writing on mescaline suggests that the forms sought by Klüver as "constants" are navigational aids for a non-local psyche, "Nissel as I."

Observation is the main threshold discussed by Fischer in his articulation of the continuum, where the self observed is treated as an imaginary construct observed by a more distributed self, one capable of treating the self as an empathetic other—"Plato, that poor fellow"—who sees itself as an empathic other, and so on. Yet observation ought not, perhaps, hold a privileged place on the hallucination-perception continuum. Instead, any algorithm that enables the inhabitation of an interconnection and reminds us of our consistency with the commons could be seen as a trope or "movement" along the hallucination-perception continuum that enables a perception of this actuality, just as the staining of a chromosome enables our visualization of heredity and its genetic mechanisms. Mandala, those nested and often geometric forms I shall return to below, offer the possibility of visualizing highly interconnected patterns of information, and have functioned as adjuncts to psychedelic experience, in one form or another, for thousands of years.

Yet it is not as a geometric or topological form that Huxley narrates his investigation, but a more intensely florid one, bursting with color:

> In this context, how significant is the enormous heightening, under mescaline, of the perception of color! For certain animals it is biologically very important to be able to distinguish certain hues. But beyond the limits of their utilitarian spectrum, most creatures are completely color blind. Bees, for example, spend most of their time "deflowering the fresh virgins of the spring"; but, as Von Frisch has shown, they can recognize only a very few colors. Man's highly developed color sense is a biological luxury inestimably precious to him as an intellectual and spiritual being, but unnecessary to his survival as an animal. To judge by the adjectives which Homer puts into their mouths, the heroes of the Trojan War hardly excelled the bees in their capacity to distinguish colors. In this respect, at least, mankind's advance has been prodigious. (A. Huxley 2004, 27)

A flower, sessile, summons an insect, tuning its broadcast toward the apian visual spectrum and vectoring bee to pollen. As co-evolved organisms, flowering plants and their pollinators emerge out of a distributed and imbricated agency: the bee, however selective its vision, is part of the reproductive system of the plant, and the flower is an aspect of a minded insect. Huxley here suggests that a quantitative understanding of color is appropriate—"they can recognize only a very few colors"—yet clearly an attentive understanding of bee ecology such as Karl Von Frisch's develops from a practiced understanding of flowers as broadcast mechanisms selected for their capacity for tuning—a quantitative difference that is qualitative. In other words, while Huxley's treatment of the exorbitance of color emphasizes its status as a dangerous supplement, an "extra" that puts survival at risk, his example indicates that the bee's visual spectrum is not excessive but precise, a siren song sung in bee-ing.

This precision, and the most efficient excess with which the flower presents itself, can remind us that as evolutionary information technologies, angiosperms work by producing the difference that makes a difference (Bateson), a qualitative tuning that relies not on the production of large *amounts* of information, but on the compressed differentiation of information. Across the background of the ecologies in which they evolve and grow, the production of scent, color, and visual imbrication stake out precise and effective domains of "bandwidth" within which the coevolution of bees and angiosperms has taken place. For, of course, it is on these channels that bees and flowers are involved in a distributed practice of plant sex, the movement and differentiation of plant alleles.

Plant sex is perhaps unusual for its object of seduction and not its methodology. Instead of beguiling or otherwise maneuvering a partner to exchange genetic information, flowers make themselves sensitive to the wind and resonant with the bee perceptual system. Charles Darwin wrote that the emergence of this beautiful co-implication was an "abominable mystery," and what was apparently abominable to Darwin is the transpecies nature of angiosperm reproduction. While Huxley treats the sense of color as a "biological luxury," it is anything but for the flower or the bee that rely on it for reproduction and food respectively.

Yet excessive color appears to be no luxury in Huxley's own account of mescaline. The innumerable colors broadcast by mescaline cluster, in Huxley's gaze and writing, into the flower. And as with the bees, flowers themselves are less objects of Huxley's gaze than zones of interconnection. Huxley, less pollinator than meditator, looks not simply at the flowers but between them:

> This was something I had seen before—seen that very morning, between
> the flowers and the furniture, when I looked down by chance, and went on
> passionately staring by choice, at my own crossed legs. Those folds in the
> trousers what a labyrinth of endlessly significant complexity! (Ibid., 30)

As with the flowers, the folds are the site of a return or a refrain, a repetition that brings the mescaline into relief. These repetitions are somewhere between action and reaction: Huxley's eyes, already on a trip, go on "passionately staring by choice." Choice it may have been, but Huxley's "passion" reminds us that it is not only "between" that Huxley gazes—he himself *is* between, perhaps between activity and passivity, being and becoming:

> And the texture of the gray flannel how rich, how deeply, mysteriously
> sumptuous! And here they were again, in Botticelli's picture. More even
> than the chair, though less perhaps than those wholly supernatural flowers,
> the folds of my gray flannel trousers were charged with "is-ness." To what
> they owed this privileged status, I cannot say. (Ibid., 33)

After this rather unusual bout of rhetorical incapacity, Huxley offers immediately to "say," a rhetorical analysis of the fold followed by a rhetorical question:

> Is it, perhaps, because the forms of folded drapery are so strange and dra-
> matic that they catch the eye and in this way force the miraculous fact of
> sheer existence upon the attention? Who knows? (Ibid.)

"Who" indeed knows? Huxley, released momentarily from the throttling embrace of the self, finds himself between flowers and folds. His analysis of the function of folds to "catch the eye" nicely foregrounds the dimensional aspect of the interaction—between the ingestion of mescaline and its departure, Huxley is irreducibly between one form of consciousness and another—something besides his self, his blessed non-self, seems to be at play, capturing his eye and his consciousness. While this division of labor or life might seem extraordinary, a veritable outsourcing of Huxley's role as narrator to plants (flowers and the peyote cactus), a look to the history of evolution reminds us that this incessant exteriority is the very essence of the natural: writing of the function of orchids to lure pollinators and the exogamy they enable, Darwin notes that "Nature tells us, in the most emphatic manner, that she abhors perpetual self-fertilisation" (1889, 293).

A twenty-first-century source discounts the fascination that is the very mechanism of the flower on humans:

> People become so obsessed with flowers it is important to remember a flower
> is nothing more than a cluster of spore-bearing leaves surrounded by whorls
> of protective and often albeit attractive leaves. (Carrington)

It is, of course, not human fascination toward which the (wild) flower is evolutionarily tuned. Instead, flowers are, as Darwin admirably put it, "well adapted for insect agency" (1896, 41). This transpecies exogamy appears to be anything but a biological luxury, and is instead an intensive necessity that renders the baroque pollen delivery system of the orchid as well as the whorls of the other angiosperms. Humans, though, share in this exogamy, and the evolutionary itinerary of the orchid is now as wedded to the human perceptual system as it is to insects' aspects of the human perceptual and even communicative system we share with bees.

Though bees arguably have language (their famous waggle dances discovered by Von Frisch), the effect of scent and form on their perceptual system requires less interpretation than attention. Huxley wisely and paradoxically avoids the interpretation of these states, preferring less the "reason for the experience" than the "experience itself":

> What is important is less the reason for the experience than the experience itself. Poring over Judith's skirts, there in the World's Biggest Drug Store, I knew that Botticelli and not Botticelli alone, but many others too—had looked at draperies with the same transfigured and transfiguring eyes as had been mine that morning. They had seen the *Istigkeit*, the Allness and Infinity of folded cloth and had done their best to render it in paint or stone. (A. Huxley 2004, 34)

Huxley's notion that mescaline afforded him access to truth of *"Istigkeit"* could be seen as a romantic identification with madness itself. Mescaline, after all, was scientifically understood to produce a scientific simulacrum of psychosis. Yet by 1964 researchers such as Lynn (Margulis) Sagan suggested that mescaline testified to another fold, that of human recursivity—their implication in a chemical ecology, what she called the "amenability of man's soul to his own researches."

> Mescaline is related to adrenaline, a known neurosecretatory hormone; and caffeine is a purine similar to the nitrogenous bases in DNA (the genetic material). If these facts do not, at best, point to physiological mechanisms, they at least attest to the knowability of consciousness, psychosis and mystical experience. The chemistry eloquently testifies to the amenability of man's soul to his own researches. (Sagan, 355)

What does Huxley see in the "Allness and Infinity of folded cloth"? He locates there a consistency, what he had earlier dubbed the "perennial philosophy"—the recursive interconnection between consciousness and reality. The flower and the fold render a state of interconnection—perhaps the fold folds our attention, bends us toward a state of non-separateness connecting at once the eye, an interior and an exterior, and offers a temporary injunction on alienation, an apprehension of the cosmos as anything but separate—who knows, even a becoming-bee. In resonating with our perceptual capacity to tune into our own ecosystems at a linguistic level distinct from interpretation, ecodelics tune us into a different segment of the hallucination-perception continuum, one that amplifies our interconnection rather than our distinction. This ecodelic insight—whatever its truth value—enables a recursive perspective on our usual egoic awareness and asks it to grapple with a universe of interconnection. While human perception tends to render any space "between" isolated objects and events as a void, mescaline (and other ecodelics) seems to call for dwelling affectively, if not cognitively, in a world where self and non-self fold into each other.

Sagan suggests that such a fluid but not liquefied identity may be a more scientifically accurate view of the world, a catalyst for breaking down illusory "defense mechanisms" and allowing for Huxley's *istigkeit*, the "world as it is."

> The drug attacks defense mechanisms built up carefully to conceal the truth of our direct sensory perceptions. One would *a priori* imagine, however, that a drug which forced us to see the world as it is would be welcomed. (Ibid., 354)

Among the "direct sensory impressions" available to humans is the impression of subjectivity—the world presents itself to us as one of individual "duration," a continuous stream of events whose integration we attempt and, at least egoically, are. Since at least the scientific "revolution," such an analog vision of the world has withered in favor of a digital vision composed of discrete, and hence reducible and repeatable, parts. "Direct sensory perceptions" seem to entail, for Sagan, a continuity between subjectivity and world, *Tat Tvam Asi*.

And yet Sagan writes precisely as a twentieth-century scientist when she suggests that LSD (and mescaline) is a visualization technology not of fantasia but of the real, and that this visualization technology amplifies "man's amenability to his own researches," highlighting the vertiginous fact that human being and becoming is caught up in a continuous recursive spiral with itself, whether reductionism likes it or not. . . . This vertigo is perhaps best navigated by the disciplined "splitting" and ongoing management of attention—the nested response-ability to and for rhetorical conditions even while one observes and plays with them—ecodelic programming. If ecodelics are visualization technologies for the apprehension of interconnection, the implication of human futures in the knowledge and tools with which they experiment, then set and setting provide a gradient for synthesizing this interconnection, the provocation and recombination that brings together, if only for a moment, self and Self.

THE DOORS OF SCIENTIFIC PERCEPTION:
A MOMENT BETWEEN HUXLEY AND HIMSELF

Drawing on contemporary research with the plant *Salvia divinorum* and meditation, even a drug warrior might grudgingly grant the capacity of entheogens to cultivate new, even sacred forms of reflection and make available the imbrication of a self with its ecosystem. Indeed, the U.S. Supreme Court, as one of the institutions that has presided over and enabled the War on Drugs, implicitly affirmed this claim in its 2006 ruling in defense of the Religious Freedom Restoration Act.[11] But from what perspectives might the visualization technology of ecodelics render a more *accurate* scientific vision of the world? As essentially plural technologies of visualization, ecodelics fold and distribute the process of visualization, one that brings the scientific gaze to a dissipative structure of vision *beyond perspective*, that "modest witness" who merely observes but does not interact with scientific truth production. Psychedelics, for Sagan and others, represented a scientific enhancement of human perception akin to the electron microscope, and the early data gathered by this new mode of scientific observation suggested that the separation between the self and the cosmos was an illusion, an artifact of egoic consciousness correctable, reflectable, by a continuously tuning and hence analog consciousness. Swiss chemist Albert Hofmann notes that LSD very specifically tunes the mind toward an awareness of itself as a weave of "cross-connections." Anticipating the argument that psychedelic effects were purely "psychopathological," Hofmann asks us to visualize the nervous system itself as a bundle of cross connections.

> If LSD acted only through a toxic effect on the brain, then LSD experiences would be entirely psychopathological in meaning, without any psychological or psychiatric interest. On the contrary, it is likely that alterations of nerve conductivity and influence on the activity of nerve connections (synapses), which have been experimentally demonstrated, play an important role. This could mean that an influence is being exerted on the extremely complex system of cross-connections and synapses between the many billions of brain cells, the system on which the higher psychic and intellectual functions depend. (Hofmann, 45)

More recently, Ralph Metzner includes the recursively situated cultures from which ayahuasca emerges and is transmitted as an aspect of an entire assemblage of visualization, or its lack thereof:

> Those who are ideologically committed to the still-prevailing Newtonian-Cartesian paradigm will at best consider the statements and descriptions of the ayahuasceros as drug-induced "hallucinations," incapable of being scientifically evaluated or verified. From the perspective of a Jamesian radical empiricism however, the phenomenological descriptions of consciousness explorers must be accorded the same reality status as observations through a microscope or a telescope. (Metzner 1991, 81)

In other words, we owe it to these reporters *to be skeptical* not toward the existence of the data, but toward its best interpretation. Skepticism toward accounts of indigenous, mestizo, and post-modern visionary experience, therefore, must itself be understood through the multiple refractions of a contemporary health state that suddenly discovers "evidence based medicine" while ignoring the evidence that other cultures have indeed coevolved with other medicines, and that those medicines appear efficacious in the context of their ecology of use, a context in which they image an ontology that is essentially layered and with which they are actively interconnected: a nested and quite tangled hierarchy or network. The entire discipline of ecology, of course, confirms this observation, as do the physics of non-locality perhaps grokked by Mitchell in his Apollo capsule. As an imaging technology for interconnection, psychedelics are indeed unusual among visualization technologies—they enjoy interdisciplinary support from multiple locales and are not often easily doubted even as their images are essentially open to multiple and even contradictory interpretations.

If programmed and tuned ecodelics induce a capacity to perceive not things but their meshed interconnection, these perceptual adjuncts may indeed prove as useful and compelling to technoscience as optical prostheses such as microscopes and telescopes, or prostheses of interference and interaction such as the electron microscope and the Scanning Tunneling Microscope respectively. Psychologist Stanislav Grof recalls that early researchers thought of LSD as a tool analogous to other visualization devices in science:

> In view of these observations, it did not seem far-fetched to see LSD as a tool comparable to a microscope or telescope. Like these devices, LSD made it possible to observe and study processes that were normally not part of our everyday experience. (125)

Recall, with philosopher of science Paul Feyerabend, that observers greeted the arrival of telescopes with anything but consensus. While telescopic observation as practiced by Galileo would seem to offer nearly demonstrative proof of the location and movements of the planets and their moons, his contemporaries appear to have been anything but convinced. In terrestrial use, the telescope, guided by a feedback loop with naked-eye observation, was duly trusted, but once the lenses were turned to the cosmos, a rhetorical crisis ensued:

> Trouble promptly arose: the telescope produced spurious and contradictory phenomena and some of its results could be refuted by a simply look with the unaided eye. Only a new theory of telescopic vision could bring order into the chaos. . . . Such a theory was developed by Kepler, first in 1604 and again in 1611. (Feyerabend, 99)

Feyerabend cites Geymonat arguing that it was only by being Corpernican that Galileo was able to view the telescopic observations as reliable.

Galileo had believed for years in the truth of Copernican theory. In Galileo's own mind faith in the reliability of the telescope and recognition of its importance were not two separate acts, rather, they were two aspects of the same process. (Ibid., 104)

This reversal of the role of observation—here observation is guided by, meshed with, premises, rather than the other way around—nicely foregrounds the role of a highly interactive observer, and it reminds us that some technoscientific observations require a new form of subjectivity—a de-centered one at that—to make the best use of a new instrument.

Our technoscientific investigations constantly browse through ideas that are far from perceptual "equilibrium," and they seem to yield insights that are not psychologically understandable by a contemporary subject formation—the way of being of a self enabled and or permitted by the infrastructure and institutions of any given historical moment. Galileo first imagines a heliocentric cosmos, dethroning human experience from its central role in nature, and then invests the telescope with the capacity to visualize a Copernican cosmos, despite the many difficulties associated with early telescopic perception.

Even concepts can require forms of subjectivity capable of affirming them before they can be validated. Evolutionary insights, for example, appear to provoke cognitive dissonance in subjects attached to notions of a static human identity. This has led evolutionary psychologist Geoffrey Miller to propose a new requirement for evolutionary theory—that it satisfy the human "hunger for self-explanation":

A theory that can't give a satisfying account of your own mind, and the minds you've loved, will never be accepted as providing a scientific account of the other six billion minds on this planet. (29)

Miller's call for a more compelling account of the evolution of mind leads him to wonder if Darwin's "Really Dangerous Idea," sexual selection, might prove more effective as a framework for Darwinian descriptions of the earth. The spread of video and information technology across the planet was driven by the very content of some very particular post-modern channels of sexual selection—namely, porn, chat, and dating—so origin stories drawing on sexual selection itself may be even better than accurate; they provoke trans-formations of the monoid self capable of grappling with a scientific vision that demands an essentially pluralized vision comfortable with fluctuation. Such organized fluctuation is helpful in rendering a more satisfying—what Miller calls a more Dionysian—vision of human intelligence and mind, one that recognizes and works with rather than against the meshed, ecosystemic nature of life and intelligence. In the next chapter, I will suggest that indeed sexual selection provides the most compelling explanation for the presence of ecodelics in the human toolkit, if only because such a theory provides useful programs for the management of ecodelic experiences themselves, encounters that can distribute the experience of a relatively encapsulated body across a field of information—ecstasy.

But it is not only a demographic skeptical of Neo-Darwinian accounts of evolution that appears attached to things and is, well, "Leary" of their inter-connection. Physicist Amit Goswami, for example, argues for an immanent interpretation of quantum mechanics, the notoriously difficult to under-stand description of the universe and its relationship to our observation of it. Here consciousness is more than a reporter on the world but an active vector in its unfolding, as crucial to our world as matter "itself." Building on the work of physicists John Von Neumann, Eugene Wigner, and Henry Stapp, Goswami claims that it is not merely measurement but consciousness that initiates the famous "collapse" of the wave function, wherein a multiplicity of itineraries and momenta of a physical system are replaced with the par-ticularity of a particle.

> Whenever we measure it, a quantum object appears at some single place, as
> a particle. . . . When we are not measuring it, the quantum object spreads
> and exists in more than one place at the same time, in the same way that a
> wave or cloud does—no less than that. (Goswami, 42)

Here consciousness becomes an "independent, causal entity" actualizing a single particle (at a single location) out of the virtual multiplicity of poten-tia: "All of a sudden the state of superposition, the multifaceted state that exists in potentia, is reduced to just one actualized facet" (ibid., 174).

Thinking this co-implication between observation and the cosmos has proved a difficult endeavor, suitable more to thought experiments than deductive reasoning. Erwin Schrödinger, who authored the wave function that bears his name as well as *What Is Life?*, an urtext of contemporary life science, famously offered his thought experiment wherein a cat is both simultaneously living and dead, a Zen-like paradox that has delighted and befuddled physicists as well as lay readers of pop physics texts since 1935. The difficulty of thinking clearly about states of mind and their apparent role in physical systems emerges at least in part from the necessary incompleteness of observation itself by any particular (interconnected) self. Psychologist Roland Fischer sums up this difficulty in a most compressed fashion, one that makes use of an example from a subset of the history of rhetoric, "literary tradition." Writing of the displacement between the referential character of a sentence and its putative subject, Fischer writes:

> It is only a comparison, following good literary tradition ("Shall I compare you to a summer's day?") and attempts to highlight the intricate problem of thought formation by self referential man. Indeed it is nothing more than a thought formed about how a thought is formed, is formed, is formed. (1994, 167)

Fischer's example—the rhetorical trope of praeritio—reminds us that this "problem" of awareness—its inability to narrate its own conditions of emergence—can be accessed only in a participatory and even experimental fashion. While mescaline may highlight the radically interconnected cosmos in which Huxley finds himself, he can narrate it only provisionally and heuristically—his tropes are experiments in re-calling and even experiencing an interconnected state as *a solo consciousness* radically divorced from the past—the one who knows the distinction on mescaline, not on mescaline. It is in the rendering of this word "I," perhaps, that Huxley's novelistic capacities become most crucial to *The Doors of Perception*.

This difficulty of observing the conditions of observation, of course, leads to a further difficulty of observing the conditions of observing the observation, and so on into an infinite regress of observation, until observation forms the entirety of both the subject and object of observation and all other objects disappear from consciousness and only a mandala or the supple folds of a skirt can orient the attention. This problem might be usefully characterized as the problem of rhetorical superposition, a situation wherein meaning itself becomes virtual, distributed over the entirety of an ecosystem and hence locatable only through a query of an entire system. Schrödinger lays out his original thought experiment on superposition as follows:

> One can even set up quite ridiculous cases. A cat is penned up in a steel chamber, along with the following diabolical device (which must be secured against direct interference by the cat): in a Geiger counter there is a tiny bit of radioactive substance, so small that perhaps in the course of one hour one of the atoms decays, but also, with equal probability, perhaps none; if it happens, the counter tube discharges and through a relay releases a hammer which shatters a small flask of hydrocyanic acid. If one has left this entire system to itself for an hour, one would say that the cat still lives if meanwhile no atom has decayed. The first atomic decay would have poisoned it. The Psi function for the entire system would express this by having in it the living and the dead cat (pardon the expression) mixed or smeared out in equal parts. (1980, 152)

In other words, from the perspective of the Psi function, the cat is both alive and dead, a superposition of corpse and cat. Readers since 1935 have treated this conundrum as the very icon of the unthinkability of the quantum. Both alive and dead, the cat differentially calls on human consciousness to respond to it. The Copenhagen Interpretation, against which Goswami argues, would suggest that the alive cat and the dead cat are mutually exclusive descriptions inseparable from the phenomenon itself. Models that articulate the interaction not only as a description but as an informational system are more general, characterizing the description itself as one state in a flow of (purposeful) bits. This informational model can yield something like a "queryable universe," where information remains identical to itself even as it is differentially actualized through different sorts of queries.

This informational theory of the universe, though, can easily forget the "it" in "bit": no doubt part of the urgency of Schrödinger's cat example for thinking the logic of the quantum—a logic itself "smeared out in equal parts"—is the call of a familiar, a cat whose very life is at stake.[12] So too might this Cheshire Cat's uncertain vitality render any recourse to the Copenhagen Interpretation, wherein scientists only have access to a description of the

world, less than persuasive to those whose attention is captured neither by
that abstraction "world" or its description, but is instead captivated by a resonance, an empathic response to an animal, "that poor fellow." As a model
organism for thinking quantum dynamics, the cat participates, albeit virtually, in an ethical conundrum as well: How to respond to a world in which
life and death become indiscernible in the context of radically unpredictable
and indiscernible matter?

Fischer sought a "multivalued logic" suitable to the hallucination-
perception continuum, and it is toward a multivalued logic that Schrödinger's cat appears to call us as well. Goswami writes that it is only when one
has overcome two-valued logic of either/or that Schrödinger's cat's status
can be understood. Jettisoning the life ring of either/or logic inevitably leads
to a bit of panic, but it is only when the control of the ego is put into disarray
that the gleam of reality can be glimpsed:

> It is one thing to talk glibly about an electron passing through two slits at
> the same time, but when we talk of a cat being half dead and half alive, the
> preposterousness of the quantum coherent superposition hits home. (80)

For Goswami, this "preposterousness" reveals the reality of quantum
mechanics' difficulties, difficulties resolved (for Goswami) only by the ide-
alist interpretation of quantum physics, wherein consciousness itself (as an
act on and with information) "comes before" the matter of its description.
This consciousness, according to Goswami, "transcends" any particular consciousness even while it is unavoidably linked to all, and owes much to the
Hindu concept of Atman, an individuated and enfolded replication of the
cosmos (Brahman) within each individual consciousness (Atman), a fractal
spatial map for *Tat Tvam Asi*. Physicist Henry Stapp describes this new multi-
valent world of quantum physics as an informational ecology:

> A world constituted not of matter, as classically conceived, but rather as an
> informational structure that causally links the two elements that combine
> to constitute actual scientific practice, namely the psychologically described
> contents of our streams of conscious experiences and the mathematically
> described objective tendencies that tie our chosen actions to experience. (38)

As a framework of higher, but not superior, logical type, an awareness of this two part informational character of the cosmos appears to be irreducible to any information abstractable from a human subject ("streams of conscious experience") and must be experienced if it is to be understood. Hence for Stapp and Fischer, it is the assay at verification in Euclidean space that initiates and is sufficient to this query; any such query or algorithm (Wolfram) must be actualized to be informative.

Yet even if we were intrigued enough to experiment with Goswami's position that there "is" a consciousness immanent to the universe whose attention and subsequent actualization composes matter, it is another matter entirely to occupy this position as a psychological subject.[13] How to respond to such a vision of an interconnected, vital cosmos, where we indeed find ourselves thoroughly "resembling material substance" as we develop and age?

Goswami's treatment of this consciousness as "transcendent" can be easily misunderstood. And clearly, following Miller and the spirit of Huxley's grandfather, whose rhetorical expertise did so much persuasive work for evolutionary theory, an account of the cosmos and life must be both satisfying and subject to high-fidelity replication: it must focus our attention if we are to learn from it. Perhaps Aldous Huxley's solicitation to imbrication is an efficacious place to begin: in order to posit that the very stuff of the cosmos is less stuff than our collective and individuated apprehension of it, one must first and foremost cease to be "one" at all, apparently dwindling identity on the incessant spiral of maximally recursive consciousness. Less an occurrence than a happening, ordinary identity dwindles and becomes implicated in the actions of an outside agency, what Huxley calls his "blessed Not-I" and, at times, the "Mind-at-Large."

Yet even if contact with this "not-I" or "Mind at Large" is achieved, how is this quantum, blessed not self to communicate with the actual self, let alone any other selves? Freed from the "throttling embrace" of the self, our narrator indeed persists, capable of navigating the transition from one domain to another—on mescaline, not on mescaline. Goswami rhetorically manages this multiplicity of consciousness and observation by ultimately locating it in a transcendence—yet that transcendence is itself, from Stapp's informational perspective, both "nowhere" and "now here." The capacity to differentiate these two statements itself requires exegesis, so much so that more entropy is produced in the process of differentiation, which itself must be differentiated, and so on. Here, a simple shift of bit renders undecidable, as per many Philip K. Dick novels, the very difference between transcendence and immanence. The very smallness of the difference makes its informational management difficult.

Huxley himself suggests that he dwells in a prison house of language, unable even to navigate the gradient or difference between two orders of representation, the linguistic and the visual:

> I am and, for as long as I can remember, I have always been a poor visualizer. Words, even the pregnant words of poets, do not evoke pictures in my mind. No hypnagogic visions greet me on the verge of sleep. When I recall something, the memory does not present itself to me as a vividly seen event or object. By an effort of the will, I can evoke a not very vivid image of what happened yesterday afternoon, of how the Lungarno used to look before the bridges were destroyed, of the Bayswater Road when the only buses were green and tiny and drawn by aged horses at three and a half miles an hour. But such images have little substance and absolutely no autonomous life of their own. They stand to real, perceived objects in the same relation as Homer's ghosts stood to the men of flesh and blood, who came to visit them in the shades. (A. Huxley 2004, 15)

While we might—with the always proleptic ghost of Huxley—be tempted to take this disjunction between words and images to be peculiar to "poor visualizers," Huxley also references the more general failure of language to *refer* to anything like the dynamics of the world, a situation Huxley punningly describes as "petrified by language" (ibid., 24). This petrification, oddly enough, is linked by Huxley not to fossilization but to life—individual consciousness it seems, can't survive the sheer vitality of the world. Distinguishing "Mind" from a more expansive and distributed "Mind-at-Large," Huxley notes that evolutionary constraints would likely select for a reduced awareness of "Mind":

> To make biological survival possible, Mind at Large has to be funneled through the reducing valve of the brain and nervous system. What comes out at the other end is a measly trickle of the kind of consciousness which

will help us to stay alive on the surface of this particular planet. To formulate and express the contents of this reduced awareness, man has invented and endlessly elaborated those symbol-systems and implicit philosophies which we call languages. (Ibid., 23)

For Huxley, then, "Life as it is" is simply incapable of being processed by the brain and the nervous system "on the surface of this particular planet" through the old plumbing biological survival has bequeathed us. One need not, with Schrödinger, Stapp, and Goswami, search in the quantum realm for evidence that our language and modes of description are inadequate to the Universe; Earth itself is more than sufficient to put our rhetorical facilities into disarray even as we are essentially interconnected with it. And Huxley's treatment of the "Mind-at-Large" is intriguingly parallel to Goswami's notion of a transcendent consciousness: "Mind-at-Large," at large, is a thoroughly distributed, "panpsychic" system and thus resists location into anything like a transcendental or serial consciousness, a consciousness out of which matter itself unfolds. And in responding to the planet as a biological entity, Huxley asks us to use our perception not to the inarticulable zone of the quantum, but to the ground beneath the reader's feet.

And the ground, of course, is anything but inert. V. I. Vernadsky, sampling and extending the formulation of nineteenth-century Austrian geologist Eduard Seuss's neologism "Biosphere" in the title of his 1926 book, *The Biosphere*, dismisses both the vitalist and mechanist understanding of life as "two reflections of related philosophical and religious ideas that are not deductions based upon scientific facts" (1998, 51). Instead, Vernadsky proposed to study the intersection between the interior milieu of the organism and the "cosmic milieu" of thermodynamics, where the earth is understood and modeled as a "region of transformers that convert cosmic radiations into active energy in electrical, chemical, mechanical, thermal, and other forms" (1998, 47), including perhaps information. As a mesh of continual transformation and entropy production,

The Living organism of the biosphere should now be studied empirically, as a particular body that cannot be entirely reduced to known physicochemical systems. Whether it can be so reduced in the future is not yet clear. (Ibid., 52)

Vernadsky's "cosmic realist" approach is now being recognized as equally fundamental to our understanding of evolution as Charles Darwin's *On the Origin of Species*, and is thoroughly complementary to it:

> What Charles Darwin did for all life through time, Vernadsky did for all
> life through space. Just as we are all connected in time through evolution
> to common ancestors, so we are all—through the atmosphere, lithosphere,
> hydrosphere, and these days even the ionosphere—connected in space. We
> are tied through Vernadskian space to Darwinian time. (Ibid., 19)

Significantly for our treatment of Huxley, Vernadsky cited Edouard Le Roy and the label "noösphere," the "thinking stratum" whose effects were transforming the planet through the effects of attention. Vernadsky, whose understanding of living systems focused on their capacities to connect and exchange energy, noted that it was difficult to recall such connection in the "thick of life."

> In the thick of life today, intense and complex as it is, a person practically
> forgets that he, and all of mankind, from which he is inseparable, are insep-
> arably connected with the biosphere—with that specific part of the planet,
> where they live. (Vernadsky 2005, 30)

For Vernadsky, this forgetting extends to the role that life has played in the transformation of the planet—geology is a life science because life does not merely occur on the planet but *irreversibly alters it*. And as the biosphere emerged out of the lithosphere, so too does the noösphere emerge and act as a vector on the twentieth-century planet, extracting mescaline sulfate from a cactus, and sending a new name through the mail from Humphry Osmond to Aldous Huxley.

For Huxley, this distributed agency of a biological, transhuman Mind-at-Large is itself recursively linked to his analogy of the "reducing valve," a rhetorical formulation that is indeed an example of itself. "Reducing valve" tropes or renders the relation between a distributed consciousness and the alienated brain as a quantitative one, and thus *compresses the information of psychedelic experience* in a way that allows the narrator to avoid taking account of the qualitative change in the observer, allowing him, perhaps fictively, to write about this character called "I": "I am and, for as long as I can remember, have always been a poor visualizer." It is perhaps little wonder, then, that the notion that consciousness is "expanded" by psychedelics followed on the heels of Huxley's trip report, an "expansionist" ethos all too amenable to egoic twentieth-century consciousness that investigated, ad infinitum, the external world, while closeting or even forgetting the investigations of subjective experience even as it constituted them through new modes of discipline and control.

But, as I hope my reading has helped amplify, there is much more to Huxley's account than "expansion" or "reduction." Huxley's articulation of *istigkeit* suggests that the antidote to the oft cited "reducing valve" is less expansion than involution—the difficult and joyful recognition that one is immanent to the system one is observing. Welcome to the noösphere, where focused attention brings forth transformations of the biosphere. Both the mechanism of a flower's enchantment and its chain of association with an exogamous system of aesthetically mediated sexual reproduction suggest a subject who is subject, even or especially in its experience of interiority, to an ecology of highly differential and thoroughly dynamic practices. As both form and content of a capacity for transpecies linkage, flowers figure and, indeed, actualize the physical connectivity necessary to any recipe that would experience the ecological collectivity of a self becoming aware that it lives on the planet described by Vernadsky.

Huxley's use of the flower as his visualization device for the difference mescaline made to his consciousness is then itself a mechanism for the alteration of consciousness—a capturing of attention and a cultivation of consciousness toward the dense imbrication of a flower—"nothing more, and nothing less, than itself." Rather than an anchoring figure that would risk re-programming egoic consciousness into a position of control, "mescalinity" emerges from a rhetorical practice, a recursive and participatory bending of language and image toward . . . recursion. While the rhetoric of the "expansion of consciousness," linked as it was to the notion that ordinary consciousness was rendered through a "reducing valve," tended to encourage a quantitative and not qualitative understanding of the difference psychedelics make, Huxley's orientation to the enthralling whorls of angiosperms insists on a volatile linkage between self and other, a linkage that undercuts the often solo drama of transcendence and reminds us of our irreducible connection to a biosphere and its emergent noösphere.

STIGMERGY AND GROUP MIND

Thus Huxley faced at least two rhetorical problems with *The Doors of Perception* and *Heaven and Hell*: communication between his non-egoic self and his egoic self, and communication between his visions and an "other." Indeed, the repetitious figure of the flower would suggest in communicating with an "other," he enabled the *reproduction* of his visions for others. All these rhetorical situations feature an apparently unbridgeable gap, a place where communication is both *a priori* impossible and necessary. Yet communicate Huxley did—to his egoic self, who continued to write about and investigate psychedelics, and to a much larger rhetorical ecology, who found in Huxley a model for 1950s lay and scientific investigations of ecodelics. This dual nature of entheogens as both technologies of communication and barriers to articulation can be seen as a symptom of the "West's" noisy reception of psychedelics: Harmine, the first candidate for the active ingredient of ayahuasca (yage), was dubbed "telepathine" in the 1920s for the brew's reputed effects on group communication, and Gordon Wasson, who first brought psilocybin mushrooms to the attention of North American readers in his 1957 *Life* article, participated in an experimental use of "the mushroom telegraph" with the CIA. Uncomfortable with group mind, twentieth-century researchers nonetheless continually sought it out. William S. Burroughs's early encounters with yage suggest one possible symptom of telepathy, an action on bodies at a distance: nausea.

Yet reports of telepathic and group cognition continually appear in the ecodelic literature, and it is in this context that Huxley emerges. In his treatment of the flower as a visualization technology for communicating the existence of a "Mind-at-Large," perhaps Huxley offers us a (now well-known) solution to a version of the Prisoner's Dilemma, that calisthenic practice of analytic philosophers and game theorists, and draws on something like the noösphere to do so. The problem is presented throughout a large literature as follows: within a six-sided cell with one concave side, you hatch a plan to escape with the help of a friend on the outside. You have no way of communicating with your friend, and your escape route is to be a tunnel dug from both directions. Your plan is likely to be successful if you and your collaborator dig at the same site. How can you ensure that the two of you dig at the same spot?

At first blush, the likelihood that you and your comrade will choose the same site is 1/12, but the form of the cell "itself" lowers the odds. According to a solution offered by Goswami, it is difference itself that enables the impossible communication. In this analysis, it is the sole convex (or concave) corner of the cell that presents the most likely digging spot, so you and your friend both begin (stealthily) digging at corner three. Why? Goswami writes in terms of the "motivation" of your friend:

> Now what is your friend's motivation to dig at this particular corner: It is you! He sees you choosing this corner for the same reason that you see him choosing it. (99)

Goswami's (unstated) interpretation is an interpretation based on imaginative difference—each guesses that the other will choose the *different* corner of the prison. It is the recognition of difference upon which both you and your collaborator alight: only corner three looks different from the inside and the outside. If there is a message that allows for this linkage of minds, then, it is a message consisting entirely of a robust and well-imagined difference. The difference between corner three and the others, in this scenario, captures attention through its divergence from imagined contexts. In this sense, Goswami's interpretation silently assumes one of the premises of information theory—that it is in the *difference or gradient* between a message and its context that information emerges.

But *how* does one perceive this difference? While the difference of corner three may enthrall with its divergence, it of course does so only because human, animal, plant, and machine perception is tuned toward difference. Indeed, "tuning" defines nothing other than altering an orientation to difference. Emergent wall following behavior in robots, wherein robots acquire orderly itineraries even while they begin with random programming, for example, occurs only via a sensitivity to the outside, a sensitivity that allows it to tune into the pattern of its world, as the floor plan of a building becomes the code with which it is programmed through competition among other

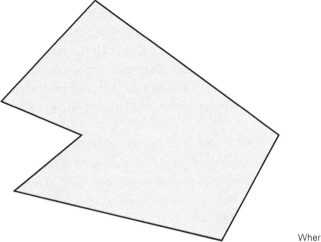

Where to dig?

robots with random distributions of (Lisp) code. Hence for the wall-following robot evolving an orderly, even lively itinerary, there must be "play" in its transduction apparatus to incorporate the signal of an orderly outside.

This sensitivity to the outside occurs only at the "cost" of opening the very control of the machine or organism to the noise of the outside, an outside where it discovers and accrues connections: the commons. Indeed, in Craig Reynolds's 1994 work, not only exteriority but noise appears to be necessary to this strategy of evolving wall-following behavior out of randomly composed Lisp code.[14]

Hence within and without the prison cell, another rhetorical practice is necessary for communication to occur: the outburst of interconnection depends upon opening the aperture of perspective toward an apprehension of the other *as other*, an opening which indeed blurs perspective itself. Psychologist Rollo May's treatment of the Delphic oracle's ecstasy is instructive on this point, as May notes that the oracle spoke in the first person, but via two others: the Pythia and Apollo.

> Apollo spoke in the first person through the Pythia. Her voice changed and
> became husky, throaty, and quavering like that of a modern medium. The
> god was said to enter her at the very moment of her seizure, or *enthusiasm*, as
> the root of the term, *en-theo* ("in-god"), literally suggests. (May, 122)

And different women became Pythia through a series of baths and draughts, a practice that transformed them into something other, and perhaps beside-themselves. May critiques Plutarch's claim that the oracles were influenced by "vapors" or other ecodelics at Delphi, but more recent evidence suggests that the geological formations at Delphi, as well as the construction of the oracle's chamber, are consistent with inspiration by ethylene (The Vaults of Erowid).

The oracle, May suggests, spoke the first person collective, the formation of a commons: Apollo speaks only if his putative opposite, Dionysus, is present, and the Pythia speaks, quavering, only on the condition of hosting Apollo by breathing in earth. "Greek vases show Apollo, presumably at Delphi, grasping Dionysus's hand" (May, 123).

> Ecstasy is a time honored method for transcending our ordinary consciousness and a way of helping us arrive at insights we could not attain otherwise . . . if we genuinely participate in the symbol or myth, we are for that moment taken "out of" and "beyond" ourselves. (Ibid., 130)

Yet if moving beyond ourselves enables the tangled hierarchy that knows where to dig, dig, so too does this imaginative, enthusiastic and even entheogenic transformation require a form if it is to render even a quavering response. The oracles, after all, had their chambers, and their repeatable rituals of ablution and inhalation, while Huxley returns to his flower. When Huxley returns to the flower, it is always on the condition of letting go of all that came before it. By analogy, to grok the noöspheric solution to the question of where to dig, one empties out the self as much as possible, making room for the viewpoint of the other. So too does one have to be an actual friend to grok the situation: *one must want the liberation of the friend* if the implicit communication is to take place—there must be an inclination to succeed which may not always be operative in such a scenario; one can imagine that many friends would find themselves reluctant to collaborate in such a situation.

And psychedelic discourse has collaborated in the opening of *The Doors of Perception*, circulating and cultivating Huxley's stigmergic gift. Huxley, sampling from evolution for the transmission of the news of our imbrication, is sampled by psychonauts Gracie and Zarkov, finally, yielding a chrysanthemum:

At the end of the "flash" of the visions you will have an after-vision of circular interlocking patterns in exquisite colors. It has been described as looking at a vaulted ceiling or dome. If you did not "breakthrough" to the levels described above, this "chrysanthemum" pattern, as we call it, is all you will see. It is worth the trip, too. (Gracie and Zarkov Productions)

Looking to practices of information compression, we can see that the chrysanthemum pattern of a DMT experience, like fractals themselves, function as techniques for squeezing more information into the same channel. Whatever its origin, the information of a DMT session becomes apprehended not as a "measly trickle," but in its compression into a complex but highly symmetrical visual form. So too does Goswami's prison offer a compressed message—so compressed it doesn't require communication, but calls only for a becoming beside oneself through the emptying of the ego and the opening to the friend, ecstasis.

In navigating this ecstasis, Huxley makes use of plant intelligence. On the one hand, Huxley deploys the flower as plant intelligence. Looking to Klüver's constants, we can see that the flower compresses all of Klüver's, however inadequate, categories, with the honeycomb operating as a kind of "major premise" to the flower, it's rhetorical and implied dimension. The very form of the flower provides a visual complexity sufficient to spatializing the itinerary of the "trip"; thus troped, bent into visual information evolved to quite haptically "grab" attention, human intelligence is oriented with a most successful rhetorical device of plant intelligence, the means by which angiosperms have covered the planet. In this sense the plant intelligence of the floral form can be understood as an "extension" of the human nervous system, a compass with which to orient ecodelic interconnection, an attention sink. On the other hand, it is the very efficacy of this floral rhetorical device that crowds out ordinary human intelligence, that mode of perception that views a world "external" to the self.

In the Buddhist mantra of *Om Mani Padme Hum*, "Mani" is the lotus, a visual focus of intention that bends the chanter toward a "turnabout in the seat of consciousness" wherein the stilled mind echoes itself in an act of (seemingly infinite) self-perception. Instead of allowing the eyes to be grabbed by the external world, vision is bent inside out, revealing not an isolated self, but the "blessed Not-I," the "Mind-at-Large." In this state of mind, Huxley does not look at the world, but instead sees with his eyes wide shut:

> Mescaline had endowed me temporarily with the power to see things with my eyes shut; but it could not, or at least on this occasion did not, reveal an inscape remotely comparable to my flowers or chair or flannels "out there." (A. Huxley 2004, 245)

Huxley's formulation resonates with the very etymology of "mystic," as it relies on a shutting off or down of the external world. Yet Huxley's scare quotes around "out there" suggest what he really shuts off is the very distinction between inner and outer.

> I was now a Not-self, simultaneously perceiving and being the Not-self of the things around me. To this new-born Not-self, the behavior, the appearance, the very thought of the self it had momentarily ceased to be, and of other selves, its one-time fellows, seemed not indeed distasteful (for distastefulness was not one of the categories in terms of which I was thinking), but enormously irrelevant. (Ibid., 35)

Inside, outside, Huxley calls the whole thing off: the "Not-self" dwells both inside (perceiving) and outside (being) the things "around" Huxley and becomes enmeshed with the parallel processing of reality. The ceaseless becoming of self becomes viewed as such, the fluidity of any particular moment announcing itself as a hyperbolic, even paradoxical quality: enormous irrelevance. Irrelevant due to its momentary and partial nature, any particular apprehension or "snapshot" of the self is nevertheless large in magnitude—it presents an "enormous" quantity that strangely lacks any measure by which it could be calibrated. Not connected to the matter at hand, *istigkeit*, the discontinuous quality of "one-time fellows" renders them qualitatively different even from each other as well as the durational, analog awareness of the "Not-self." Unbroken in its consistency even as any particular self comes into and out of existence, the "Not-self" enables an apprehension of the "Mind-at-Large" (McFarlane). Here Huxley, with mescaline, seems to notice a structure larger than ego, and he does so via a mobius vision taking in both inside and outside. He becomes aware of performing acts of austerity, a "refraining" of awareness that seems to intensify an awareness of awareness through that glyphic snapshot of a mobius strip, "infinity."

> Compelled by the investigator to analyze and report on what I was doing (and how I longed to be left alone with Eternity in a flower, Infinity in four chair legs and the Absolute in the folds of a pair of flannel trousers!), I realized that I was deliberately avoiding the eyes of those who were with me in the room, deliberately refraining from being too much aware of them. (A. Huxley 2004, 36)

It is of course absurd to treat any single trip report as exemplary, if for no other reason than the graphomania associated with ecodelics suggests that trip reports are essentially plural, with each trip recursively effecting the next. I seek here only to teach a way of reading the trip report that makes sense of what the trip report does for psychonautic practice, and such a teaching itself seeks to transform the reader into a psychonaut, one navigating and hence manifesting the ubiquitous and sometimes barely livable fluctuations of psyche—life. In short, I seek to amplify what trip reports do rather than what they say, since for Huxley, language itself harbors a mistake, even is a mistake. For Huxley, the encounter with language is the site of a most particular failure: a failure of words to be talismanic for such "infinite" visions even as they induce a petrified illusion:

> Every individual is at once the beneficiary and the victim of the linguistic tradition into which he has been born the beneficiary inasmuch as language gives access to the accumulated records of other people's experience, the victim insofar as it confirms him in the belief that reduced awareness is the only awareness and as it bedevils his sense of reality, so that he is all too apt to take his concepts for data, his words for actual things. That which, in the language of religion, is called "this world" is the universe of reduced awareness, expressed, and, as it were, petrified by language. (Ibid., 23)

Here language is understood as both uncannily powerful and inert: insufficient to the news of the recursively looped immanence of the Mind-at-Large, language is nonetheless granted the status of a human invention, and an "implicit philosophy." But more than an achievement, these implicit philosophies function as robust amnesia software: they induce forgetfulness not only of the expansive reality of the "Mind-at-Large," but also the very difference between concepts and things. Huxley suggests by negation that this forgetfulness is biologically useful: "As Mind at Large seeps past the no longer watertight valve, all kinds of biologically useless things start to happen." (Ibid., 26)

> In some cases there may be extra-sensory perceptions. Other persons discover a world of visionary beauty. To others again is revealed the glory, the infinite value and meaningfulness of naked existence, of the given, unconceptualized event. In the final stage of egolessness there is an "obscure knowledge" that All is in all that All is actually each. This is as near, I take it, as a finite mind can ever come to "perceiving everything that is happening everywhere in the universe." (Ibid., 26)

This substitution of a separated self for an immanent cosmos where "All is in all that All is actually each," while, for Huxley, "biologically" adaptive, renders beings strangely incapable of connection:

> Embraced, the lovers desperately try to fuse their insulated ecstasies into a single self-transcendence; in vain. By its very nature every embodied spirit is doomed to suffer and enjoy in solitude. Sensations, feelings, insights, fancies all these are private and, except through symbols and at second hand, incommunicable. We can pool information about experiences, but never the experiences themselves. From family to nation, every human group is a society of island universes. (Ibid., 12)

Hence the impossibility of describing the psychedelic state in words is, for Huxley, only a subset of a more general problem—the inability of humans to escape the hold of linguistic forms and transmit the particularity of experience in the formation of a commons. Language is both closed—incapable, by virtue of its status as a medium, of referring to any particular thing—and enclosing, as it produces the simulacrum of autonomy, an individuality that senses but cannot articulate or share, difference. It was this gulf between sensation and articulation—even the connection between one moment and another—that led Karl Popper to define science in terms of its capacity to be falsified. If induction, since David Hume, could not connect even a cause and an effect, science must then, for Popper, become an ordeal, acts of the imagination subjected to repeated assays or tests. John C. Lilly noted that "What I believe to be true is unbelievable."[15] Scientific knowledge is, in this view, best pursued as an enterprise of deconstruction—a nonstop effort to undo convention in the service of inquiry. Science in this view does not seek proof, but instead selects for the resonating survival of paradox—"What I believe to be true is unbelievable (Lilly qtd. in Brown 1993)"—before the infinitely diverse challenges available to the scientific imagination.

And, yet, of course, *The Doors of Perception* is Huxley's attempt to precisely share a difference—the difference of mescaline, the difference of a human perception, cleansed and capable of at least a glimpse of the infinite. This is the realm, for Huxley, of those domains not describable in language for good reason: they are nonsymbolic.

> The non symbolic of the mind's antipodes exist in their own right, . . .
> our perceptions of the external world are habitually clouded by the verbal
> notions in terms of which we do our thinking. (Ibid., 92)

For Huxley, writing in *Heaven and Hell*, the strangeness of these non-symbolic domains is aligned with their distance—these "antipodes" are the "Australia" of the mind—freakishly odd, yet actual. Huxley's colonialist vision of otherness is part of a larger debt psychonautics and psychedelics owe to the rhetoric of "exploration," but it also rather nicely exemplifies the viability of Huxley's modeled mismatch between language and mescaline experience. The very failure of Huxley's comparison to render the strangeness of those "antipodes of mind" is a measure of his success, and it is this failure that, shared, calls forth further experimentation, a script for further investigations of the allegedly biologically useless "Mind-at-Large."

Huxley's treatment of ecodelic experiences, that down under of the mind, also works to foreground the nonsymbolic evolution of these states of mind. For if there are strangely pouched and billed creatures to be found in the antipodes, these organisms are highly visible evidence for evolutionary differentiation so persuasively defended by Huxley's grandfather. Novel and yet vital, these creatures are perhaps even more spectacularly divergent than those strangely quiescent creatures of the Galapagos observed by Darwin and feasted on by generations of ravenous seamen.

And if Huxley despaired of a description sufficiently other to render these antipodes of the mind, he nonetheless returns, again and again, to the flowers and their color:

> Their color (that hallmark of givenness) shines forth with a brilliance which seems to us preternatural, because it is in fact entirely natural—entirely natural in the sense of being entirely unsophisticated by language or the scientific. (Ibid., 92)

Here Huxley suggests that color—the very thing which, between and among the petals of the nosegay, seizes his attention—is untainted by language, is literally not affected by the "sophistic." Yet, as I will argue in the next chapter, patterns of color and sound likely trace the very evolution of rhetorical function in human culture, and it is perhaps only an inclination toward language that induces us to forget persuasion's debt to rhythm, color, and sound. If Huxley's lovers, embracing, asymptotically seek ecstatic transcendence in futility, they do so within the natural techniques of ecstasy quite literally evolved through sexual selection. Here Huxley seems to forget that such exorbitance—such as the attention attraction at work in a fan of peacock feathers—is hardly biologically useless, but is instead the difference that makes the difference when it comes to many of the "natural" evolutionary processes that have sculpted us.

If language induces a forgetting, it is a repetitious one, those common places associated with rhetoric. In working with the *istigkeit* somehow rendered by a fold of drapery, Huxley treats these sartorial tropings as a curious admixture of abstract and embodied:

> Torn between fact and wish, between cynicism and idealism, Bernini tempers the all but caricatural verisimilitude of his faces with enormous sartorial abstractions, which are the embodiment, in stone or bronze, of the everlasting commonplaces of rhetoric. (Ibid., 31)

For Huxley, the commonplaces of rhetoric can function to induce fits of abstraction that foster the production of entropy—the loss of attention—

a breakdown in the order tempering the aesthetic tendency toward banal repetition by bending representation not toward the actual but toward the fantastic. They can also create frameworks for an all too orderly and serial consciousness, the habitual self that channels the "measly trickle" that substitutes for the "Mind-at-Large."

If such an inheritance of repetitious language is an unbearable failure, it is the very medium of Huxley's complaint and the paradoxical vector of psychedelic knowledge—language both shapes psychedelic experience in practice even as it is insufficient to communicate the experiences. What, then, does the transcription of psychedelic states into language achieve? If language is incapable of achieving any degree of reference to the psychedelic state except through its breakdown or break up, this failure nonetheless achieves something like Socratic success, as ecodelics, and the attempt to put their effects into language, seem to be potent tools of transformation precisely as they remind us of language's insufficiency. The mescaline eater, Huxley writes, is educated by the plant intelligence crowding out his self:

> He will be wiser but less cocksure, happier but less self-satisfied, humbler
> in acknowledging his ignorance yet better equipped to understand the
> relationship of words to things, of systematic reasoning to the unfathomable
> Mystery which it tries, forever vainly, to comprehend. (Ibid., 79)

Yet this failure serves to increase the reflexivity of the post-psychedelic observer, whose failure to articulate the conditions of her own insight serve as a useful heuristic for blocking truly transcendental claims: psychedelics may sometimes seem to reveal a theory of everything, but no such theory has included an adequate description of psychedelics or of the states they induce such that they substitute for the state itself. This blockage can be more than negative: once released from the throttling embrace of the premise that language is a high-fidelity information protocol between nature and the brain, the psychonaut can begin to search for other forms of (nonsymbolic) order on which to hang the complexities of ecodelic experience. Trolling for, finding, and deploying such pre-given templates for order is known as "stigmergy," and it is for their stigmergic function in compressing the information transmissions known as trip reports that Huxley deploys his flowers. Angiosperm evolution has provided Huxley with a means of seductively orienting the eye and consciousness of a viewer, and he samples from it no less than he samples William Blake. While at Harvard, Timothy Leary would begin to sample from Burroughs and Brion Gysin's method of the cut up to transmit the ineffable yet repeatable effects of manifesting the psyche through plant reagents. Huxley practices a botanical mode of rhetoric, one that grafts plant-insect communication into another, symbolic, order of plant reproduction. Huxley's flowers indeed enabled the replication of mescaline experiences, and they did so by turning, "troping" the attention of readers for whom *The Doors of Perception* became a crucial software of nascent psychedelic culture. This culture, in noticing patterns other than egoic ones, paradoxically amplified the effect of thought on the biosphere and hence could be seen to be noöspheric in its effects. Vernadsky defined the noösphere as the transformative effects of thought on the planet:

> As for the coming of the noösphere, we see around us at every step the empirical results of that "incomprehensible" process. That mineralogical rarity, native iron, is now being produced by the billions of tons. Native aluminum, which never before existed on our planet, is now produced in any quantity. The same is true with regard to the countless number of artificial chemical combinations (biogenic "cultural" minerals) newly created on our planet. The number of such artificial minerals is constantly increasing. All of the strategic raw materials belong here. Chemically, the face of our planet, the biosphere, is being sharply changed by man, consciously, and even more so, unconsciously. (Samson and Pitt 1999, 99)

I have tried to make the case elsewhere that one must look at technical communication through a tropical lens—as if it, too, included mantra—if one is to understand its function, but this demand is unavoidable in the case of psychedelics, a discourse in which ordinary premises of communicative rationality become non sequiturs and mescaline, LSD, and other "cultural minerals" begin to circulate on the planet, transforming the "thinking stratum" which begins to become aware, however fitfully, of its own interconnected existence, "Mind-at-Large." In communicating about one of these "cultural minerals," mescaline sulfate, Huxley offers rhetorical "flower power" to orient his investigation and that of his readers.

If it appears that norms of referentially cannot survive the psychedelic experience, the last two hundred years of psychonautic discourse is nonetheless constant in its ongoing desire to articulate the ecodelic experience into language. The premise that a stable identity serves as an ideal but improbable referent for the statement, "I am tripping," is a case in point: a consistent element of psychedelic and shamanic discourse is the dismemberment of identity itself: ego death, Huxley's "release from the throttling embrace of the self." Perhaps Terence McKenna went the furthest in his association of psychedelics and language in his deployment of the gerund "Englishing," and McKenna's locution nicely captures the way in which the "autonomous power" of plants provokes "unpredictable spins"—the loopy (Hofstadter) event of the psychedelic encounter, and the ordeal of articulation that follows. Thus an emphasis on the tropes associated with psychedelic reports seeks not a universal meaning or message at the heart of ecodelic experiences, but instead looks at texts and their sheer quantity and facility as symptoms of a fundamentally corporeal, and hence biological, response, a response deeply implicated with its "set and setting": programs for survival through collective awareness of awareness and the production of enormous amounts of discourse that "fail" to capture this cross-connective nature of human consciousness.

And yet the distinction between set, setting, and compound is itself rendered uncertain both in the psychedelic experience and after. In chapter 4, I will reenact the continual re-discovery of the fundamentally programmable nature of psychedelic experience by Leary and others, but for now it is enough to remark that since set and setting—the context of the ecodelic experience—plays such a causal role, rhetorical practices are a crucial vector in ecodelic experience.

This relation between ecodelic experience and its description is no doubt recursive. Writer William Burroughs, whose encounter with peyote is the subject of only an aside in *Junkie*, treats his letters to Allen Ginsberg as fictional tools with which to unravel the psychedelic noir of his ayahuasca journeys. Burroughs, whose hand had recently fired the bullet that took his wife Joan's life during their first and only game of William Tell, headed south in search of "yage." In *Queer*, Burroughs writes that Amazon curanderos used the brew to determine the identity of a murderer, and Burrough's letters to the young Allen Ginsberg read like a noir quest where a man goes stoically in search of his wife's murderer, expecting to find himself, or at least to find out who he is *now.* Instead, the noir mutates into science fiction, as "Bill Lee," Burroughs's alter ego, finds himself plugged into an extraterrestrial network: "This is what occurs to me. Yage is space time travel" (1975, 47). Perhaps only fiction can be a sustainable "program" for navigating the often loopy and synesthetic multiplicity of psychedelic experience, a parallel transduction of information breaking the narrative barrier—only the cut up can render the "space time travel" without an incessant linearity—"the throttling embrace of the self"—that would convert the attention vortex of an ayahuasca journey into distinct segments of time, what Burroughs dubs "excisors of telepathic sensitivity, osteopaths of the spirit" (ibid., 46). Burroughs's cut up itself begins to resonate, as the cadences and gaps between segments of coherence engage the reader with their rhythm and transmit something of the paratactic "travel" sometimes induced by the drinking of yage.

Hence there can be no question of understanding psychedelics through an analysis of psychedelic compounds themselves, but instead we must grapple with the complex of relations implicated in psychedelic experience and the strange ontologies and rhetorical antics they seem to summon. Our question has thus moved from one mystery to another: what are ecodelic rhetorical practices, and what can they do?

Yes.

I had fallen through the earth itself. Alan Shoemaker, a self-described "gringo shaman" from North America living in exile in Iquitos, received us in his courtyard with his wife Mariella and his two children. Alan, in an interview before we headed out to the jungle, had taught me a highly compressed but thoroughly effective methodology for dealing with ayahuasca: "Just let go!" My usual response would have been: "Yes, but how?" Coming from this man, whose calm in the face of his persecution and jailing by U.S. and Peruvian authorities I admired, though, I understood immediately that if I followed his recipe with a still mind, it could be done.

I had meditated to surprising effect while waiting for Norma to show on the first night. The effect was surprising because I was only a casual meditator, having gone to it for help after my mother's death and during my subsequent regime of prescription serotonin re-uptake inhibitors, dopamine agonists, and lap swimming. Yet there I was, sitting on the thin foam mat set out for the night's ceremony, breathing in and breathing out, trying not to notice that I was in the archetypical situation of, as the Velvet Underground might have had it, waiting on the (Wo)man.

There was quite a bit of breathing, since while Norma had instructed John—an independent radio producer with whom I was making an audio documentary about ayahuasca tourism—and I to be ready to drink by 7:30, she didn't arrive in her feathered crown with Adonai, her apprentice, and Pepito, her green parrot familiar, until nearly nine. Waiting was no doubt part of the notorious ayahuasca diet, which involves abstinence from sugar, salt, alcohol, and sex, and a diet of plantains, fish, and rice. The breathing wasn't happening through my nose, since the diet—which I had taken fairly seriously for the previous two weeks in North America—called for the ayahuasca drinker to refrain from taking pharmaceuticals, including the Claritin-D which had been my daily antihistamine-Sudafed habit. So too had I avoided my habitual use of the Albuterol inhaler to manage my persistent asthma, so I probably made for a fairly noisy meditator there on the banks of the Yanayacu River, if I can be forgiven for introducing a paradox where one is perhaps not strictly necessary.

The previous two nights, there had been little sleep. On the flight from Miami to Lima, my now meditating about to drink ayahuasca self practiced poor Spanish by seeking to decode the tirades of passengers seated across the aisle from me throughout the entirety of the overnight flight. A man in his thirties seemed to be disagreeing with a slightly younger woman about the best means of filling out their papers in order to fool the customs agents, but that served only to convince me that my Spanish was worse than I had feared.

After Norma appeared in her crown of iridescent feathers and blew-whistled tobacco smoke over the liter bottle, I drank down the bittersweet orange liquid and returned to my breathing. The ayahuasca seemed to come on very quickly, and the drinker panicked as he struggled to find a glass of water to quell the immense heat and quench the thirst in which I was now immersed. The ayahuasca drinker stumbled in the dark toward a cooler of water he knew to be there, and miraculously found it. Adonai, the apprentice ayahuascero, who had led me on a short ethnobotanical walk earlier that day and played his guitar throughout the hot afternoon while we lounged in hammocks, was at my side in an instant. "You must sit down, yes."

Adonai gave me his shoulder to lean on as I stumbled with my glass of water. I had found Adonai's Peruvian folk songs comforting earlier in the day, until he had begun strumming the opening bars to "Hotel California," and my first bout of dread began, inducing the lengthy meditation.

This was much, much stronger than the ayahuasca drinker could have anticipated, so he leaned into Adonai's shoulder and let go, falling through the earth. In an oral and visual fractal, he heard and saw himself fall, "I am letting go."

2

RHETORICAL MYCELIUM

Psychedelics as Eloquence Adjuncts?

I cure with Language. Nothing else. I am a Wise Woman. Nothing else.
   —Maria Sabina, Mazatec curandera

Each word becoming more and more dense, too dense to be uttered from now
on, word complete in itself, word in a nest, while the noise of the wood-fire
in the fireplace becomes the only presence, becomes important, strange and
absorbing in its movements. . . . In a state of expectancy, an expectancy that
becomes with each minute more pregnant, more vigilant, more indescribable,
more painful to endure . . . and to what point can it be endured?
   —Henri Michaux, *Miserable Miracle*

Kekule's dream here's being routed now past points
which may arc through the silence.
   —Thomas Pynchon, *Gravity's Rainbow*

BETWEEN THE "BARDO THODOL" AND THE HUXLEYS

AN ENTIRE GENERATION REMEMBERS WHERE IT WAS WHEN LEE OSWALD became "Lee Harvey Oswald"; something like rhetorical exergy seemed to be at work, as if the enormity of the sacrifice called forth a new word, a middle name to resonate with "Fitzgerald."[1] The media burst that accompanied the assassination of John F. Kennedy was spiking toward infinity when Laura Huxley administered one hundred micrograms of LSD-25 to the nearly departed author and psychonaut Aldous Huxley.

Rhetorical practices are practically unavoidable on the occasion of death,

100

and scholars in the history of rhetoric and linguistics have both opined that it was as a practice of mourning that rhetoric emerged as a recognizable and repeatable practice in the "West." Yet if indeed rhetorical practice and death seem unavoidably linked, the oratory that accompanied Aldous Huxley's departure focused on stillness rather than loss. Here Laura Huxley's anticipatory grief became the occasion for the production of guidance and recipes, orientation toward the Clear Light that occasions an awareness of death as a passage to be navigated in the "Bardo Thodol". In this sense it would be easy to see that what W. Y. Evans-Wentz called *The Tibetan Book of the Dead* functions, for the Huxleys, as a map and a script, a map of orientation and a script for interconnection through the ordeal and disarray of grief. Laura Huxley's reading of, and Aldous's listening to, the "Bardo Thodol" serves as a psychedelic "program" or script that would convert the serial and temporal domain of life and death into the diffused, parallel, and distributed spatial domain of the bardos. It is perhaps this capacity of some rhetorical practices to induce and manage the breakdown of borders—such as those between male and female, life and death, silence and talk—that deserves the name "eloquence." Indeed, the *Oxford English Dictionary* reminds us that it is the very difference between silence and speech that eloquence manages: a. Fr. *éloquent*, ad. L. *loquent-em*, pr. pple., f. *loqui* to speak out.[2]

In Evans-Wentz's translation, the nearly departed navigates the bardo (literally, "between") through recourse to a repetition, a repetition of the void:

> O nobly-born Aldous, the time hath now come for thee to seek the Path (in reality). Thy breathing is about to cease. Thy guru hath set thee face to face before with the Clear Light; and now thou art about to experience in its Reality in the Bardo state, wherein all things are like the void and cloudless sky, and the naked, spotless intellect is like unto a transparent vacuum without circumference or centre. At this moment, know thou thyself—, and abide in that state. I, too, at this time, am setting thee face to face.
>
> Having read this, repeat it many times in the ear of the person dying, even before the expiration hath ceased, so as to impress it on the mind (of the dying one). (91)

As a repetition and an utterance, this algorithm of navigational efficacy is that of a mantra. By tuning Merlin Donald's "literacy" brain toward repetition, the semantic content of an utterance gradually decreases, while the purely "sonic" share of the rhetorical transaction grows, grows, grows, focusing the attention on emptiness as semantics are crowded out, the Clear Light of open mind. Philosopher and adept Lama Anagarika Govinda gave the most succinct analysis of the "mantric word" in his investigations of OM, the "unstruck" sound out of which all other sonic landscapes emerge in the Vedic and Upanishadic traditions that form the historical archive for the "Bardo Thodol."

> *Repeat thou* these (verses) clearly, and remembering their significance as thou repeatest them, go forwards, (O nobly-born). Thereby, whatever visions of awe or terror appear, recognition is certain; and forget not this vital secret art lying therein. (Ibid., 103; emphasis mine)

Harvard psychologists Timothy Leary, Richard Alpert, and Ralph Metzner remixed Huxley's psychedelic appropriation of the "Bardo Thodol" with their own "practical translation" of the text for an emerging demographic of psychonauts catalyzed by the very existence of the book. Here, rhetorical practices of repetition are the crucial adjunct to the psychedelic adjuncts themselves:

> The drug is only one component of a psychedelic session. Equally important is the mental and spiritual preparation. . . . The authors find no need to invent new mental and spiritual materials for this purpose. The great literature of meditation lends itself very well to this use. (Leary et al., cover)

Yet for Huxley, in the light of our last chapter, this reading aloud of the "Bardo Thodol" could seem like so much flapping jaw. For, as we saw, his texts made a shibboleth of the incapacity of language: for him, we are beings "petrified" by language, imprisoned by its formulations accrued from the past, continuously mistaking its fossilized imprints for reality. Yet as clusters of mantra, Laura Huxley's words are not measured by their capacity to refer adequately to her grief or to Aldous's departure, but instead function as resonance with a prior reading. Such resonance can perhaps most easily be inhabited through the repetition of OM, which less transmits a message than creates a commons through eloquence, a "speaking out" that both joins

Laura to her husband and sends him on his way.

And ecodelics do, as The Vaults of Erowid database teaches us every day, induce humans to "speak out" at great length. We become prolix, graphomaniacal beings intent on transmitting the nature of our psychedelic experiences. Perhaps only dreams have induced as much rhetorical and hermeneutical activity in humans. I hypothesize that ecodelics enter human evolution via their role as adjuncts to eloquence. Causing us to speak out about the ineffable until we exhaust the capacities of our language unto silence, ecodelics take language to its limit and encourage its innovation. In contact with widespread ecodelic practice, even fonts begin to alter in the infinite, joyful labor of psychedelic testimony. Perhaps our oldest psychedelic text, the Rigveda, likely emerges from versification contests, skaldic "duels" of languaging that were in part induced by soma, the entheogenic adjunct whose exact identity is elusive but whose existence is certain (Easwaran 1987, 252).

And despite Huxley's concern that such an opening of the doors of (rhetorical) perception would be biologically "useless," properly Darwinian treatments of such ordeals of signification would place them squarely within the purview of sexual selection—the competition for mates. If psychedelics such as the west African plant Iboga are revered for "breaking open the head," it may be because we are rather more like stags butting heads than we are ordinarily comfortable putting into language (Pinchbeck 2004, cover). And our discomfort and fascination ensues, because sexual selection is precisely where sexual difference is at stake rather than determined. A gradient, sexuality is, of course, not a binary form but is instead an enmeshed involutionary zone of recombination: human reproduction takes place in a "bardo" or between space that is neither male nor female nor even, especially, human. Indeed, sex probably emerged as a technique for exploring the space of all possible genotypes, breaking the symmetry of an asexual reproduction and introducing the generative "noise" of sexuality with which Aldous Huxley's flowers resonated. In this context, psychedelics become a way of altering the context of discursive signaling within which human reproduction likely evolved, a sensory rather than "extra-sensory" sharing of information about fitness.

DOCTORS OF THE WORD

In an ecstatic treatment of Mazatec mushroom intoxication, Henry Munn casts the curandera as veritable Sophists whose inebriation is marked by an incessant speaking:

> The shamans who eat them, their function is to speak, they are the speakers who chant and sing the truth, they are the oral poets of their people, the doctors of the word, they who tell what is wrong and how to remedy it, the seers and oracles, the ones possessed by the voice. (Munn, 88)

Given the contingency of psychedelic states on the rhetorical conditions under which they are used, it is perhaps not surprising that the Mazatec, who have used the "little children" of psilocybin for millennia, have figured out how to modulate and even program psilocybin experience with rhetorical practices. But the central role enjoyed by rhetoricians here—those doctors of the word—should not obscure the difficulty of the shaman/rhetorician's task: "possessed by the voice," such curanderas less control psychedelic experience than consistently give themselves over to it. They do not wield ecstasy, but are taught by it. Munn's mushroom Sophists are athletes of "negative capability," nineteenth-century poet John Keats's term for the capacity to endure uncertainty. Hence the programming of ecodelic experience enables not control but a practiced flexibility within ritual, a "jungle gym" for traversing the transhuman interpolation.

The necessity of programming ecodelic events only underscores the irreducibility built into psychedelic encounters. Psychonauts often report, for example, that a given compound or plant can lose its "magic," and that subsequent LSD or MDMA sessions do not capture the intensity or transformative power of the initial experiment. One possibility is that the self is heavily "programmed" by the first experience itself, and is hence not as open to the vicissitudes of the plant or compound. The first trip itself becomes a script for further sessions. The self that survived the first session brings that knowledge recursively to subsequent sessions, perhaps vastly reducing a still infinite array of experiences actualized from the same conjunction of compound, set, and setting. Working on and through identities is, then, for the psychonaut, as crucial to learning from psychonautic experiments as learning about plants and chemistry, and rhetorical practices are potent transducers, as the name "I" might continually remind us.

Fundamental to shamanic rhetoric is the *uncertainty* clustering around the possibility of being an "I," an uncertainty that becomes the very medium in which shamanic medicine emerges. While nothing could appear more straightforward than the relationship between the one who speaks and the subject of the sentence "I speak," Munn writes, sampling Heraclitus, "It is not I who speak . . . it is the logos." This sense of being less in dialogue with a voice than a conduit for language itself leads Munn toward the concept of "ecstatic signification."

> Language is an ecstatic activity of signification. . . . Intoxicated by the mushrooms, the fluency, the ease, the aptness of expression one becomes capable of are such that one is astounded by the words that issue forth from the contact of the intention of articulation with the matter of experience. At times it is as if one were being told what to say, for the words leap to mind, one after another, of themselves without having to be searched for: a phenomenon similar to the automatic dictation of the surrealists except that here the flow of consciousness, rather than being disconnected, tends to be coherent: a rational enunciation of meanings. Message fields of communication with the world, others, and one's self are disclosed by the mushrooms. (Ibid., 88–89)

If these practices are "ecstatic," they are so in the strictest of fashions. While recent usage tends to conjoin the "ecstatic" with enjoyment, its etymology suggests an ontological bifurcation—a "being beside oneself" in which the very location, if not existence, of a self is put into disarray and language takes on an unpredictable and lively agency: "words leap to mind, one after another."[3] This displacement suggests that the shaman hardly governs the speech and song she seemingly produces, but is instead astonished by its fluent arrival. Yet this surprise does not give way to panic, and the intoxication increases rather than retards fluency—if anything, Munn's description suggests that for the Mazatec (and, perhaps, for Munn) psilocybin is a rhetorical adjunct that gives the speaker, singer, listener, eater access to "message fields of communication." How might we make sense of this remarkable claim? What mechanisms would allow a speaker to deploy intoxication for eloquence?

Classically speaking, rhetoric has treated human discourse as a tripartite affair, a threefold mixture of ethos, an appeal based on character; logos, an appeal based on the word; and pathos, an appeal to or from the body.[4] Numerous philosophers and literary critics since Jacques Derrida have decried the Western fascination with the logos, and many scholars have looked to the rich traditions of rhetoric for modalities associated with other offices of persuasion, deliberation, and transformation. But Munn's account asks us to recall yet another forgotten rhetorical practice—a pharmacopeia of rhetorical adjuncts drawn from plant, fungus, and geological sources. In the context of the Mazatec, the deliberate and highly practiced ingestion of mushrooms serves to give the rhetor access not to individually created statements or acts of persuasion, but to "fields" of communication where rhetorical practice calls less for a "subject position" than it does a capacity to abide multiplicity—the combination and interaction, at the very least, of human and plant.

Writer, philosopher, and pioneering psychonaut Walter Benjamin noted that his experiments with hashish seemed to induce a "speaking out," a lengthening of his sentences: "One is very much struck by how long one's sentences are" (20). Longer sentences, of course, are not necessarily more eloquent in any ordinary sense than short ones, since scholars, readers, and listeners find that eloquence inheres in a response to any given rhetorical context. Indeed, Benjamin's own telegraphic style in his hashish protocols becomes extraordinary, rare, and paradoxical given his own claim for long sentences in a short note. Yet Benjamin's account does remind us that ecodelics often work on and with the etymological sense of "eloquence," a "speaking out," an outburst *of* language, a provocation *to* language. Benjamin reported that it was through language that material forms could be momentarily transformed: "The word 'ginger' is uttered and suddenly in place of the desk there is a fruit stand" (ibid., 21).

And yet if language and, indeed, the writing table, is the space where hashish begins to resonate for Benjamin, it does so only by making itself available to continual lacunae, openings and closings where, among other things, laughter occurs. For precisely as they are telegraphic, the hashish protocols of Benjamin create a series of non sequiturs:

> I would draw a connection between the laughter and the extraordinary
> mental vacillation. Stated more precisely: the laughter is connected with,

among other things, the great detachment. Moreover, this irresolution—
which contains a potential for affectation—is, to a certain extent, an external projection of the sensation of internal ticklishness. It is remarkable that
one speaks freely and rather impulsively—without strong resistance—about
sources of inhibition that lie in superstition and other such things. . . . An
elegy by Schiller contains the phrase: "the hesitant wings of the butterfly."
This in regard to the coexistence of wingedness with the feeling of indeci-
sion. You follow the same paths of thought as before. Only, they appear
strewn with roses. (Ibid., 21–22)

Hashish, then, is an assassin of referentiality, inducing a butterfly effect in
thought. In Benjamin, cannabis induces a parataxis wherein sentences less
connect to each other through an explicit semantics than resonate together
and summon coherence in the bardos between one statement and another. *It
is the silent murmur between sentences that is consistent while the sentences continually differentiate until, through repetition, an order appears:* "You follow the same
paths of thought as before. Only, they appear strewn with roses."

For a comparable practice in classical rhetoric linking "intoxication" with
eloquence, we return to Delphi, where the oracles made predictions persuasive even to the always skeptical Socrates, predictions whose oracular
ecodelic speech was rendered through the invisible but inebriating "atmosphere" of ethylene gases—a geological rhetoric. Chemist Albert Hofmann,
classicist Carl Ruck, ethnobotanist Jonathan Ott, and others have made a
compelling case that at Eleusis, where Socrates, well before Bartleby, "pre-
ferred not" to go, the Greek Mysteries were delivered in the context of an
ecodelic beverage, perhaps one derived from fermented grain or the ergot-
laden sacrament *kykeon,* chemically analogous to LSD.[5] These Mystery rites
occasioned a very specific rhetorical practice—silence—since participants
were forbidden from describing the *kykeon* or its effects. But silence, too, is
a rhetorical practice, and one can notice that such a prohibition functions
rhetorically not only to repress but also to intensify a desire to "speak out" of
the silence that must come before and after Eleusis.

And Mazatec curandera Maria Sabina is explicit that indeed it is not language or even its putative absence, silence, that is an adjunct or "set and
setting" for the mushrooms. Rather, the mushrooms themselves are a languaging, eloquence itself, a book that presents itself and speaks out:

> At other times, God is not like a man: He is the Book. A Book that is born
> from the earth, a sacred Book whose birth makes the world shake. It is the
> Book of God that speaks to me in order for me to speak. It counsels me, it
> teaches me, it tells me what I have to say to men, to the sick, to life. The
> Book appears and I learn new words.[6]

Crucial to this "speaking" is the way in which Maria Sabina puts it.
Densely interactive and composed of repetition, the rhetorical encounter
with the mushroom is more than informative it is pedagogical and transformative: "The Book appears and I learn new words." The earth shakes
with vitality, manifesting the mushroom orator.[7] Like any good teacher, the
mushrooms work with rhythms, repetitions that not only reinforce prior
knowledge but induce one to take leave of it. "It counsels me, it teaches me."
The repetition of which and through which Maria Sabina speaks commu-
nicates more than knowledge, but allows for its gradual arrival, a rhythm of
coming into being consonant and perhaps even resonant with the vibrations
of the Earth, that scene of continual evolutionary transformation.

More than a supplement or adjunct to the rhetor, the mushroom is a
transformer. Mary Barnard maps out a puppetry of flesh that entails becoming a transducer of the mushroom itself: "The mushroom-deity takes possession of the shaman's body and speaks with the shaman's lips. The shaman
does not say whether the sick child will live or die; *the mushroom says*" (248).

Nor are reports of psilocybin's effects as a rhetorical adjunct peculiar to
Munn or even the Mazatec tradition. Over a span of ten years, psychologist
Roland Fischer and his colleagues at Ohio State University tested the effects
of psilocybin on linguistic function. Fischer articulated "the hallucination-
perception continuum," wherein hallucinations would be understood less
as failed images of the real than virtual aspects of reality not verifiable in
the "Euclidean" space projected by the human sensorium. Fischer, working
with the literary critic Colin Martindale, located in the human metabolism
of psilocybin (and its consequent rendering into psilocin) linguistic symptoms isomorphic to the epics of world literature. Psilocybin, Fischer and
Martindale argued, provoked an increase in the "primary process content"
of writing composed under the influence of psilocybin. Repetitious and yet
corresponding to the very rhetorical structure of epics, psilocybin can thus
be seen to be *prima facie* adjuncts to an epic eloquence, a "speaking out" that

leaves rhetorical patterns consistent with the epic journey (Martindale and Fisher).

And in this journey, it is often language itself that is exhausted—there is a rhythm in the epic structure between the prolix production of primary process content and its interruption. Sage Ramana Maharshi described *mouna*, a "state which transcends speech and thought," as the state that emerges only when "silence prevails."

> It is meditation without mental activity. Subjugation of the mind is meditation; deep meditation is eternal speech. Silence is ever-speaking; it is the perennial flow of "language." It is interrupted by speaking; for words obstruct this mute language. . . . By silence, eloquence is meant. Oral lectures are not so eloquent as silence. Silence is unceasing eloquence. It is the best language. There is a state when words cease and silence prevails. (Maharshi, 48)

A more recent study conducted of high-dose psilocybin experience among international psychonauts suggested that over 35 percent of subjects heard what they called "the logos" after consuming psilocybin mushrooms.

> Based on the responses to the question of the number of times psilocybin was taken, the study examined approximately 3,427 reported psilocybin experiences (n = 118). Of the total questionnaire responses (n = 128), 35.9% (n = 46) of the participants reported having heard a voice(s) with psilocybin use, while 64.0% (n = 82) of the participants stated that they had not. (Beach)

It may seem obvious that respondents who hear voices are, *prima facie*, unreliable witnesses, but the frequency of these reports of "eternal speech," of "the logos," must be understood as reliable symptoms of psilocybin ingestion although their meaning remains elusive. It is difficult, for example, to avoid the notion that psilocybin ingestion caused the voices, especially because the voices occurred with much more frequency among subjects consuming five or more grams of dried psilocybin. Philosopher Karl Popper defined science as the set of hypotheses that can be tested, and, in Popperian fashion, this account would seem to be worthy of repetition, a controlled experiment in which this apprehension of the logos would be falsifiable. And yet no such falsifiable test can control for the possible transformation of the self investigating the logos. The cause, if it is one, is specific not only to psilocybin but to the self; fundamental to accounts of "logos" experiences on psilocybin is the report that the voices offer information with specific purchase for the psychonaut. This *information risks altering the subject*, and is thus the risk one must take when testing the "Listening for the Logos" hypothesis:

> Though there are a number of different types of voice experiences, the common thread running through them all is the imparting of information to the listener. This is the crucial importance of voices. In traditional usage, the mushroom voices give healing information. (Ibid.)

Thus any attempt to falsify "listening for the logos" would require subjects willing to be healed, perhaps even subjects willing to be healed of *being subjects*. What are the practices associated with being capable of receiving such (often transformative) information? While eating psilocybin mushrooms itself is clearly implicated by the participants of the study, its author, and common sense, it is also clear that, if these sacraments were necessary, they were not sufficient to the experience of the logos. Subjects were much more likely to report the information imparting voices if they combined the mushrooms with an environment of total darkness, and the psychonautic demographic selected for by the study suggests that subjects affirmed the risk that they might be healed of ordinary subjectivity. Implicit to this understanding of a "listening" to the logos, too, is the eloquent silence of which Sage Ramana Maharshi speaks.

As with Leary, Alpert and Metzner's understanding of the adjunct as but one "component" of the psychedelic session, "listening to the logos," listens in to the tuning of the logos and (re)mixes the mushrooms with darkness and an inclination to hear. So too must the original impetus of the study itself be noted here—the late "shamanologist" and psychonaut Terence McKenna wrote persuasively that human beings were involved in an evolutionary dance with psilocybin. For McKenna, consciousness itself is the result of the habitual encounter with the Transcendental Other often provoked by the ingestion of this *serotonin in drag*.[8] Given the contingency of psychedelic experience on the rhetorical contexts in which it is used, it is difficult to avoid the impression that some subjects heard the voice precisely because they had read, or heard, the always persuasive McKenna. *The Psychedelic Experience* dubs this capacity for psychedelic experience to mirror any self beholding it the "cosmic ink-blot test."

> He converts the life flow into a cosmic ink-blot test—attempts to label each form. "Now I see a peacock's tail. Now Muslim knights in colored armor. Oh, now a waterfall of jewels. . . ." Verbalizations of this sort dull the light, stop the flow and should not be encouraged. (Leary et al., 56)

If it is a rhetorical adjunct, then, psilocybin does not intensify an "interior" eloquence that is subsequently amplified, but instead enables a practically infinite resonance with set and setting, an inclination to be affected by the outside sometimes associated with healing.

And despite the rhetoric of expansion associated with psychedelics—those tools of "mind expansion" and "expanded consciousness"—"listening to the logos" also requires paradoxical but actual practices of blockage. The ego and its cataloging of experience must give way to listening if silence is to prevail. By dialing down the ego toward a vanishing point, human subjectivity is pulled inside out and the transhuman other broadcasts on that track usually labeled "interior" but whose very experience is itself contingent on an opening to a state in which inside and outside dissolve: the ego, that static aspect of the self that knows what is possible and impossible, must lose its sound track, which dwindles to nothing but the "chattering monkey" if the auditory response to psilocybin is to be investigated. "Verbalizations of this sort dull the light, stop the flow and should not be encouraged."

Listening, of course, is anything but passive. But nor is it a doing of the usual sort, and often involves the blockage of the self rather than its cultivation, a shutting up that is not a shutting down.[9] This "shutting up," though, is more than verbal and is often an ordeal. If silence "prevails," this victory is hardly the insignificant *accomplishment* of the rhetorical practice of *mouna;* silence does not prevail by accident. If voices snare the attention of the psilocybin eater, they do so only to the extent that she has ears for it. This practice takes place on the very precipice of agency—the logos listener is intensely aware of an utter demand to respond, to "speak out," and this response requires a recognition that threatens the contours of human identity itself—a focusing not on the self but on those nonhuman entities troubling the very distinction between the inner world and the outer world, entities whose appearance is linked to the disappearance of the self through its gradual or sudden dwindling, ego death.

The focusing of attention would usually be understood as an intensification of the self, a shutting down of exterior distractions that allow the solitary self its connection to the matter at hand. Novices squint when they focus their vision down the barrel of a rifle. Yet here it is the action of an external agency distinctly other than the self that enables the individual's "fixing" of attention on the self in its actual and interconnected nature. Aldous Huxley's rhetorical challenge, both as he lay dying and as the flowers sent him on a trip, was to focus on the "Clear Light" of *istigkeit*. And the yoga of diverse traditions teaches one to focus on the "brief cessation of the breath," particularly "the still point at the end of the out-breath, before you start breathing in again" (Tolle, 245). This point of stillness or void is the "blessed Non-self" where "silence prevails," and enables an apprehension of interconnection, if only between one breath and another.[10]

In the "Bardo Thodol" remixed by Leary et al., it is only by avoiding attachment and its attendant panic that the dying and, therefore, the living, might avoid catastrophic rebirth, and in psychonautic practice a focus on the Clear Light—the visual correlative of the mute eternal speech of the logos?—opens the psychonaut to bliss rather than distress. Both as an outburst and as a subsequent silence, then, eloquence can serve as an attention management device. Leary et al. offered the alterity and enormity of the evolving universe itself as a conceptual mandala with which to focus the attention:

> O nobly-born, listen carefully:
> The Radiant Energy of the Seed
> From which come all living forms,
> Shoots forth and strikes against you
> With a light so brilliant that you will scarcely be able to look at it.
> Do not be frightened.
> This is the Source Energy which has been radiating for billions of years,
> Ever manifesting itself in different forms.
> Accept it.

Do not try to intellectualize it.
Do not play games with it.
Merge with it.
Let it flow through you.
Lose yourself in it. (Leary et al., 123)

When the focused self can only emerge through its own disappearance into a radiating "Source Energy," the message is clear: ecodelically finding your "center" affects radical transformation, a deliberate becoming-other that emerges from accepting a radiant and transhuman alterity in silence. "Now I see a peacock's tail. Now Muslim knights in colored armor. Oh, now a waterfall of jewels." While the content and nature of any particular transformation is practically infinite, this silencing and dwindling of the self through the investigation of ecodelics affects a figure-ground transformation in which the self emerges as an effect of an Other, an exocentric transformation of any neurologic attached to Merlin Donald's myth of the isolated mind. Leary's words apply to this ecodelic hypothesis:

> If you take L.S.D. to check out these hypotheses, you will automatically alter your consciousness. . . . Like the telescope to astronomy and the microscope to biology, L.S.D. compels a dramatic revision of our neurologic. L.S.D. does alter brain function, change personality, create vulnerable suggestibilities totally determined by set and setting. (Leary and Leary, 11)

Yet, just as Galileo's deployment of the telescope toward the cosmos rather than the earth required an observer capable of seeing a Copernican solar system, so too did LSD's visions challenge the credulity, imagination, premises, and rhetorical capacities of their (now ecologically enmeshed) beholder. This "automatic" transformation of the logos listener's consciousness with the combination of an adjunct and *a desire to test a hypothesis*, though, need not be human or even vocal in the conventional sense. Alan Watts writes of an organ that fluctuates between the solidity of matter and the moistness of flesh:

> I am listening to the music of an organ. . . . The organ seems quite literally to speak. There is no use of the vox humana stop, but every sound seems to issue from a vast human throat, moist with saliva. (39)

Here Watts, in his masterful narration of his own psychedelic experiences, deploys the figure of the voice in order to depict an entire cosmos of sound, suggesting that even in silence, the voice (in its "mute language") and its anatomy provides psychonauts with a template for navigating the sometimes overwhelming flow. Combined with the testimony of Maria Sabina and other doctors of the word, it also suggests that the "cosmic ink-blot test" is particularly attracted not only to the peacock tail but also to its human metonym, the voice, in our human management of this transhuman interpellation.

Inevitably, this flow fluctuates between silence and discourse. Michaux's experiments with psychedelics rendered the now recognizable symptoms of graphomania, silence, and rhetorical amplification. In *Miserable Miracle*, one of the three books Michaux wrote "with mescaline," Michaux testifies to a strange transformation into a Sophist:

> For the first time I understood from within that animal, till now so strange and false, that is called an orator. I seemed to feel how irresistible must be the propensity for eloquence in certain people. Mesc. acted in such a way that it gave me the desire to make proclamations. On what? On anything at all. (81)[11]

Hence, while their spectrum of effects is wide ranging and extraordinarily sensitive to initial rhetorical conditions, psychedelics are involved in an intense inclination to speak unto silence, to write and sing in a time not limited to the physical duration of the sacramental effect, and this involvement with rhetorical practice—the management of the plume, the voice, and the breath—appears to be essential to the nature of psychedelics; they are compounds whose most persistent symptoms are rhetorical. Aldous Huxley makes the analogy between oratory and "hallucinogens" explicit:

> Art and religion, carnivals and saturnalia, dancing and listening to oratory, all these have served, in H. G. Wells's phrase, as Doors in the Wall. And for private, for everyday use there have always been chemical intoxicants. All the vegetable sedatives and narcotics, all the euphorics that grow on trees, the hallucinogens that ripen in berries or can be squeezed from roots all, without exception, have been known and systematically used by human beings from time immemorial. (2004, 62)

That psychedelics leave a prolix discursive record should now be clear, and that the record is likely to select precisely for such verbosity is also suggestive; technologies of media production and rhetorical adjuncts, psychedelics never stop telling us what they are once we start asking. Yet how are we to understand this sometimes mute, sometimes chattering, and always incessant logos once we begin to listen for it?

The more distributed causal context for a rhetorical practice associated with psychedelics—a listening for the logos—only amplifies the importance of thinking of rhetorical ecologies as networked, indeed mycelial, ecologies. In this case, listening to the logos involves an identity practice isomorphic to the deployment of a blindfold or darkened room. As a silencing self that suddenly, but unmistakably, finds itself anything but alone, the challenge of navigating an experience of *ecstasis* often hinges on the capacity to endure a transhuman interpellation, a hailing by and hospitality toward an entity that is anything but human, "the Source Energy which has been radiating for billions of years, Ever manifesting itself in different forms" (Leary et al., 123). And as in the mantric tradition of the "Bardo Thodol" with which we began, the repetition of language can provide the self with the serenity to endure the strangeness of psychedelic experience. Language, then, becomes the occasion for a feedback loop, where utterances and writings that seem to enable the endurance and enjoyment of psychedelic experience are replicated, programming further ecodelic investigations, and so on, perhaps unto *mouna*.

The psychologist Stanley Krippner compiled perhaps the most complete survey of the effects of psychedelics on linguistic function. For Krippner, as for Aldous Huxley, psychedelics intensify the usual "failures" of language to refer:

> A. J. Ayer (1946, 65) once claimed, ". . . we are unable, in our everyday language, to describe the properties of sense-contents with any great precision, for lack of the requisite symbols. . . ." Ayer's statement about normal, everyday perception has even greater application when chemically altered perception is considered. The difficulties subjects have in describing their experiences are compounded by the difficulties experimenters often have in interpreting these reports in terms of some organizational structure. (214)

Yet if Krippner returns to this refrain detailing the incapacity of language, he also recognizes that such a failure involves the paradoxical agency of submission. Listening is the action through which Krippner makes this point most succinctly:

> Listening is the receptive process by which aural language assumes meaning. As listening involves attending to a stimulus, the act often includes a commitment to respond in some way to the messages that are received. The Native American Church communicants commit themselves to a positive experience while the unfortunate subjects of poorly handled LSD experiments commit themselves to a negative experience. In both cases, language plays a key role in determining which way the commitment will turn. (217)

Alan Watts, listening, hears the sound of one species evolving. No longer a "voice" in any usual sense, the logos becomes all the names in history, as "stratagems" form evolutionary strata as clear as the rings on a tree:

> As I listen, then, I can hear in that one voice the simultaneous presence of all the levels of man's history, as of all the stages of life before man. Every step in the game becomes as clear as the rings in a severed tree. But this is an ascending hierarchy of maneuvers, of stratagems capping stratagems, all symbolized in the overlays of refinement beneath which the original howl is still sounding. Sometimes the howl shifts from the mating call of the adult animal to the helpless crying of the baby, and I feel all man's music—its pomp and circumstance, its gaiety, its awe, its confident solemnity—as just so much complication and concealment of baby wailing for mother. And as I want to cry with pity, I know I am sorry for myself. I, as an adult, am also back there alone in the dark, just as the primordial howl is still present beneath the sublime modulations of the chant. (45)

Note that what is represented in these layers of language is precisely evolution itself recapitulated by human development, including the evolution of language. Is this a kind of primal scene for the evolution of stratagems upon stratagems, all enabled by another stratagem—the use of a psychedelic? Crucial to Krippner's analysis, though, is the efficacy of psychedelics in peeling

away these strata of rhetorical practice. By withering some layers of perception, others are amplified:

> In one experiment (Jarvik et al. 1955), subjects ingested one hundred micro-
> grams of LSD and demonstrated an increase in their ability to quickly cancel
> out words on a page of standardized material, but a decreased ability to
> cancel out individual letters. The drug seemed to facilitate the perceptions of
> meaningful language units while it interfered with the visual perception of
> non-meaningful ones. (Krippner, 220)

Krippner notes that the LSD functioned here as a perceptual adjunct,
somehow tuning the visual perception toward increased semantic and hence
rhetorical efficacy. This intensified visual perception of language no doubt
yielded the familiar swelling of font most associated with psychedelic art and
pioneered by the psychedelic underground press (such as the *San Francisco
Oracle*.) By amplifying the visual aspect of font—whose medium is the psychedelic message—this psychedelic innovation remixes the alphabet itself,
as more information (the visual, often highly sensory swelling of font) is
embedded in a given sequence of (otherwise syntactic and semantic) symbols. More information is compressed into font precisely by working with
the larger-scale context of any given message rather than its content. This
apprehension of larger-scale contexts for any given data may be the very signature of ecodelic experience. Krippner reports that this sensory amplification even reached dimensional thresholds, transforming texts:

> Earlier, I had tasted an orange and found it the most intense, delightful taste
> sensation I had ever experienced. I tried reading a magazine as I was "com-
> ing down," and felt the same sensual delight in moving my eye over the
> printed page as I had experienced when eating the orange. The words stood
> out in three dimensions. Reading had never been such a sheer delight and
> such a complete joy. My comprehension was excellent. I quickly grasped the
> intent of the author and felt that I knew exactly what meaning he had tried
> to convey. (221)

Rather than a cognitive modulation, then, psychedelics in Krippner's
analysis seem to affect language function through an intensification

of sensory attention on and through language, "a complete joy." One of Krippner's reports concerned a student attempting to learn German. The student reported becoming fascinated with the language in a most sensory fashion, noting that it was the "delicacy" of the language that allowed him to, well, "make sense" of it and indulge his desire to "string" together language:

> The thing that impressed me at first was the delicacy of the language. . . .
> Before long, I was catching on even to the umlauts. Things were speeding
> up like mad, and there were floods of associations. . . . Memory, of course, is
> a matter of association and boy was I ever linking up to things! I had no dif-
> ficulty recalling words he had given me—in fact, I was eager to string them
> together. In a couple of hours after that, I was even reading some simple
> German, and it all made sense. (Ibid.)

Krippner reports that by the end of his LSD session, the student "had fallen in love with German" (222). Krippner rightly notes that this "falling" is anything but purely verbal, and hypothesizes that psychedelics are adjuncts to "non-verbal training": "The psychedelic session as non-verbal training represents a method by which an individual can attain a higher level of linguistic maturity and sophistication" (225).

What could be the mechanism of such a "non-verbal" training? The motor-control theory of language suggests that language is bootstrapped and developed out of the nonlinguistic rhythms of the ventral premotor system, whose orderly patterns provided the substrate of differential repetition necessary to the arbitrary configuration and reconfiguration of linguistic units. Neuroscientist V. S. Ramachandran describes the discovery of "mirror neurons" by Giaccamo Rizzolati. Rizzolati

> recorded from the ventral premotor area of the frontal lobes of monkeys and
> found that certain cells will fire when a monkey performs a single, highly
> specific action with its hand: pulling, pushing, tugging, grasping, picking up
> and putting a peanut in the mouth etc. different neurons fire in response to
> different actions. One might be tempted to think that these are motor "com-
> mand" neurons, making muscles do certain things; however, the astonish-
> ing truth is that any given mirror neuron will also fire when the monkey in

question observes another monkey (or even the experimenter) performing
the same action, e.g. tasting a peanut! (Ramachandran)

Here the distinction between observing and performing an action are
confused, as watching a primate pick up a peanut becomes indistinguishable
from picking up the peanut, at least from the perspective of an EEG. Such
neurological patterns are not arbitrary, linked as they are to the isomorphic
patterns that are the developmentally articulated motor control system of the
body. This may explain how psychedelics can, according to Krippner, allow
for the *perceptual* discernment of *meaningful* units. By releasing the attention
from the cognitive self or ego, human subjects can focus their attention on
the orderly structures "below" conscious awareness and distributed across
their embodiment and environments. Robin Allot has been arguing for the
motor theory of language evolution since the 1980s:

> In the evolution of language, shapes or objects seen, sounds heard, and
> actions perceived or performed, generated neural motor programs which, on
> transfer to the vocal apparatus, produced words structurally correlated with
> the perceived shapes, objects, sounds and actions. (1989)

These perceived shapes, objects, sounds, and actions, of course, include
the sounds, smells, visions, and actions continually transmitted by ecosystems and the human body itself, and by focusing the attention on them, we
browse for patterns not yet articulated by our embodiment. Significantly, as
neuroscientist Ramachandran points out, this "mirror neuron" effect seems
to occur only when other living systems are involved:

> When people move their hands a brain wave called the MU wave gets blocked
> and disappears completely. Eric Altschuller, Jamie Pineda, and I suggested at
> the Society for Neurosciences in 1998 that this suppression was caused by Riz-
> zolati's mirror neuron system. Consistent with this theory we found that such
> a suppression also occurs when a person watches someone else moving his
> hand but not if he watches a similar movement by an inanimate object.

Hence, in this view, language evolves and develops precisely by nonverbal means in interaction with other living systems, as the repetitions proper

to language iterate on the basis of a prior repetition—the coordinated movements necessary to survival that are coupled to neurological patterns and linked to an animate environment. By blocking the "throttling embrace of the self," ecodelics perhaps enable a resonance between the mind and nature not usually available to the attention. This resonance creates a continuum between words and things even as it appears to enable the differentiation between meaningful and nonmeaningful units:

> One subject, while smoking marijuana, looked at a magazine cover and reported a concretization experience: The magazine featured a picture story about Mexico, and the cover featured large letters spelling out the name of that country. As I looked at the letters, they turned into Aztec men and women. They retained their shape as letters, but subtle shades and shadows became eyes, heads, arms, and legs. That part wasn't so bad, but when Aztecs began to move across the page, I quickly turned the magazine over! (Krippner, 225)

Language itself seems to be activated by ecodelics, amplifying the abstract symbolization of alphabetic font to take on more explicit sensory content even as numerous writers attempt to compose any response adequate to the visions until they stop in recognition of the paradoxically "hopeless" task of transcribing the sensations into language. Krippner notes that the novelist S. Weir Mitchell became unusually eloquent about mescaline in 1896 even as he, like Aldous Huxley, found language wanting:

> Written language attempts to convey meaning through printed symbols. Although S. Weir Mitchell (1896), one of the first to write a description of a psychedelic experience, stated that his peyote experience was " . . . hopeless to describe in language," he later managed to describe " . . . stars, delicate floating films of color, then an abrupt rush of countless points of white light [that] swept across the field of view, as if the unseen millions of the Milky Way were to flow in a sparkling river before my eyes." His account was sufficiently vivid . . . to be able to suggest that he substituted primitive thinking in the form of visual images for conceptual thought. (Ibid., 233)

This continuum between the abstract character of language and its motor control system is consistent with Krippner's observation that "at the sensory level, words are encoded and decoded in highly unusual ways" (238). This differential interaction with the sensory attributes of language includes an interaction with rhythms and puns common to psychedelic experience, a capacity to become aware of a previously unobserved difference and connection. Puns are often denounced as, er, *punishing* a reader's sense of taste, but in fact they set up a field of resonance and association between previously distinct terms, a nonverbal connection of words. In a highly compressed fashion, puns transmit novel information in the form of a meshed relation between terms that would otherwise remain, often for cultural or taboo reasons, radically distinct.[12] This punning involves a tuning of a word toward another meaning, a "troping" or bending of language toward increased information through nonsemantic means such as rhyming. This induction of eloquence and its sensory perception becomes synesthetic as an oral utterance becomes visual:

> The guide asked me how I felt, and I responded "Good." As I uttered the word "Good," I could see it form visually in the air. It was pink and fluffy, like a cloud. The word looked "good" in its appearance and so it had to be "Good." The word and the thing I was trying to express were one, and "Good" was floating around in the air. (Ibid., 235)

By engaging with language as a sensory-motor interaction, one subject reportedly rapidly became a skilled typist:

> In one first-person report (Roseman 1966), a subject claimed that he learned how to become a skilled typist by means of psychedelic experience. Instead of emphasizing the more ideational aspects of the writing process, the subject concentrated on sheer motor activity. (Ibid., 236)

Hence, if it is fair to characterize some psychedelic experiences as episodes of rhetorical augmentation, it is nonetheless necessary to understand rhetoric as an ecological practice, one which truly works with all available means of persuasion (Aristotle), human or otherwise, to increase the overall dissipation of energy in any given ecology. One "goes for broke," attempting the hopeless task of articulating psychedelics in language until exhausting language of any possible referential meaning and becoming silent. By locating "new" information only implicit in a given segment of language and not semantically available to awareness, a pun increases the informational output of an ecosystem featuring humans. This seems to feedback, as with the daily growth of The Vaults of Erowid trip report archive, onto the search for new techniques and technologies for further increasing the amount of information extractable and shareable from psychedelic experiences. Timothy Leary, in his attempts to adequately record psychedelic experiences for scientific representation despite the great speed at which the information was experienced, deployed the "experiential typewriter," a keyboard with various macros for inscribing a rapid shorthand of psychedelic states, linking psychedelic experience to the invention of at least one novel information technology, and chapter 4 will review more thoroughly the evidence that ecodelics are implicated in the emergence of paradigm-breaking information technologies such as Polymerase Chain Reaction. Mycologist Paul Stamets suggests that it is the very form of the mycelium from which the Internet was "bootstrapped" in an act of cybernetically stigmergic eloquence, a silicon testimony to Gaia:

> Waves of mycelial networks intersect and permeate through one another. This interspersing of mycelial colonies is the foundation of soils worldwide. Although seemingly undifferentiated under the microscope, the ability of fungi to respond to natural disasters and sudden changes in the environment are a testimonial to their inherent intelligence. I believe that mycelia are Earth's natural Internet, the essential wiring of the Gaian consciousness. The recent creation of the computer Internet is merely an extension of a successful biological model that has evolved on this planet for billions of years. The timing of the computer Internet should not be construed as a happenstance occurrence. *Sharing intelligence might be the only way to save an endangered ecosystem.* The planet is calling out to us. Will we listen in time? The lessons are around us. Will we learn? (Stamets)

THE TRANSHUMAN INTERPELLATION

The ringing sound in all the senses
of everything that has ever been Created
all the combinations recurring over and
over again as before—
Every possible Combination of Being—all the old ones:—all the old Hindu
Sabahadadie-pluralic universes
ringing in Grandiloquent
Bearded Juxtaposition
    —Allen Ginsberg

To be ordered to say your piece of the puzzle, that is the fathomless mystery.
    —Gus

Visions of the sea come before me, of life struggling on coastal rocks,
of birds overhead. The present is the moving summation of all that
has gone before; all things are interrelated. I understand myself as
the continuing culmination of the evolutionary process.
    —Navi

He said, I am the voice of one crying in the wilderness.
    —John 1:23[13]

Besides a fluid multitasking that allows more than one incessant voice to consciously traverse a body, this transhuman interpellation demands above all the attention of the listener. The urgency of Stamets's question—"Will we listen in time?"—reminds us that for the psychonaut, response-ability is not optional, and that ecodelics are, pace Aldous Huxley, attention attractors and distractors. Listening to the logos requires the psychonaut to block, and then, perhaps, release any understanding of themselves as individuals alone in the cosmos, hacking their evolutionary destiny by everywhere and always strengthening and amplifying their autonomy and individual freedom while forgetting the radical interconnection through which they come into being. "Dropping" an ecodelic entails a commitment, knowingly or not, to "drop" the ego. Both trip manuals and trip reports are pedagogical on this point: one must loosen the grip of identity on autonomy and allow oneself to become a veritable field, to cease to "be" in any given location.

For Huxley, writing in *Heaven and Hell*, this "transporting" effect, the distribution of psyche, is achieved through radical shifts in scale:

> The most transporting landscapes are, first, those which represent natural objects a very long way off, and, second, those which represent them at close range. (2004, 127)

Here both the molecular and the astronomical obey the Hermetic dictum, "As above, so below"; artists break scale by shifting to what Huxley describes as a "non-human point of view":

> The same non-human point of view must be adopted by any artist who tries to render the distant scene. (Ibid., 129)

One poster to the Internet took on this transhuman point of view both from space and from the planet itself:

> In the heightened state of awareness induced by 5 dried grams of *Stropharia cubensis*, the author has had the experience of looking down upon the planet from deep space and actually seeing these social organisms as gigantic creatures. They bore a remarkable resemblance to the formal structure of single-celled organisms, such as paramecia, amoeba, etc. The Teacher was narrating this vision with the following comments:
>
> Human beings are the neural structures of these emergent life forms. Although they have always been here, only recently have any of them begun to gain sufficient neural cells (human beings) to begin to become self-reflectively aware. (Elfstone)

Paired with an apprehension of the logos, this tuning in to ecodelia suggests that in "ego death," many psychonauts experience a perceived awareness of what Vernadsky called the noösphere, the effects of their own consciousness on their ecosystem, about which they incessantly cry out: "Will we listen in time?"

In the introduction, I noted that the ecodelic adoption of this non-local and hence distributed perspective of the biosphere was associated with the apprehension of the cosmos as an interconnected whole, and with the language of "interpellation" I want to suggest that this sense of interconnection often appears in psychonautic testimony as a "calling out" by our evolutionary context. Ethnobotanist Dennis McKenna was induced to cry out for the planet during an ayahuasca ceremony:

> Suddenly I was wracked by a sense of overwhelming sadness, sadness mixed with fear for the delicate balance of life on this planet . . . " What will happen if we destroy the Amazon," I thought to myself, "what will become of us, what will become of life itself . . . ?" (2003, 157)

The philosopher Louis Althusser used the language of "interpellation" to describe the function of ideology and its purchase on an individual subject to it, and he treats interpellation as precisely such a "calling out." Rather than a vague overall system involving the repression of content or the production of illusion, ideology for Althusser functions through its ability to become an "interior" rhetorical force that is the very stuff of identity, at least any identity subject to being "hailed" by any authority it finds itself response-able to. I turn to that code commons Wikipedia for Althusser's most memorable treatment of this concept:

> Memorably, Althusser illustrates this with the concept of "hailing" or "interpellation." He uses the example of an individual walking in a street: upon hearing a policeman shout "Hey you there!", the individual responds by turning round and in this simple movement of his body she is transformed into a subject. The person being hailed recognizes himself as the subject of the hail, and knows to respond.[14]

This sense of "hailing" and unconscious "turning" is appropriate to the experience of ecodelic interconnection I am calling "the transhuman interpellation." Shifting back and forth between the nonhuman perspectives of the macro and the micro, one is hailed by the tiniest of details or largest of overarching structures as reminders of the way we are always already linked to the "evolutionary heritage that bonds all living things genetically and behaviorally to the biosphere" (Roszak et al., 14). And when we find, again and again, that such an interpellation by a "teacher" or other plant entity (à la the logos) is associated not only with eloquence but also with healing,[15] we perhaps aren't surprised by a close-up view of the etymology of "healing." *The Oxford English Dictionary* traces it from the Teutonic "heilen," which links it to "helig" or "holy." And the alluvial flow of etymology connects "hailing" and "healing" in something more than a pun:

> A Com. Teut. vb.: OE. *hlan* = OFris. *hêla*, OS. *hêlian* (MDu. *hêlen, heilen*, Du. *heelen*, LG. *helen*), OHG. *heilan* (Ger. *heilen*), ON. *heil* (Sw. *hela*, Da. *hele*), Goth. *hailjan*, deriv. of *hail-s*, OTeut. **hailo-z*, OS. *Hál* <HALE><WHOLE>[16]

Hailed by the whole, one can become healed through ecodelic practice precisely because the subject turns back on who they thought they were, becoming aware of the existence of a whole, a system in which everything "really is" connected—the noösphere. Such a vision can be discouraging and even frightening to the phantasmically self-birthed ego, who feels not guilt but a horror of exocentricity. It appears impossible to many of us that anything hierarchically distinct, and larger and more complex than Homo sapiens—such as Gaia—could exist, and so we often cry out as one in the wilderness, in amazement and repetition.

Yes, this time was more deliberate, and the triplet sequence unfolded in three separate domains or aspects of the same fundamental experience. The pixelated doorway and its Labyrinthine Way itself gave way to a bird entity I think I recalled from the first night. There he was, distinct from the voice that graced my gourd, presenting himself with little ado and wielding a wand that sounded and felt like a Theroman as he passed it over my ribosomes.

"I'm here to heal you. Your asthma."

"Gentle, please, it was difficult. The other time."

"You must learn how to breathe, from here."

3

RHETORICAL ADJUNCTS AND THE EVOLUTION OF RHETORIC

Darwin's Impassioned Speech

In the vigils I clap and whistle; at that time I am transformed into God.
   —Maria Sabina

It's true you can access any circuit in your brain and change your
mind. But it's time you faced the facts, Timothy. We're turning on
the most powerful sexual organ in the universe! The brain.
   —Timothy Leary, *Flashbacks*

Among the most noteworthy phenomena, the true wonders of our planet, is
the mystery of the bees, which is at the same time a mystery of flowers. The
love-duet between two creatures so immensely far removed from one another
in their form and development must have once been attested, as if by a stroke
of magic, through innumerable acts of caring. The blossoms are reshaped
into sex organs which adapt themselves in a wondrous fashion to completely
foreign creatures—flies, hawk moths and butterflies, but also sunbirds
and hummingbirds. At one time, they were pollinated by the wind.
   —Ernst Jünger, "The Plant as Autonomous Power"

Standing on the bare ground, my head bathed by the blithe air, and
uplifted into infinite space, all mean egotism vanishes. I become a
transparent eyeball—I am nothing; I see all; the currents of the Universal
Being circulate through me—I am part or particle of God.
   —Ralph Waldo Emerson, "Nature"

I F MUNN'S MAZATEC SOPHISTS OFFER US AN ACCOUNT OF A TRULY transhuman rhetorical practice taking in both Homo sapiens and *Psilocybe cubensis,* it remains to be seen if this is anything but an oddity and/or gift of the Mazatecs. Is rhetoric to be understood as a fundamentally human endeavor, with the Mazatec mushroom adjunct and our logos listeners merely proving odd exceptions to an overwhelming rule?

In short, no. Much has been written on the historical emergence of rhetoric, but neither scholars nor biologists have paid much heed to an evolutionary account of our most treasured faculty.[1] But with recent scientific research locating animal, insect, and bacterial communication at every level of our ecosystems, it is time to place rhetorical practice and facility into an evolutionary and ecological context, a context foregrounded in the work of psychologist Timothy Leary.

> Suddenly
> I begin
> to feel
> Strange. (Leary 1968, 19)

The verbosity, musicality, graphomania, and oft reported fluency induced by ecodelics is often paired with a despair for reference and control until the "hopeless" attempt at articulation is, however momentarily, exhausted. The graphomania seems to be particularly intensive in the context of this transhuman interpellation suggesting that, as the philosopher Martin Heidegger might put it, something other "reigns" (156). For Leary, this "other" is nothing other than evolution. Nothing could be clearer from Leary's corpus than the fact that he is an evolutionary psychologist, yet his extraordinary writings of "Psy Fi" did not seem to find ears capable of hearing his call toward a radical response-ability to and for evolution. In *High Priest,* Leary faces the rhetorical burden of narrating his first trip and immediately begins to sample and remix, grafting together the *I Ching,* the *Book of John,* and *Sgt. Pepper's Lonely Hearts Club Band* in the attempt to testify to the otherness that so evidently reigned for him when first he ingested psilocybin in Cuernavaca, Mexico, at the age of 35. The strangeness of the feeling immediately induced a crying out in Leary, as he asked for the help of those eloquent ones that came before him, such as Dante:

O muses, O great Genius, aid me now. O memory that wrote down what I saw, here shall your noble character be shown. (1968, 19; quoting Dante's *Inferno*, Canto II)

As with Huxley before him, the ineffability of psychedelic experience seemed to induce the need to sample and remix as Leary sought to tune even his sub-audible eloquence toward an experience that would be both navigable and shareable. At first, the samples or grafts respect the boundary between themselves and the "trip report." After citing Dante, Leary remixes the Beatles' "A Day in Life" (itself a weave) and the *Book of Changes*, weaving them into his text that becomes multiply voiced in the margins, breaking up the linearity of the alphabetic text until finally, iconically, sampling, with a difference, Emerson's eyeball in a visually self-referential pun:

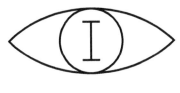

Eyecon

Here Leary echoes not only Emerson, but mycologist Gordon Wasson's echo of Emerson in his 1957 introduction of psilocybin to mass awareness in *Life*: "There I was, poised in space, a disembodied eye, invisible, incorporeal, seeing but not seen" (Wasson).

Leary then intensifies the mix, and begins to cut-up his poetic stream of consciousness narration of the trip, weaving other CAPITALIZED TEXTS directly into his description of the scene. Which muse answers his call?

Laughter in dark room. IT IS INTERESTING TO CONTEMPLATE A TAN-GLED BANK CLOTHED WITH MANY PLANTS OF DIFFERENT KINDS. Gearhart sitting up in dark shouting WITH BIRDS SINGING ON THE BUSHES WITH VARIOUS INSECTS FLITTING ABOUT. Oh God don't let this end AND WITH WORMS CRAWLING THROUGH THE DAMP EARTH. Gearhart goatee bobbing AND TO REFLECT THAT THESE ELABORATELY CONSTRUCTED FORMS SO DIFFERENT FROM EACH OTHER. Gearhart

gone in ecstasy AND DEPENDENT UPON EACH OTHER IN SO COMPLEX
A MANNER I know his ecstasy HAVE ALL BEEN PRODUCED BY LAWS
ACTING AROUND US. We are high. High Priests. (1968, 28)

Here in the eponymous sample from *High Priest*—the book's title is a
(singular) sample from itself—Leary enacts not a neurological but a plant
intelligence, as Darwin's contemplation of the sublime evolutionary immanence of mind and nature from *On the Origin of Species* is *grafted through
interruption* into the ecstatic rant that forms a commons of diverse texts
and experiences:

> It is interesting to contemplate a tangled bank, clothed with many plants of
> many kinds, with birds singing on the bushes, with various insects flitting
> about, and with worms crawling through the damp earth, and to reflect
> that these elaborately constructed forms, so different from each other, and
> dependent on each other in so complex a manner, have all been produced by
> laws acting around us. (Darwin 1909b, 528)

Yes it is "a tangled bank *clothed* with many plants of different kinds" that
captures Charles Darwin's contemplative attention, as Darwin looks beyond
the surface appearance of *separation* and toward the actual *entangled* order
not always recognized by our senses but which, instead, call us to contemplate the interconnection of all things. Darwin promises that this suggested
contemplation is "interesting," that it will hold our attention:

> with birds singing on the bushes, with various insects flitting about, and
> with worms crawling through the damp earth, and to reflect that these
> elaborately constructed forms, so different from each other, and dependent
> on each other in so complex a manner, have all been produced by laws act-
> ing around us. (Ibid.)

Darwin, his attention entangled by the bank, leads us rhetorically (sha-
manically?) with three connected repetitions, each leading to a multiplicity;
"With birds singing . . . with various insects flitting . . . and with worms
crawling" and "reflects" these simple repetitions toward a vision of difference, dependence, and alliteratively "constructed . . . complexity"; "elabo-

rately constructed forms, so different from each other, and dependent on each other in so complex a manner."

Having introduced this baroque vision whose sheer number of activities already challenges the mind attempting to imagine the scene—contemplating, singing, flitting about, crawling, tangling—Darwin also introduces the scalar and spatial differences across which this entangling bank now spreads into his suggested contemplation. The bird sings in the symbolic, aesthetic, informational realm of birdsong and its habitat, sky, while Darwin's implicit exploration of the space of visible ecosystems continues down through which insects, perhaps grasshoppers, vaguely "flit," and ending finally, as it must, with the worms and the earth beneath the feet of each and every mortal reader. It may seem that the archetypical (and, of course, evolutionary) realm of water has been overlooked in this passage, well sculpted to challenge even the most enthusiastic imagination, until we recall that our bank is itself, of course, already "entangled" with the sometimes turbulent flow which it hosts and from which it cannot be reliably separated.

Leary's sample reminds us that in these lines Darwin was precisely a psychonaut, as the coevolutionary interconnections of ecosystems are beheld by Darwin with more than simple interest; this is a fascination manifested by a psyche looking beyond the distorted appearance of separation—the bank itself—and toward its interconnection. This entangled bank entangles not only Darwin's reader, but Darwin himself; he too must be understood as part of this set of "elaborately constructed forms, so different from each other, and dependent on each other in so complex a manner" (ibid.).

And yet Darwin's rhetorical engineering of this highly intertwingled vision of nature is abruptly funneled into the univocality and declarative simplicity of "have all been produced by laws acting around us." Darwin introduces the complexity of his vision and benefits from the sublime aesthetic responses induced by it precisely in order to answer a proleptic *need for order* opened up by the tangled dynamism of the scene. Manifested by Leary, this typographical interruption of his own voice departs from the confessional style and becomes part of the tangle, the formation of an evolutionary commons response-able to his psilocybin experience. This "cutting-up" of Leary's text parallels a cutting up of the ego that enables an observation of the tangled bank of reality, Huxley's "Mind-at-Large," Vernadsky's noösphere. While egoic perception insists on the radical discontinuity between self and other, material and divine, psychedelics often enable the apprehension of these categories as a continuum, a differential (often holographic) whole. Dr. Hilarius, the LSD-prescribing psychiatrist of Thomas Pynchon's *The Crying of Lot 49,* sums it up as well as can be expected: "There is me, there are the others. You know, with the LSD, we're finding, the distinction begins to vanish. Egos lose their sharp edges" (111). At the post-WWII cybernetic gatherings, the Macy Conferences, psychophysicist Heinrich Klüver opines that a similar egoic dissolution was at work in Kekule's famous vision of the benzene ring, which Klüver termed "a direct symbolic representation of thoughts." Klüver suggests that Kekule had an "entoptic" vision of the "Mind-at-Large"—a vision induced within his own eyes:

> This hexagonal pattern was seen entoptically by Purkinje, Konig, and many other observers. I have seen it on several occasions, not with closed eyes, but on the ceiling after awakening. It is highly probable that Kekule, who was used to watching repeated visions and configurations gamboling before his eyes, also saw hexagonal patterns either entoptically or in hypnagogic hallucinations. Unfortunately, my hypothesis that the benzene ring was psychologically derived from the hexagons of this well-known honeycomb design can never be checked. I should mention that hexagonal patterns also appear in mescaline hallucinations, in the visual phenomena of insulin hypoglycemia and when viewing flickering fields under certain conditions. (Josiah Macy Jr. Foundation, 233)

The transhuman interpellation need not, then, be understood as emerging from a distant evolutionary other, but is instead also stigmergically useful feedback from an aspect of ourselves not available to ordinary consciousness. It often requires nothing but a literal contemplative look "within" observable only through a rhetorical manipulation of that wily character, the ego. It can involve, for example, a turning or tuning of our attention to language, a troping of our vision isomorphic to the head twitch characteristic of reindeer that have ingested, as reindeer do, *Amanita muscaria* mushrooms: a double take.

Leary's grafting, entangling, and doubling of the narrative voice of his own "confessional" text reminds us that this double take entails interconnection as well as interruption. Anthropologist and yage drinker Michael

Taussig exemplifies this breakdown and reconfiguration of serially continuous consciousness with the rhetorical strategy of montage:

> The "mystical insights" given by visions and tumbling fragments of memory pictures oscillating in polyphonic discursive room full of leaping shadows and sensory pandemonium are not insights granted by depths mysterious and other. Rather, they are made, not granted, in the ability of montage to provoke sudden and infinite connections between dissimilars in an endless or almost endless process of connection-making and connection-breaking. (Taussig qtd. in Luna and White 2000, 123)

Note that Taussig's subsequent use of montage as a mode of ethnographic description—"MONTAGE: the manner of the interruptedness; the sudden scene changing which breaks up any attempt at narrative ordering and which trips up sensationalism"—itself focuses on the creation of gaps or lacunae by language in language. The quotation marks around "mystical insights," for example, does the (insufficient) rhetorical work of focusing on the formulations we have inherited for these states of consciousness, a "strategic zone of vacuity" that thwarts any final treatment of what the yage session "is." Yage drinkers, in this view, become search engines for eloquence, and Taussig's transcription suggests that yage's effect emerges by interrupting habitual or "sensational" forms of thought in montage through the creation of a commons, the space of interconnection through which montage (such as that used by Leary) unfolds in an "endless or almost endless process of connection-making and connection-breaking." Such a commons demands participation, and exists as such only through the processes of its interaction. *The commons is less a place than a developmental and evolutionary process.* For Taussig this commons is formed fundamentally through a recursive relation to language, although, with Kekule, we might also be capable of recursive (entopic) visions of our own eyes.

The recursive bending of the reader's attention toward the failures of its own transmission is, for Taussig, a mode of experimenting with and on one's "experience" without recuperating it into any final form. "Hence the very grounds of representation itself are raked over" (ibid.). "Hence" for Taussig any ethnographic description of yage must itself be interrupted or divided from itself in Brechtian alienation, the "A-effect" that nicely puns on "ayahuasca." "Perhaps that is the formula for the profoundest possible A-effect, standing within and standing without in quick oscillation" (ibid., 124).

This oscillation is itself Taussig's "image of truth that flickers through the yage nights of which I write," *not an image but its differential doubling*, and being truthful to yage involves a double take precisely on language and the rhetorical practices with which it is ordered. Taussig looks to Walter Benjamin's treatment of German tragic drama for such an interruption that creates a kind of "double consciousness" (Dubois) about language, what he calls "Benjamin's preoccupation with montage":

> For fruitful antecedents, [Benjamin] looked back beyond German baroque
> to those forms of drama where the montage principle first made its appearance. He finds it wherever a critical intelligence intervenes to comment
> upon the representation. (Ibid., 125)

Montage would seem to have made its arrival much earlier through the graphic and sophistic(ated) arts, not to mention the transduction of bacterial DNA. Indeed this "critical intelligence" can sometimes be nothing but the (rhythmic) intervention of noise breaking up a pattern, as in the croak of a frog. Writer William S. Burroughs begins cutting up his final "Yage letter" to Allen Ginsberg, a method that transmits the flicker of "space-time travel" induced by his own drinking of yage.

> Larval beings passed before my eyes in a blue haze, each one giving an
> obscene, mocking squawk (I later identified this squawking as the croaking
> of frogs). (Burroughs 1975, 24)

Note here that Burroughs samples and integrates the (from the perspective of Burroughs) noisy iteration of the croaking frogs by narrating their appearance to his sensorium: a jittery, croaking noise generates not disorder but order. Timothy Leary even sought to use such montage as a way of generating worthy scientific reports of psychedelic experience (Leary 1989). Yet if Taussig, Leary, and Burroughs seem to paradoxically agree that it is only this "ability of montage" to create the conditions necessary to "mystical insights," we might still wonder where this "ability" might come from. As an interruption of attention, montage is a troping or "turning" of attention resonant with Huxley's flowers, a noisy silencing that enables transformation. Burroughs writes in his "Atrophied Preface" to *Naked Lunch*, itself a cut-up emerging after his yage experiences, that even reading the cut-up requires a sudden silence, a bardo in "an endless or almost endless process of connection-making and connection-breaking" given between ellipses:

> How-to extend levels of experience by opening the door at the end of a long
> hall. . . . Doors that open only in *Silence*. . . . *Naked Lunch* demands Silence
> from The Reader. Otherwise he is taking his own pulse. (1959, 203)

Such silence, as any beginning meditator knows, sometimes comes and goes, and thus itself becomes an oscillation. Recall Roland Fischer's description of "self-referential man":

> It is only a comparison, following good literary tradition ("Shall I compare
> you to a summer's day?") and attempts to highlight the intricate problem of
> thought formation by self-referential man. Indeed it is nothing more than a
> thought formed about how a thought is formed, is formed, is formed. (1994,
> 167)

Such "standing within and standing without in quick oscillation" is necessary if not sufficient to an apprehension of self-reference, and it would seem to be a characteristic feature of at least some ecodelic experience, as Taussig both experiences yage and experiences himself experiencing yage in an "image of truth that flickers through the yage nights of which I write." Roland Fischer, searching for a common signature of "psychotomimetic" compounds, suggested that along with their effects on handwriting pressure and spacing, it is the effect of psychedelics on the frequency and amplitude of saccadic vision—the involuntary "rapid flick-like movement" of the eyes necessary to vision—that best characterizes the common action of psychedelics (Hebbard and Fischer). In the last chapter, we saw that ecodelics are intensely implicated in the production of the variable cadences of eloquence, and I suggest that we look to the evolution of oscillatory behavior in mate pairing and group bonding to understand why this image of truth might, of all things, flicker.

THE GRASSHOPPER LIES HEAVY

Insult not the divine grasshoppers, the sweet singers!
—Edmond Rostand

*Please accept references to "global consciousness" as a convenient metaphor.
It is a label for a possible source for effects and correlations that remain
essentially mysterious. All this is subtle, and we can at this point only report
the data, hoping to understand its meaning better as we go along.*
—Roger Nelson, Global Consciousness Project

*The males in the three saltatorial families in this Order are remarkable for their
musical powers, namely the Achetidae or crickets, the Locustidae for which there is
no equivalent English name, and the Acridiidae or grasshoppers. The stridulation
produced by some of the Locustidae is so loud that it can be heard during the night
at the distance of a mile . . . and that made by certain species is not unmusical even
to the human ear, so that the Indians on the Amazons keep them in wicker cages.
All observers agree that the sounds serve either to call or excite the mute females.*
—Charles Darwin, *The Descent of Man*[2]

Charles Darwin was hardly charmed by the results or processes of sexual
selection. In Darwin's presentation of natural selection, it is danger and not
beauty or eloquence that drives evolutionary change:

> Although natural selection can act only through and for the good of each
> being, yet characters and structures, which we are apt to consider as of very
> trifling importance, may thus be acted on. When we see leaf-eating insects
> green, and bark-feeders mottled-grey; the alpine ptarmigan white in winter,
> the red grouse the colour of heather, we must believe that these tints are of
> service to these birds and insects in preserving them from danger. (1909b, 98)

This is the "survival of the fittest" described by Spencer, a vector of evolution offering a kind of utilitarian satisfaction to readers, where Nature,
however red in tooth and claw, fundamentally *makes sense*. Programmed
repetitiously by a dangerous environment, organisms accrue protection
through the most "trifling" of details. In this context, there is no such thing

as a "trifle." And yet sexual selection introduces a fluctuation into this referentiality between the environment and the very color of organisms. In both *On the Origin of Species* and *The Descent of Man*, Darwin's most extensive treatment of sexual selection, Darwin oscillates between "the law of battle" and "charm." The former fits comfortably within the treatment of natural selection, while the latter threatens the natural selective order with the mechanisms of seduction.

And Darwin himself oscillated between these positions when engaging in his own practice of sexual selection. Entertaining the possibility of his marriage to Emma Wedgewood, Darwin was pulled to and fro, pro and con. Assuming the conventions of double-entry bookkeeping, Darwin worked through the decision "on a blue scrap of paper," where he "filled up 'for' and 'against' columns, rambling in a self centered way" (Desmond and Moore, 257). The columns are revealing—"Marry" includes "better than a dog anyhow" and "Not Marry" features "degradation into indolent, idle fool." But even more instructive is the scrambling of the categories themselves, as Darwin begins "rambling" in the "Marry" column about the imagined horrors of bachelor life and is haunted by marriage's "loss of time"—in the "Not Marry" column. "Marry" won the day when this oscillation ended in a vision:

> Only picture to yourself a nice soft wife on a sofa with good fire, & books & music perhaps—Compare this vision with the dingy reality of Grt. Marlboro St. (Ibid.)

Here Darwin seems to convince himself in an act of self-referential *ekphrasis*, the rendering into words of a visual scene. Indeed, Darwin charms himself into marriage here precisely through the induction of a vision, the erosion of the distinction between illusion and reality *induced by a rhetorical algorithm*, "picture to yourself." It is, even, a montaged, dashing vision of Mary: "Marry-Mary-Marry Q.E.D." Darwin's oscillation on that blue scrap of paper is illustrative of a core component of sexual selection itself, the production of a "fluttering" to and fro that scrambles the very categories of before and after, here and there, illusion and reality, inside and outside. In *The Descent of Man*, Darwin focused his attention on the minutiae of such vibrations of all kinds:

With birds of paradise a dozen or more full-plumaged males congregate in
a tree to hold a dancing-party, as it is called by the natives: and here they
fly about, raise their wings, elevate their exquisite plumes, and make them
vibrate, and the whole tree seems, as Mr. Wallace remarks, to be filled with
waving plumes. When thus engaged, they become so absorbed that a skillful
archer may shoot nearly the whole party. (1909a, 403)

As the end of this segment suggests, it is sexual selection's purchase on
attention that so disturbs Darwin's vision of natural selection. While the
actions of these birds of paradise can erode even the distinction between
vegetable and, perhaps rhetorical, animal ("and the whole tree seems . . . to
be filled with waving plumes"), most troubling to Darwin's vision of natural
selection is the spectacle of a single hunter shooting the lot, the erosion of
the boundary between organismic life and death. Freud wrote:

At the height of being in love, the boundary between ego and object threatens to melt away. Against all of the evidence of the senses, a man who is in
love declares that "I" and "you" are one, and is prepared to behave as if it
were a fact. (Roszak et al., 16)

For Darwin, the danger of even temporarily eroding the boundaries
between individuals who are everywhere in competition for survival is
acute. In his first treatment of this "other" vector of evolution, Darwin shut-
tles back and forth between two different modalities of sexual selection as
he weaves his argument: the "law of battle" and "charm." The "law of battle"
with which Darwin begins offers a comfortable analogy to natural selection. But the wide evidence for the effects of "charm" in nature undercuts
the analogy between the struggle for survival and the struggle to produce
progeny. In his very first publication on evolution, a joint 1858 paper published with the aforementioned Alfred Wallace, Darwin writes of this "second agency" in this writing of multiple plumes:

Besides this natural means of selection, by which those individuals are
preserved, whether in their egg, or larval, or mature state, which are best
adapted to the place they fill in nature, there is a second agency at work in
most unisexual animals, tending to produce the same effect, namely, the

struggle of the males for the females. These struggles are generally decided by the law of battle, but in the case of birds, apparently, by the charms of their song, by their beauty or their power of courtship, as in the dancing rock-thrush of Guiana. (1909a, 474)

The "law of battle" is hence no law at all, but instead only the general (and not absolute) condition for the sorting of traits. With "charm" and the "dancing rock-thrush of Guiana," Darwin introduces the possibility that survival comes not to the fittest but the sexiest, those who are adepts of attention gathering:

> The most vigorous and healthy males, implying perfect adaptation, must generally gain the victory in their contests. This kind of selection, however, is less rigorous than the other; it does not require the death of the less successful, but gives to them fewer descendants. The struggle falls, moreover, at a time of year when food is generally abundant, and perhaps the effect chiefly produced would be the modification of the secondary sexual characters, which are not related to the power of obtaining food, or to defence from enemies, but to fighting with or rivalling other males. The result of this struggle amongst the males may be compared in some respects to that produced by those agriculturists *who pay less attention* to the careful selection of all their young animals, and more to the occasional use of a choice mate. (Ibid., emphasis mine)

Whatever the respective values for the vectors of sexual and natural selection in any given ecosystemic niche, Darwin treats them as distinct modes by which the diversity of life has come into being through the focusing of "attention." And birds are "rhetorical familiars" for Darwin's argument, model organisms that best model for humans the fascination and attention capture at work in sexual selection. Darwin manages the rhetorical dissonance between natural and sexual selection by moving gently between languages of struggle and charm. From Darwin's treatment of the issue, it is clear that "charm" was paradoxically less persuasive than "battle" to his (imagined) reader. Here, Darwin segues between battle and song through the figure of a contest, one featuring "strange antics":

> Amongst birds, the contest is often of a more peaceful character. All those
> who have attended to the subject, believe that there is the severest rivalry
> between the males of many species to attract, by singing, the females. The
> rock-thrush of Guiana, birds of paradise, and some others, congregate; and
> successive males display with the most elaborate care, and show off in the
> best manner, their gorgeous plumage; they likewise perform strange antics
> before the females, which, standing by as spectators, at last choose the most
> attractive partner. (1909b, 102)

Here Darwin manages the *collective effect of song* with the *dividing language*
of "severest rivalry" and "contest." Lest Darwin's reader find insufficient
struggle in the mechanisms of evolution, melody itself becomes the outcome
of this "severest rivalry." Again, Darwin has recourse to the language of
attention management:

> Those who have closely attended to birds in confinement well know that
> they often take individual preferences and dislikes: thus Sir R. Heron has
> described how a pied peacock was eminently attractive to all his hen birds.
> (Ibid.)

Note that breeders need to pay as close attention as the peahens themselves. It is only by forming a commons, "closely attending," that selective
effects may be perceived. Yet even as he focuses his attention on what he
calls the "mystery of mysteries" via natural selection, Darwin does not want
to get too close to another subject:

> I cannot here enter on the necessary details; but if man can in a short
> time give beauty and an elegant carriage to his bantams, according to his
> standard of beauty, I can see no good reason to doubt that female birds, by
> selecting, during thousands of generations, the most melodious or beautiful
> males, according to their standard of beauty, might produce a marked effect.
> (Ibid.)

While Darwin elsewhere leverages "marked effect" out of the analogy
between domestic breeding and the ubiquitous but sometimes difficult to
perceive workings of natural selection, Darwin pleads for another space and

time to focus on the "necessary details." Here his analogy attends more to the efficacy of attention in transforming traits than to any homology with natural selection. It is in this sense that Darwin's model of sexual selection can be seen to be acting through what Vernadsky later dubbed the "noösphere." While the biosphere irreversibly and undeniably altered the lithosphere from which it emerged, the noösphere transforms the biosphere via the gathering and application of attention. For the attention of the hen, focused as it is on the beauty of the males, is, according to contemporary evolutionary psychologists, precisely the attention of a breeder guessing at the future of her progeny, and in so doing she brings forth the future of her species. Darwin links the law like connection between past and future through the action of this "second agency":

> Some well-known laws, with respect to the plumage of male and female birds, in comparison with the plumage of the young, can partly be explained through the action of sexual selection on variations occurring at different ages, and transmitted to the males alone or to both sexes at corresponding ages; but I have not space here to enter on this subject. (Ibid., 150)

Any threat sexual selection might pose to the rigors of natural selection is rhetorically balanced by the appeal to the laws that these exorbitant, even garish birds may brilliantly exemplify and articulate. Disruptive to our attention because of their purchase on it, birds (and their songs) nonetheless focus our attention on the law-like order of scientific knowledge. Birdsong and plumage are the veritable audio and visual media with which Darwin gathers and renders his scientific description of the world. What the diverse imaging technologies such as scanning tunneling microscopes, nuclear magnetic imagining, and biofluorescent sequencers are to the "frontiers" of contemporary science and medicine, bird plumage and song were for Darwin. It was Darwin's rhetorical burden and gift to translate these imaging devices by which he was visualizing evolution into the linguistic and visual idiom of his "one long argument," an argument in which he sometimes had recourse to the aforementioned *ekphrasis*. Consider, for example, Darwin's careful, elaborate, and collective visualization in his later work devoted almost entirely to sexual selection, *The Descent of Man*:

> If we picture to ourselves a progenitor of the peacock in an almost exactly
> intermediate condition between the existing peacock, with his enormously
> elongated tail-coverts, ornamented with single ocelli, and an ordinary gal-
> linaceous bird with short tail-coverts, merely spotted with some colour, we
> shall see a bird allied to Polyplectron—that is, with tail-coverts, capable of
> erection and expansion, ornamented with two partially confluent ocelli,
> and long enough almost to conceal the tail-feathers, the latter having
> already partially lost their ocelli. The indentation of the central disc and
> of the surrounding zones of the ocellus, in both species of peacock, speaks
> plainly in favour of this view, and is otherwise inexplicable. (Ibid., 443)

This ekphratic signification asks us to "picture to ourselves" this visual
display, relying implicitly on the very visionary activity induced by the feathers themselves.[3] But while this act of deliberate visualization relies upon the
conscious attention and intention of a reader, this same activity is treated as
an example of the peahen unconscious in an act of "appreciation":

> The males of Polyplectron are no doubt beautiful birds, but their beauty,
> when viewed from a little distance, cannot be compared with that of the
> peacock. Many female progenitors of the peacock must, during a long line of
> descent, have appreciated this superiority; for they have unconsciously, by
> the continued preference for the most beautiful males, rendered the peacock
> the most splendid of living birds. (Ibid.)

Sufficient "space" for a discussion of sexual selection indeed had to wait
for *The Descent of Man*, and here again Darwin oscillates between the "strange
antics" of charm and the more familiar law of battle. Darwin assures us that
at least some of the time, male birds are weaponized, introducing a long
discussion of charm with the attraction of a fight and shifting our attention
toward a fluttering list and arriving finally at "naked skin . . . gorgeously
colored":

> Male birds sometimes, though rarely, possess special weapons for fighting
> with each other. They charm the female by vocal or instrumental music of
> the most varied kinds. They are ornamented by all sorts of combs, wattles,
> protuberances, horns, air-distended sacks, top-knots, naked shafts, plumes

and lengthened feathers gracefully springing from all parts of the body. The beak and naked skin about the head, and the feathers, are often gorgeously coloured. (Ibid., 365)

Hence if natural selection seems to operate, for Darwin, as a rigorous sorting of traits for survival, sexual selection offers a diverse array of attention management devices "gracefully springing from all parts of the body." With sound and vision, sexual selection also captures attention through the olfactory channel:

> The males sometimes pay their court by dancing, or by fantastic antics performed either on the ground or in the air. In one instance, at least, the male emits a musky odour, which we may suppose serves to charm or excite the female; for that excellent observer, Mr. Ramsay, says of the Australian musk-duck (*Biziura lobata*) that "the smell which the male emits during the summer months is confined to that sex, and in some individuals is retained throughout the year; I have never, even in the breeding-season, shot a female which had any smell of musk." (Ibid., 366)

Hence well before Vernadsky's discussion of the noösphere, Charles Darwin was following the trace of the effects of attention on the biosphere. And while the aforementioned musk "can be detected long before it can be seen" (Ibid.), such competition for attention tends to produce collectives. Common to all of Darwin's analyses of sexual selection on everything from insects to humans is the commons itself. Bowerbirds, locusts, and orators all seek attention and the production of a collective. Here Darwin offers a veritable lek about lekking, a gathering of large numbers of organisms for courtship and mating:

> Thus in Germany and Scandinavia the balzen or leks of the black-cocks last from the middle of March, all through April into May. As many as forty or fifty, or even more birds congregate at the leks; and the same place is often frequented during successive years. The lek of the capercailize lasts from the end of March to the middle or even end of May. In North America "the partridge dances" of the *Tetrao phasianellus* "last for a month or more." Other kinds of grouse, both in North America and Eastern Siberia, follow nearly

the same habits. The fowlers discover the hillocks where the ruffs congregate by the grass being trampled bare, and this shews that the same spot is long frequented. The Indians of Guiana are well acquainted with the cleared arenas, where they expect to find the beautiful cocks of the rock; and the natives of New Guinea know the trees where from ten to twenty malebirds of paradise in full plumage congregate. . . . Small parties of an African weaver (*Ploceus*) congregate, during the breeding-season, and perform for hours their graceful evolutions. Large numbers of the solitary snipe (*Scolopax major*) assemble during dusk in a morass; and the same place is frequented for the same purpose during successive years; here they may be seen running about "like so many rats," puffing out their feathers, flapping their wings, and uttering the strangest cries. (Ibid., 413)

Perhaps thinking of his own rhetorical activities, Darwin connects the very cadence of language with these strange cries that excite a gathering:

The impassioned orator, bard, or musician, when with his varied tones and cadences he excites the strongest emotions in his hearers, little suspects that he uses the same means by which his half-human ancestors long ago aroused each other's ardent passions, during their courtship and rivalry. (Ibid., 586)

Note that it is to the differentiation of sound—"varied tones and cadences"—that Darwin links the efficacy of the excitation. Here, along with his lengthy analysis of the variable tones composing the tail feathers of a peacock, Darwin is in accord with the definitions cyberneticist Gregory Bateson later composed for "information," that primal stuff of the noösphere:

Any aggregate of events or objects . . . shall be said to contain "redundancy" or "pattern" if the aggregate can be divided in any way by a "slash mark," such that an observer perceiving only what is on one side of the slash mark can guess, with better than random success, what is on the other side of the slash mark. We may say that what is on one side of the slash contains information or has meaning about what is on the other side. (Ibid., 131)

Paradoxically for Bateson's definition here, it follows that a surprising entry on the other side of the "slash mark" contains more information for an implied observer than a simply redundant one. This generation of more information through divergence, though, relies on a prior redundancy or pattern from which the "difference that makes a difference" can emerge. Even as Bateson's slash mark seeks to rigorously separate one side from the other, we see that the "difference that makes a difference" relies upon a prior pattern, an implicate order or habit from the past through which the present differentially comes into being. Music—and hence birdsong—is exemplary in this regard, demonstrating the always parallel operation even of serial consciousness and the efficacy of the oscillations seemingly proper to sexual selection. Physicist David Bohm uses music as his basic analogy for articulating this parallel operation of consciousness:

> Consider, for example, what takes place when one is listening to music. At a given moment a certain note is being played but a number of the previous notes are still "reverberating" in consciousness. Close attention will show that it is the simultaneous presence and activity of all these reverberations that is responsible for the direct and immediately felt sense of a whole, unbroken, living movement that gives meaning and force to what is heard. (2003, 106)

Bohm goes so far as to argue that this phenomenological investigation of music reveals a fundamental cosmological fact, the mutual entanglement of consciousness and cosmos: "In listening to music, one is therefore directly perceiving an implicate order." In this "immediately felt sense of a whole, unbroken, living movement," Bohm offers an immanent view of consciousness in which it is "implicit" or enfolded within the only apparently separate order of matter. The "reverberations" of past notes remain implicit to the explicit order of sound and the temporal domain of matter. An observer, in this view, *is* itself the interference patterns emerging out of these reverberations with an always already prior template of vibration. Emerging as it does from an implicit and hence non-conscious but not unminded order, consciousness is capable of being rendered "explicit" only through the application of another informing layer: "close attention." Nobel laureate and heterodox physicist Brian Josephson was compelled by this analogy as well, with his view and attention drawn to it:

We seem virtually forced to view music in a similar fashion that is to say as the outward expression of more fundamental phenomena occurring at deep levels of the mind, or of consciousness. (Josephson and Carpenter, 281)[4]

Of course Bohm's arguments for the efficacy of attention in rendering explicit the implicit or immanent order of things begs the question of *how* such "close attention" is to be mustered. Sexual selection, which has (in an extended sense) fashioned the flowers of perception as well as those other rhetorical attractors, the insect stridulation with which this section began, explores the space of all such effective capturing of attention. The peahen, beholding the highly symmetrical patterns of a peacock in full display, attends an oscillation between redundancy and novelty, and in so doing she likely interacts with an implicate order constituted by, but not exhausted by, fitness for "survival." By focusing on the symmetry of an organism, for example, another organism can make a good heuristic guess about the developmental past, and evolutionary future, of that organism. The symmetry of a swallowtail's feathers are a "speaking out" about that organism's developmental and genetic past, an evolutionary eloquence (Ridley, 152).

Yet this movement between redundancy and novelty—as in the refrain and then melody of a cardinal's song, symmetry and its divergence—induces an oscillation for which "true" and "false" are often insufficient categories; the cardinal sings on the hallucination-perception continuum. For this movement is precisely the movement of creation, a non-referential emergence of novelty: Surprise! (Something happened!) While "true" and "false" are categories that can be repetitiously brought to bear on already existing statements, creativity involves an imaginative register where such categories *can only retroactively apply*. As testable processes requiring existence as well as falsifiability, true and false discourses become true and false only as the result of a laborious process and network of interaction. Yet those interactions and statements themselves emerge from acts of creation which themselves can be neither true nor false but simply and totally novel for any given context, "once upon a time called right now!" And just as an extensive tradition since Plato has worried over the possible fakery of eloquence, Darwin reports through a "good observer" that creative and successful sexual selection contests were only miming and not achieving naturally selective effects compatible with the "survival of the fittest" or any other strictly referential system:

> In the case of *Tetrao umbellus*, a good observer goes so far as to believe that
> the battles of the male "are all a sham," performed to show themselves to
> the greatest advantage before the admiring females who assemble around;
> for I have never been able to find a maimed hero, and seldom more than a
> broken feather. (1909a, 374)

But this "sham" induces an *actual* attraction or focusing of close attention. Darwin suggests that this "sham" is not the only such contest to diverge from the alleged rigor of natural selection:

> Even with the most pugnacious species it is probable that the pairing does
> not depend exclusively on the mere strength and courage of the male; for
> such males are generally decorated with various ornaments, which often
> become more brilliant during the breeding-season, and which are sedu-
> lously displayed before the females. The males also endeavour to charm or
> excite their mates by love-notes, songs, and antics; and the courtship is, in
> many instances, a prolonged affair. Hence it is not probable that the females
> are indifferent to the charms of the opposite sex, or that they are invari-
> ably compelled to yield to the victorious males. It is more probable that the
> females are excited, either before or after the conflict, by certain males, and
> thus unconsciously prefer them. (Ibid.)

As many scholars have noted, it is this specter of female *choice* that alarms both Darwin and his imagined reader, and Darwin is careful here to attribute an "unconscious" preference to female birds for the ornaments and eloquence of the males. Yet "consciousness" itself, as Roland Fischer has pointed out, means "thinking with" until at least the sixteenth century in Europe, and it is with this more distributed activity of consciousness that such birds can be understood to be engaged. Becoming tuned to the slightest of differences among the "love-notes," such female bird practices transform species through the application of attention. The capacity to be affected by a differential of color, sound, or symmetry is selected for, as an example, among peahens. Yet, so too does the male aspect of the sexual selection gradient seek out novel ways of affecting hens that do not indicate in any strictly referential way, survival fitness:

> A bird that merely shows its feathers, such as a peacock or a bird of paradise, might be a cheat whose strength has been sapped by bad habits since he grew the plumes. After all, peacock feathers still shine brightly when their owner is dead and stuffed. (Ridley, 157)

While it is continually subject precisely to this drift away from the rigors of "natural selection," the no less natural vector of evolution Darwin dubs "sexual selection" effects remarkable and speedy change on the populations of organisms caught up in their sometimes "runaway" logic (Hall, Kirkpatrick, and West 2000, 1862). Such attention tuned to the likely success and survival of progeny slowly transforms the biosphere into a fluttering, musky swirl of color. The cardinal, chirping and trilling from his perch, wants to be heard, and the peacock, in full vibratory display, wants beholding. Observed, these performances bring forth highly variable, increasingly brilliant colors and intricate patterns differential to a given ecosystem through the selection of mates whose DNA, diet, developmental strategy, and ecosystem can develop such exorbitance. Margulis and Sagan write:

> We suspect that the sheer variety of plant, animal and fungal sexuality, where each individual belongs to a different population of potential mates, increases the rate of energy degradation and entropy production. Ecosystems rich in sexual species, such as the Amazon rain forest, more efficiently reduce the solar gradient than the less sexual bacterial ecosystems beneath the ice of Antarctica. (1997, 150)

Feeding back onto the evolution of novel forms like the human brain itself, these effects of sexual selection spectacularly display the effects of the "thinking stratum" on the biosphere, the commons through which an implicate order of consciousness (and its economies of attention) creatively emerges. Human consciousness, in this view, emerges out of the thermodynamic push to produce ever greater amounts of entropy, and evolves out of the high-resolution attempt to focus close attention and guess at the future reproductive, as well as survival success, of progeny from any given mate. Bowerbirds, with their elaborate nests, generate elaborate and local order while enabling the dissipation of yet more energy (entropy) through reproduction.

Grasshopper stridulation, then, becomes something like montage, a constant interruption of itself that creates the conditions for symmetrical breakdown and linkage through the telecommunicative "slashing" or cutting up of the sonic commons. Through the rhythmic slashing of the sounds of, for example, an Amazonian night, a stridulating insect fulfills Bateson's definition for the transmission of information—a combination of pattern and surprise for the attention focused on the moment before and after a chirp—in a most compressed fashion. Unlike the birdsongs and amphibian or primate calls such insects interrupt, the stridulators need not create intricate patterns or melodies, but instead can "remix" them through the application of a variable rhythm. That grasshopper stridulation is regulated through the muscarine complex suggests that its eloquence is mediated by compounds contained in what is perhaps the oldest continuously used human entheogen, the *Amanita muscaria* mushroom. Indeed, through the application of a pharmacological dose of muscarine, grasshoppers begin to simulate courtship behavior in the lab:

> Repeated injections of small volumes of muscarine elicited stridulation of increasing duration associated with decreased latencies. This suggested an accumulation of excitation over time that is consistent with the suggested role of mAChRs in controlling courtship behavior: to provide increasing arousal leading to higher intensity of stridulation and finally initiating a mating attempt. (Wenzel, Elsner, and Heinrich, 876)

To charm a commons into being, then, organisms need not produce sounds that are melodious, only ones that are informative within any given ecosystem. The grasshopper can then both transmit information indicating location and create one of Darwin's most deployed rhetorical strategies: *gradatio*. Through a continuous and gradual accretion emerging out of repetition, evidence and sound piles up until the "heaps" produce rhetorical conviction and a change of degree becomes a change in kind. This accretion, and its dynamics, is perhaps analogous to what physicist Per Bak called "self organized criticality," as in the sudden collapse of a sand dune that inscribes patterns in the swirling winds of the Sahara or the additional feedback loop of "increasing arousal leading to higher intensity of stridulation and finally initiating a mating attempt." In the case of the grasshopper and, as we shall see below, many other organisms, ecodelic plants and compounds are highly implicated in this rhetorical and evolutionary strategy, as the endogenous production of an entheogen (in the muscarine complex) induces an outburst of highly structured and repetitious eloquence, yielding, perhaps, another iteration of grasshoppers, all the energy they will dissipate, as well as an ancient rhetorical trope: *repetitio.* The Rigveda, for example, likely emerged from "skaldic" contests in versification (Easwaran, 252) competitions whose very content concerned the use of Soma, an entheogenic drink that some scholars have argued is likely to be *Amanita muscaria* (Wasson 1968, 10). Divine singers indeed!

BORROWED FROM THE BIRDS? CYRANO'S PANACHE

And because the montage of stridulation does not take place against a background of pure silence any more than it consists of absolute repetition, the sonic engineering of the kairotic (properly timed) organism could "play" off of the sonic production of another organism and hence "outsource" the evolutionary (and therefore thermodynamic) complexity of song in an act of compression. Here we see a possible rhetorical strategy in common with Christian de Neuvillette, the face for Cyrano de Bergerac's more properly linguistic eloquence in Edmond Rostand's play, *Cyrano de Bergerac.* Evolutionary psychologist Geoffrey Miller notes that Rostand's play is a compelling allegory of sexual selection, where eloquence can be sampled as well as rendered and the display of fitness is hacked or "cheated" through a distributed logic as a network is mobilized for the production of an allegedly binary bond—male/female (Miller, 378). *Cyrano de Bergerac* imagines how eloquence might function to create a group bond as well as a mate-pair bond in human culture and reminds us of the role of lengthy discourse for courtship as well as survival. Grasshoppers form a commons for survival as well as mating. It is clear their use of sound keeps them close together which is essential for species development and survival.[5]

The eponymous Cyrano demonstrates that systems of eloquence are always subject to a sampling precisely because they involve a primordial sharing. As "content provider" for the physically handsome Christian and the beautiful Roxanne, Cyrano is entangled in a love triangle though his "generous counterfeit" of poesy:

ROXANE:
I see through all the generous counterfeit—
The letters—you! (Rostand 2008, 96)

Yet this "counterfeit" simulation wherein Cyrano provides Christian with rhetorical instruction and, eventually, content, is hardly devoid of reality. Cyrano—"The letters—you!"—is, if anything, more lively and real in Rostand's play than the flesh and blood of Christian. It is ironically Cyrano described here by the Duke:

Pity him not! He has lived out his vows,
Free in his thoughts, as in his actions free!
True! I have all, and he has naught;
Yet I were proud to take his hand! (Ibid., 91)

So too does the peacock butterfly play such fakery to its advantage: its ocelli aflutter, the *Lepidopter* throws images of a predator into the sensorium of *its* predator in an act of "intimidation."[6] And the eloquence of plants permits sampling, as in an orchid's imitation of a female wasp, the Internet's recapitulation of mycelia, or indeed Darwin's own sampling of the "Tree of Life" or "that entangled bank" in his well-nigh Kabbalistic visualization of "one long argument" for evolution.[7] Geoffrey Miller has pointed out that human sexual selection is hardly a one-way affair, with choice vectoring from both males and females through the medium of discourse—gestural, oral, and written; "rides" and "cribs" pimped and made over—suggesting that such discourse is also a way to evaluate fitness. As such, Miller argues that sexual selection is a fundamentally discursive "game" that requires at least two players, and when Christian first speaks to Roxanne, he stumbles and merely repeats himself:

ROXANE:

Ay, speak to me of love.

CHRISTIAN:

I love thee!

ROXANE:

That's

The theme! But vary it.

CHRISTIAN:

I . . .

ROXANE:

Vary it!

CHRISTIAN:

I love you so!—(Ibid., 53)

Able only to iterate and re-iterate his egoic consciousness "I . . . I love you so!," Christian fails to give any rhetorical pleasure to Roxanne, as his speech is devoid of the "varied tones and cadences" of (Darwinian) oratory, "That's the theme. But vary it."[8] But Christian can render only repetition, and his seduction comes up short for Roxanne: "I hoped for cream—you give me gruel!" (Ibid.). Miller suggests that the evolution of this "Robin's" behavior, a searching for variations on a theme, is a search that has driven the process by which the big brains of Homo sapiens have evolved.[9] This search for a theme—symmetry—and variety is perhaps a way of evaluating fitness and performing fitness for another.

In *Cyrano*, it is only by forming a commons that all the characters can experience (paradoxically individuated) love. With Roxanne, Cyrano borrows from the rhetoric of generosity when first the distributed scheme for sexual selection—wherein it is from Cyrano's plume that Christian's love-notes are produced—is hatched:

CHRISTIAN (with despair):

Eloquence! Where to find it?

CYRANO (abruptly):

That I lend,

If you lend me your handsome victor-charms;

Blended, we make a hero of romance! (Ibid., 45)

By forming such an (admixture) commons of eloquence, an implicit and implicate order emerges in which a distributed relation is articulated and interconnected, "blended," and "lended." To put it even more redundantly, in *Cyrano*, it is only by forming a commons that a commons can be formed. And it is only by breaking with conventional practices of identity and authorship that this distributed composition can be shared. As two aspects of a coevolutionary system, the male and female attractors seek out fitness effects, e.g., eloquence and the capacity to analyze discourse. Cyrano, with his nose, "my visage's center ornament," functions as a substitute for the discursive phallus of Christian, and "Robin . . . so-called Roxanne"[10] meets eloquence with eloquence and its capacity for the focused attention on analysis and its production.

In these contexts, as eloquence adjuncts, ecodelics would appear to be something like rhetorical Viagra. And Christian, waiting on the whispers of Cyrano, begins to come into relief as a "logos listener." Now Cyrano becomes an adept of attention, enacting spontaneous, skaldic versification while deploying that other appendage, the sword, with panache:

Oh, for a rhyme, a rhyme in o?—
You wriggle, starch-white, my eel?
A rhyme! a rhyme! The white feather you SHOW!
Tac! I parry the point of your steel;
—The point you hoped to make me feel;
I open the line, now clutch
Your spit, Sir Scullion—slow your zeal!
At the envoi's end, I touch.
(He declaims solemnly):
Envoi.
Prince, pray Heaven for your soul's weal!
I move a pace—lo, such! and such!
Cut over—feint!
(Thrusting):
What ho! You reel?
(The viscount staggers. Cyrano salutes):
At the envoi's end, I touch! (Ibid., 21)

This double performance of (s)wordplay involves the discovery of intervals, distinctions between smaller and smaller differences involved in the continuous reflection of the lover on the loved, a word on its rhyming double, all in the doubled service of an "envoi" or letter that most definitely has a "point." Although acts of reflection are deemed untrustworthy "navel gazing" or, worse, the immaterial ponderings of a Sophist (Socrates in Aristophanes), such iterated reflection is hardly bereft of fertility. For the commons depends upon the exorbitant production of this commons of eloquence, one described in iridescence by Cyrano, "Dream-fancies blown into soap-bubbles!":

CYRANO:
Oh! We have our pockets full,
We poets, of love-letters, writ to Chloes,
Daphnes—creations of our noddle-heads.
Our lady-loves,—phantasms of our brains,
—Dream-fancies blown into soap-bubbles! Come!
Take it, and change feigned love-words into true;
I breathed my sighs and moans haphazard-wise;
Call all these wandering love-birds home to nest.
You'll see that I was in these lettered lines,
—Eloquent all the more, the less sincere!
—Take it, and make an end! (Ibid., 45–46)

The fundamentally shared nature of this exorbitant and iridescent production is what most astonishes those who, devoid of love, are not able to even imagine the ecstasis described with such caution by Freud, who described it as a threat:

The boundary between ego and object threatens to melt away. Against all of the evidence of the senses, a man who is in love declares that "I" and "you" are one, and is prepared to behave as if it were a fact. (qtd. in Roszak et al., 16)

This melting of boundaries is characteristic of a sudden capture of attention, an inability to look away, a double take that is nearly emblematic of love. Christian's very first action testifies to a capture of attention by "Magdaleine Robin—Roxane, so called!": Christian, who is dressed elegantly, but rather behind the fashion, seems preoccupied, and keeps looking at the boxes (Rostand 1898, 7).

Christian's attention is focused on the possible arrival of Roxanne, and so his gaze moves to and fro, between the boxes and that other spectacle competing (unsuccessfully) for his attention, the stage. Darwin, in a text originally to be included in *The Descent of Man*, describes the veritable face of attention, a visage subject to sudden transformation:

> ATTENTION, if sudden and close, graduates into surprise; and this into astonishment; and this into stupefied amazement. The latter frame of mind is closely akin to terror. Attention is shown by the eyebrows being slightly raised; and as this state increases into surprise, they are raised to a much greater extent, with the eyes and mouth widely open. (1916, 278)

Crucial to Darwin's treatment of the issue is his recognition of the bodily envelope as a surface of continual implication and explication, as an unfolding of the face in one direction becomes folded in the other, transversal direction:

> The raising of the eyebrows is necessary in order that the eyes should be opened quickly and widely; and this movement produces transverse wrinkles across the forehead. The degree to which the eyes and mouth are opened corresponds with the degree of surprise felt; but these movements must be coordinated; for a widely opened mouth with eyebrows only slightly raised results in a meaningless grimace, as Dr. Duchenne has shown in one of his photographs. On the other hand, a person may often be seen to pretend surprise by merely raising his eyebrows. (Ibid.)

Darwin discovers in the foldings and unfoldings of the face as it is captured by attention an analog (continuous) mechanism for the production of digital (discrete but interconnected) states. It is perhaps this continual involution and evolution of the face by which its attention gathering proceeds—a simple symmetry from which it can literally be troped, moved. This immanent movement unfolds in Rostand's play when Christian, that "pretty fellow," catches a pickpocket in the theater:

CHRISTIAN (holding him tightly):

I was looking for a glove.

THE PICKPOCKET (smiling piteously):

And you find a hand. (Rostand 2008, 12)

Cyrano articulates a scene of implication, wherein an order is activated in a given context by *unfolding it*. Cyrano is an *explicator* both of Roxanne's beauty and Christian's desire. Here the resources by which pairing takes place are themselves distributed and not easily confined to the binary system of one male/one female pair. This distributed logic suggests an "exocentric" system of courtship trolling for an implicate order, a search engine for solutions to the "problem" of sexual reproduction. Recent studies of the neurology of birdsong suggest that even the production of birdsong, that "charming" mode of sexual selection, is itself distributed among two neurological systems which require noise to be truly explicated and actualized:

Thus, while the HVC→RA projection carries the learned song, the LMAN→RA projection carries the jitter that induces the variability in motor output necessary for the imitation of a model. This jitter, presumably, is imposed on the firing of the same RA neurons that receive the more orderly output from HVC. When the LMAN neurons are silent (or absent), the HVC→RA pathway produces a stereotyped pattern; when the LMAN→RA neurons are firing, song is more variable. (Nottebohm)

Cyrano lends his wit and panache so that three might effectively bond two. In the distributed eloquence of the neurology of birdsong, it is a "jitter" rather than any "jabber" whose addition is required, the "lending" that enables a "blending." And what is (b)lended into the mixture is noise. Nottebohm discovers the multiple components necessary to the commons engineering of birdsong, and his map of birdsong eloquence ought also to inform a more networked or mycelial model of rhetoric than our usual brain in a box, authorial model of eloquence or its neatly dissected tripartite model of the rhetorical triangle: audience, writer, text; pathos, ethos, logos. Even Nottebohm's model (learned song "bassline," plus noise jitter funk) is at least implicitly "brain-centric," looking for the, however multiple, components of his experimental variable eloquence system within the functional anatomy of the brain. Yet because noise itself becomes a crucial component in the generation of the novel and beautiful order of birdsong, might organisms also seek out *exogenous* sources of noise just as Christian outsources his eloquence? Indeed, Cyrano himself had some help. In his last words, Cyrano claims that there is one thing that cannot be taken from him—his "panache"—and panache is of course an effect of style and wit, and its literal and etymological translation is itself a pun on pen, and perhaps the effects of sexual selection in general: a panache is a "tuft of feathers" upon a man's hat, deployed to capture the attention, and borrowed from a male bird's attempt to do the same.

FROM NOSE TO NOISE TO NOÖSPHERE: ATTENTION!

> *"Attention," a voice began to call, and it was as though an oboe had suddenly become articulate. "Attention," it repeated in the same high, nasal monotone. "Attention."*
> —Aldous Huxley, *Island*[11]

Noise proves to be a strange attractor indeed for the emergence of order. The Global Consciousness Project, for example, recapitulates the twin oscillator model of birdsong in its attempt "to study direct effects of consciousness on physical systems." The GCP has distributed over 50 "EGG" (Electrogaiagram) devices around the planet. Each of the geographically dispersed EGG devices

> usually produce a continuous sequence of completely unpredictable numbers which can be recorded in computer files. Experiments have shown that human consciousness can make the string of numbers slightly non-random when people hold intentions to do so, or when there is a special state of coherent group consciousness. The difference is very small, but statistical analysis demonstrates that this correlation of the REG behavior with something about consciousness is real. It is as if our wishes could change the 50/50 odds of a coin flip ever so slightly. (Global Consciousness Project, "Procedures")

Crucial to the GCP project is the production of noise and the application of attention to it.[12] Before it is to become song, birdsong is a duet with noise; *noise is the very medium upon which the order is played.* While common sense suggests that noise abounds and is easily produced, the production of pure noise is in fact a constantly evolving technoscientific frontier that begins with ancient oracle bones and dice and unfolds toward quantum tunneling. The GCP project, for example, use three distinct laboratory-grade random event generators:

> The Global Consciousness Project uses three different random-event genera-
> tors (REG or RNG). All three use quantum-indeterminate electronic noise.
> They are designed for research applications and are widely used in laboratory
> experiments. They are subjected to calibration procedures based on large
> samples, typically a million or more trials, each the sum of 200 bits. In the
> GCP application, an unbiased mean is guaranteed by XOR logic. Although
> they have different fundamental noise sources, they all provide high-quality
> random sequences that are functionally equivalent. (Ibid., "REG Design")

In the neurology of birdsong, the addition of a jitter enables the produc-
tion of variable song, and this jitter is induced through the LMAN pathway, itself modulated by the GABA agonist and the entheogenic compound mus-
cimol. Evolution often works by making endogenous what had been exog-
enous and vice versa, so the noise necessary to variable song (and hence successful mating) could come from an organisms' diet as well as an inter-
nalized regime. In his "Poem on Hashish," translated into English by Aleis-
ter Crowley, French poet Charles Baudelaire treated this strategy of dietetic supplement (with cannabis seeds) as a truism:

> Who does not know the extravagant behavior of hens which have eaten
> grains of hemp-seed, and the wild enthusiasm of the horses which the
> peasants, at weddings and on the feasts of their patron saints, prepare for a
> steeplechase by a ration of hemp-seed, sometimes sprinkled with wine? (5)

Whether endogenously or exogenously produced, Nottebohm suggests two different interpretations of the role that this "jitter" plays in the development of the varied cadences of song:

> One possibility is that birds trying to imitate a model succeed by retaining,
> from the variability generated, those patterns that more closely approxi-

mate the model and discard the rest, thus, over a period of time, achieving a perfect imitation. A second possibility is that the variable mismatch between a model and the attempted imitation drives output modification, so that patterns that had not occurred before now first appear. Both mechanisms would depend on auditory feedback. It is the first time we are so close to a mechanism for vocal learning.

Nottebohm's oscillation between the two interpretations, of course, is not itself mutually exclusive—both approximation and novelty would occur as evolutionary transformations under the influence of symmetry breaking noise. At its limit, this noise may be nothing but a shift in collective attention, altering the focus of attention and hence by extension attention itself. One of the peculiarities of consciousness is the effect it can have on itself—it cannot be observed without changing in kind; consciousness can attend to itself, "feedback," and in attending, it is altered. This capacity to *respond to information* is understood by Bohm to be the very hallmark of mindedness, which he attributes even to that aspect of the implicate order we model as the "electron."

> The quantum theory, which is now basic, implies that the particles of physics have certain primitive mind-like qualities which are not possible in terms of Newtonian concepts (though, of course, they do not have consciousness). (Bohm 1990, 272)

Roger Nelson of the Global Consciousness Project links his own understanding of that "hard problem," consciousness, to Bohm's concept of "active information."

> Consciousness is a source of active information, and . . . the objects of attention for consciousness can be sinks that attract and hence actualize the information. (Global Consciousness Project)

Here Nelson and Bohm are in accord with rhetorician, economist, and Nobel Laureate Herbert Simon, who articulated this uncannily "active" nature of information in terms of its capacity to "consume" the attention of observers:

> What information consumes is rather obvious: it consumes the attention of
> its recipients. Hence a wealth of information creates a poverty of attention,
> and a need to allocate that attention efficiently among the overabundance of
> information sources that might consume it. (73)

Nelson's language of "sinks" and Simon's image of information as an attention "consumer" are both congruent with Bohm's continual recourse to the language of attention in describing the implicate order. The "sinks" and "consumers" can be said to become "implicated" in the sense that they fold information and its apprehension into a single space and give it the form of that space. In becoming affected by information, the attention can change in kind becoming an archetype, an ideology, a poem. For Bohm, the capacity to be affected by information ("in-formed") is the very definition of mind.

> A new notion of active information . . . plays a key role in this context.
> The word in-form is here taken in its literal meaning, i.e., to put form into
> (rather than in its technical meaning in information theory as negentropy).
> One may think of the electron as moving under its own energy. The quan-
> tum potential then acts to put form into its motion, and this form is related
> to the form of the wave from which the quantum potential is derived.
> (Bohm 1990, 279)

If this "wholeness that informs" is fundamental to our reality for Bohm, Goswami, Vernadsky, and Margulis, why, we might ask, do we not perceive its "holomovement"? For Bohm, the evolution of ordinary human consciousness has been informed by memory's nature: a temporal "recording" device that turns our attention toward the "static and fragmented."

> One reason we do not generally notice the primacy of the implicate order
> is that we have become so habituated to the explicate order . . . the activa-
> tion of memory recordings whose content is mainly that which is recurrent,
> stable, and separable, must evidently focus our attention on what is static
> and fragmented. . . . The more subtle features of the unbroken flow (e.g., the
> "transformations" of musical notes) generally tend to pale into such seem-
> ing insignificance that one is, at best, only dimly conscious of them. (Bohm
> 2003, 113)

Bohm's concept of the implicate order in which all of apparent reality ("the explicate order") manifests from a potential, implicate, or enmeshed order was itself manifested as a response to the puzzling results of quantum mechanics, but it also points to a biotic component of all order, the ecosystemic conditions under which anything like an observation can emerge.

> A more detailed analysis shows that the quantum potential for the whole system then constitutes a non-local connection that brings about the above described organized and orderly pattern of electrons moving together without scattering. We may here make an analogy to a ballet dance, in which all the dancers, guided by a common pool of information in the form of a score, are able to move together in a similar organized and orderly way, to go around an obstacle and re-form their pattern of movement. (Bohm 1990, 280–81)

Bohm's analogy, by focusing on similarities, also magnifies differences within the "common pool of information." In Bohm's treatment, the difference between the ballet and the cosmos resides in the practical identity between the cosmos and the score; unlike the score described above, the implicate order manifesting in the patterned behavior of electrons is actualized by Everything That Is: the electron's "score" has not been compressed by anything smaller than the unfolding evolving cosmos itself. Yet within the cosmos, as Aldous Huxley's flowers teach us, there are patterns that allow for precisely such compression. Sexual selection would appear to be a search engine for such patterns in the implicate order, an ecstatic browsing for patterns that will increase the capacity of any system to produce ever greater amounts of entropy.

TROLLING FOR IMPLICATE ORDER

> *And when I stumble onto a psilocybin site already claimed by goats, I make*
> *no attempt to drive them off—partly out of respect for them, but partly*
> *from fear of being attacked by animals already under the influence of this*
> *powerful drug and therefore doubly recalcitrant and dangerous.*
> —Giorgio Samorini, *Animals and Psychedelics*[13]

Architect Christopher Alexander seeks to amplify the patterns formed by this implicate order as they in-form (in Bohm's sense) the human experience of subjectivity. The press kit promises veritably ecodelic effects from the reading of Alexander's text, a four-volume collection entitled *The Nature of Order*:

> When we are finished reading, we see everything (not many things, *everything*) differently. We see human relations differently; we see science differently; we see architecture and art differently. We see the idea of God or Buddha differently. We enter a domain, in which, at last, we see ourselves as related to the world, and in which the objective structure of the world, and the internal structure of our own selves have somehow merged and been healed. (Alexander 2004, press kit; emphasis in original)

This ability to see *"everything . . . differently"* can be seen, pace Huxley, to emerge only when the smaller scale structure of the human ego ceases to be the primary attention attractor (figure) and other larger scale forms (ground, implicate order) come into relief. Psychologist Edward de Bono suggests that the process of human development involves precisely the breakdown in such familiar and established patterns:

> de Bono talks of the value of depatterning in evolutionary development—forcing the mind to break with established routines and habits, and so discover new approaches to important matters of survival. (Pilkington)

The recent work of Italian ethnobotanist Giorgio Samorini has gone furthest in an investigation into the evolutionary context of psychedelics as pattern breakers, and Samorini's findings are extremely suggestive in the context of sexual selection and its evident deployment of noise (Nottebohm). Indeed, while Samorini does not himself conclude that sexual selection is the likely evolutionary vector for ecodelics, his evidence and analysis continually allude to it. It is as if there were a Darwinian and ecological unconscious implicit to Samorini's text testifying to the likely sexually selected trait of psychedelic use in human and animal evolution.

Samorini's work focuses on the unavoidable evidence that animals and insects habitually seek out inebriants, and wonders over the evolutionary implications of such habits and habitats:

Could it be that animals that consume various plant inebriants develop enhanced senses and perceptual acuity that confer an adaptive advantage in evolution? (xi)

And Samorini extends this notion from perception and sensation to "awareness":

Awareness-enhancing plant drugs are indeed sought out by certain animals. And behavior which increases mating such as eating prosexual or libido-stimulant plant drugs (the so called aphrodisiacs), means disproportionate breeding by that savvy individual who thus breeds more of its gene type into the species. (xii)

Timothy Leary flashed backed to mandrake root early in his conceptualization of LSD experience:

The mandrake root was apparently the first sex stimulus. It was mentioned twice in the bible. Pythagoras "advocated" it. Machiavelli wrote a comedy about it. (Leary 1994, 32)

More recently, Olivia Judson as well as Lynn Margulis and Dorion Sagan provided keen observations on the likely role of aphrodisia in nature. Margulis and Sagan speculate, as this book argues, that sex itself has emerged in response to the planetary telos of increasing the dissipation of information and the sun derived energy it both helps locate and squander.

The entire unfolding of evolution is a response to an unexportable excess, a growing surplus of sun-derived energy. Both the sex act and the tiger are complexities of the biosphere. (Margulis and Sagan 2000, 199)

While the language of purpose runs counter to the contemporary taboo against teleological argument in biology, purpose has been one of the defining characteristics of biological inquiry since its inception. Aristotle famously equated the very difference between the animate and the inanimate in terms of "telos." Ethnobotany has discovered indigenous studies of living systems that begin with the question of purpose: "What is this plant *for*?" In a psychological context, biological inquiry is a continual meditation on the distinction between living and non-living systems, and in so doing it continually provokes questions about what is often called life's "meaning" but is perhaps more usefully mapped as evolutionary desire—*What does evolution want?* Contemporary thermodynamics teaches us that evolution systemically wants to dissipate greater and greater amounts of exergy or "available energy." By converting energy into information, new (formerly implicate) energy gradients are explicated, increasing overall entropy production.

Artificial life grows as an example for this response to an "exhortation to excess"—the need to dissipate greater and greater amounts of energy: representing life for a living, artificial life-forms convert energy—electricity, programmer time, hobbyist attention, scientific publications, computer hardware construction and maintenance, and everything else that composes artificial life—into differentially self-replicating patterns of information seductive to users looking to enliven their world of very real dwindling biodiversity, producing more entropy than their absence. Indeed, artificial life could be seen as demonstrative proof for the theses of Margulis, Sagan, Kay, Salthe, Schneider, Swenson, and others that living systems emerge such that more entropy is produced, as the silicon "phylum" becomes host to organisms neither more nor less queer than deep-vent cyanobacteria.

> It is interesting to contemplate an entangled screen, PLUGGED AND CORDED, SURROUNDED BY plants of many kinds, with COFFEE CUPS VARIOUSLY FILLED BY UNKNOWN LIQUIDS PERCHED UPON THE DESK, a USER singing, with various insects flitting about, AND THE FLOOR LITTERED WITH THE FLOTSAM AND JETSOM OF THOUGHT, and to reflect that these elaborately constructed ARTIFICIAL LIFE forms, so different from each other, and dependent on each other in so complex a manner, have all been produced by laws acting around us. (Darwin, via Leary, remixed by mobius)

Viewed thermodynamically, this entangled screen, with new ways to live and die in the Milky Way, lives as much off human attention as electricity. Human attention, seduced by the digital critters on the screen, boots the program, consuming resources in the system, slowing down the processing of a lengthy book manuscript, proliferating alife organisms. . . . And alife creatures can be uncannily psychedelic, interrupting our attention, creating more entropy. And perhaps the death of artificial life-forms, as manifestations of our consciousness, are an expression of the cosmos' need to creatively destroy information to avoid what systems scientist Stanley Salthe has characterized as the "senescence" of information overload in such thermodynamic systems trolling for increased entropy production (Salthe, 153).

Groping for a way to understand this sort of exorbitance among the widespread animal use of inebriants, Ronald Siegel posited the existence of a "fourth drive" found in some animals: the instinct to alter consciousness. Siegel had the chance to observe birds engaged in some fascinating, even romantic behavior as he was watching a pair of cedar waxwings drunk on firethorn berries:

> Despite their reputation for sleek plumage, never seeming to have a feather out of place, the waxwings were left rumpled and tipsy by the intoxications. Yet they still had the ability to engage in a unique courtship display. The male fluffed his feathers and turned his head away from the female, who did the same. Then the male passed a firethorn berry as a "present." He offered it to his partner at the tip of the beak and she accepted it. The berry was passed back and forth several times and, eventually, it was eaten by one of the birds at the end of the display. (1989, 59)

While Siegel perceives an apparent contradiction between the courtship display and the birds' obvious intoxication, the context of Darwin's analysis suggests that this breakdown in the autopoietic function—"the waxwings were left rumpled and tipsy"—is in fact fundamental to the process of sexual reproduction, manifesting the seeming paradox of sexual selection: the production of a complex of order—"sleek plumage"—such that it might be put into disarray.

> In 1926 J. Grinnel observed the behavior of these birds in his garden: "There were dozens of robins on the bushes and everywhere on the ground. They appeared tame and dazed. Some lay on the earth in the dirt with their wings awry, I regretted the fact that their condition rendered the birds unusually easy to catch by our cat." (Samorini, 52)

And, as per my claim last chapter that, sometimes, the sexually selective vector searches for eloquence in intoxication, Samorini notes that for birds, too, plant adjuncts can induce highly context contingent and selective eloquence:

> Sparrows have been observed entering storehouses to feed on hemp seeds, which seem to excite and stimulate them. In fact many kinds of birds show this partiality for marijuana seeds, and it is related in many different regions of the world that eating them modifies their behavior—that, for example, they "sing" at greater length and with more ardor and are more inclined toward amorous displays. People who raise parrots add a certain percentage of hemp seeds to their animals diets to make them more talkative. In Italy, many canary owners do the same thing, to stimulate their birds to sweeter song. (Ibid., 54)

And why must this song become, synesthetically, "sweeter"? As eloquence adjuncts, hemp seeds here can be seen to offer a competitive advantage to a bird who has sampled Darwin's pharmacy in "amorous display." For it is precisely such "eloquence" that, in many species, correlates with increased fitness in the form of increased reproductive success. Mandrills too extend their phenotypes with alkaloid adjuncts during courtship:

> When a male mandrill must engage in combat with another, either to establish his claim to a female or to climb a rung of the hierarchical ladder, he does not begin to fight without forethought. Instead, he first finds and digs up an iboga bush, eating its root; next, he waits for its effect to hit him full force (which can take from one to two hours); only then does he approach and attack the other male he wants to engage in battle." (Ibid., 58)[14]

TRIPPING TROPED: IS SEXUAL SELECTION
A CONTENT PROVIDER FOR PSYCHEDELICS?

A CAPTCHA

*When I shut my eyes I saw blackness, but sometimes a random image
would appear. When I laid next to and hugged my girlfriend, I shut
my eyes and saw a forest, and then a group of vivid flowers.*
   —Mr. Nice Guy

In the nuptial embrace personality is melted down.
   —Aldous Huxley, *Island*

*Clouds from left to right through optical field. Tail of a pheasant (in centre
of field) turns into a bright yellow star; star into sparks. Moving scintillating
screw; "hundreds" of screws. A sequence of rapidly changing objects in
agreeable colors. A rotating wheel (diameter about 1 cm.) in the centre of a
silvery ground. Suddenly in the wheel a picture of God as represented in old
Christian paintings.—Intention to see a homogenous dark field of vision: red
and green shoes appear. Most phenomena much nearer than reading distance.*
   —Heinrich Klüver, *Mescal: The "Divine" Plant and Its Psychological Effects*

*Outside was the most beautiful scene. A peacock stood in a circle of stone, feathers
spread wide, projecting its energy into my direction. . . . Running through the house,
a race against the purging urging me on, I was consumed with desire to be outside.
Once there, I ran across the deck and vomited over the side. Below me, awaiting a feast
was a large group of peacocks. As I puked they fed. Not on the food that came from
my stomach, but the ooze from my head. They felt the energy in the house growing
and knew we were coming to give them a gift. As they fed, I realized they were
receiving my consciousness. They were accessing altered states via me. As I realized
this, the group of birds began to let their DNA flow. All around the now birds, were
the eternal birds, flowing in brilliant blue patterns of life tide. I was amazed. . . . My
now self became aware of my eternal self. The me recycling through history. All of
nature rejoiced and vibrated at a higher level, aware that another was conscious of
the true resonance. Layers of information began to flow into my field of awareness.*
   —Souldier

Once I managed to behold a bright ascending multitude of sparks, but at half-altitude it transformed itself into a group of silently moving spots from a peacock's tail. During the experiment I was very impressed that my state of mind and the type of hallucinations harmonized so consistently and uninterruptedly.
    —W. A. Stoll quoted in *LSD: My Problem Child*

Finally, we note among the visions of the Saints one called the Universal Peacock, in which the totality is perceived thus royally appareled.
    —Aleister Crowley, "Absinthe: The Green Goddess"

There are no lights on, yet I am blinded by patterns that look like beautiful peacock feathers. . . . The trees look as if they're all connected. I cannot distinguish the branches of one tree from another. It looks so gorgeous. Again, it looks like a giant peacock spreading its magnificent, beautiful feathers in the morning sun.
    —thechubbygoblin

Then, on cue, the experience took off into another direction. I was shown very interesting and complex imagery: Hindu gods and eyes of Horus patterns like peacock feathers swirling ad infinitum. My forehead felt like it was opening. When realized that I was thinking about the nature of this substance, 2c-t-2, a name suddenly popped into my head with insistence: Shiva. Flames and eyes everywhere. Wow!
    —Gandolf

If you start to see peacocks hop around the room, bounce off your head, the ceiling has giant pimples, gnomes start to peek out and tease you and your arms stretch out through walls then a few miles down the street while they are changing liquid crystal colors, then it's a sure sign that it's not a good time to make up with your parents, especially don't start philosophical conversations with them about God, politics etc. or you will be locked up for life.
    —"Trip Toys"

The image is that of the peacock who ingests poisonous plants . . . transmuting it into an elixir responsible for his beauty.
    —Lhundup Sopa, *Peacock in the Poison Grove* [15]

If psychedelics leave undeniable traces on the rhetorical fossil record even as, or because, they refuse individual interpretation, they do so because they are fundamentally troped. Indeed, the older coinage for (some) of these compounds, "psychotropic," nicely captures this work of psychedelics without confining our mechanism to those of "consciousness." Among its many other meanings, "Psyche" is breath, the fundamental unit of movement in yogic rhetorics. A trope is a pattern of information that "turns" the attention, a repeatable response in response to response, one version of which is known as the "flowers" of rhetoric. A logic of replication is involved here; teachers of rhetoric who sought to write down and "store" the many recipes of the rhetorical tradition for capturing and altering human attention rightly saw in the flower an easily replicable mnemonic device. The initial "alert" of a psychedelic—that first unmistakable but nonetheless stray sign that Something Is About to Happen—often arrives precisely as a tropism, a movement or recoil that can later iterate in the closed-eye visuals of double helical or fractal arrays. . . . If fractal sets themselves often take the trope of the patronym—e.g., the Mandelbrot set, the Lorenz attractor—psychedelics as a whole offer a more diverse toolbox of response. Perhaps, even, diversification is itself the response we can link most coherently to the feeling and epistemology of imbrication with the transhuman—the evolutionary imperative to become that is a common place of psychedelic discourse. For contemporary psychonauts enmeshed by the accrued psychedelic discourse I have treated this far as well as the informational web of the Internet, the persistently ecodelic symptom of graphomania is given an outlet and even indulged by a culture of the "blogosphere" whose leading edge was The Vaults of Erowid. The result is an online compendium of contemporary trip reports whose content provides us with a remarkable archive of data with which to study ecodelic discourse and program our experiences with awareness, health, and safety.

It is true, though, that our data, our "rhetorical fossil record," faces a remarkable and perhaps singular evidentiary challenge: psychonautic discourse continually clusters around protests of ineffability—however troped, the psychedelic experience is representable only, apparently, by itself. In chapter 4, we will watch with some compassion and humor as Albert Hofmann struggles to speak persuasively under the influence of LSD-25 or even about LSD until finally offering the compound itself as an act of communication beyond words.

And the ordeal of speaking is only one of the panoply of rhetorical difficulties instigated by psychedelics. We will watch with near identificatory zeal as Hofmann attempts to explain his unprecedented findings to his boss. But given these difficulties in responding to psychedelic experience, it is nonetheless difficult to avoid repetition and consistency across accounts. *Sexual selection would appear to be a content provider for psychedelic experience, as if their evolutionary use were allegorized or imprinted by some of its content as well as its form* (the breakdown of male/female polarity, a breakdown in symmetry). Peacock feathers appear in both the evolution of peacock reproduction and the images of ecodelic experience precisely because both processes involve the search for attention sinks. That mandrills use "virtual" rather than "actual" peacock feathers does not matter to the mandrill; sexual selection includes the focusing of attention on tiny differentials of fitness and aphrodisiac effects. The capacity of sexual selection to browse for new forms of attention sinks and symmetry breaking continues with the hacking of "captchas," security frameworks that seek to capture human attention through characters readable only by actual human wetware. Author Cory Doctorow details this scheme:

> Someone told me about an ingenious way that spammers were cracking "captchas"—the distorted graphic words that a human being has to key into a box before Yahoo and Hotmail and similar services will give her a free email account. The idea is to require a human being and so prevent spammers from automatically generating millions of free email accounts. The ingenious crack is to offer a free porn site which requires that you key in the solution to a captcha—which has been inlined from Yahoo or Hotmail—before you can gain access. Free porn sites attract lots of users around the clock, and the spammers were able to generate captcha solutions fast enough to create as many throw-away email accounts as they wanted. . . . Which suggests a curious future, where commodity pornography, in great quantities, is used to incent human actors to generate and solve Turing tests like captchas in similarly great quantities.

Here I want to juxtapose the evidence for sexual selection's role in the evolution of human consciousness with writer Jeremy Narby's recent arguments concerning shamanic knowledge of DNA:

It no longer seemed unreasonable to me to consider that the information about the molecular content of plants could come from the plants themselves, just as the ayahuascero's claimed. However, I failed to see how this could work concretely. (2001, 52.)

Given the often compelling character of ecodelic visions, it is indeed important to state that nothing ought to be thought (in advance) to be an unreasonable attribute of them. If nothing else, such visions are persuasive, inducing further discourse. As with the bird familiars that frequently accompany ayahuasca visions, it is often through song and discourse that the representation of fitness and recombination of DNA unfolds in human culture. Sexual selection is perhaps precisely the attempted broadcast and evaluation of information about likely fitness, so if Narby's arguments are correct, then in one sense the shamans are "merely" recapitulating bird practices in their management of ecstatic states and the transduction and evaluation of genetic as well as developmental information. As with bird-song, it is music that provides a crucial medium for this intensive act of evolutionary communication.

According to the shamans of the entire world, one establishes communication with spirits via music. For the ayahuascero's, it is almost inconceivable to enter the world of the spirits and remain silent. Angelika Gebhart-Sayer discusses the visual music projected by the spirits in front of the shaman's eyes: it is made up of three dimensional images that coalesce into sound and that the shaman imitates by emitting corresponding melodies. (Ibid., 68)

So too does the characteristic symmetry breaking of some sexually selected traits—the intensification of sexual differentiation—render self and other into a duplex: "it makes light of the sexes, and of the opposition of contraries; it is male and female too, *a twin to itself*" (ibid., 66).

Years earlier, pioneering psychonaut Gordon Wasson embarked on his mycological and entheogenic quest precisely as a result of his wedding vows, in some ways launching the psychedelic revolution. As an experiment with captchas as a tool for the distributed study of the ecodelic hypothesis, I will leave it to the reader to pursue their own rhetorical analysis of this allegory of sexual selection, aided and abetted by my use of italics.

It was a walk in the woods, many years ago, that launched my wife and me on our quest of the mysterious mushroom. We were married in London in 1926, she being Russian, born and brought up in Moscow. She had lately qualified as a physician at the University of London. I am from Great Falls, Montana, of Anglo-Saxon origins. In the late summer of 1927, recently married, we spent our holiday in the Catskill Mountains in New York State. In the afternoon of the first day we went strolling along a lovely mountain path, through woods criss-crossed by the slanting rays of a descending sun. We were young, carefree and in love. Suddenly my bride abandoned my side. She had spied wild mushrooms in the forest, and racing over the carpet of dried leaves in the woods, she knelt in poses of adoration before first one cluster and then another of these growths. *In ecstasy she called each kind of by an endearing Russian name. She caressed the toadstools, savored their earthy perfume.* Like all good Anglo-Saxons, I knew nothing about the fungal world and felt that the less I knew about those putrid, treacherous excrescences the better. For her they were things of grace, infinitely inviting to the perceptive mind. She insisted on gathering them, laughing at my protests, mocking my horror. *She brought a skirtful back to the lodge. She cleaned and cooked them. That evening she ate them, alone. Not long married, I thought to wake up the next morning a widower.*

These dramatic circumstances, puzzling and painful for me, made a lasting impression on us both. From that day on we sought an explanation for this strange *cultural cleavage* separating us in a minor area of our lives. Our method was to gather all the information we could on the attitude toward wild mushrooms of the Indo-European and adjacent peoples. We tried to determine the kinds of mushrooms that each people knows, the uses to which these kinds are put, the vernacular names for them. We dug into the etymology of those names, to arrive at the metaphors hidden in their roots. We looked for mushrooms in myths, legends, ballads, proverbs, in the writers who drew their inspiration from folklore, in the clichés of daily conversation, in slang and the *telltale recesses of obscene vocabularies.* We sought them in the pages of history, in art, in Holy Writ. We were not interested in what people learn about mushrooms from books, but what untutored country folk know from childhood, the folk legacy of the family circle. It turned out that we had happened on a novel field of inquiry. (Wasson 1957, 113; emphasis mine)

The bird deity is back. This time there is a different mood. His blue-crested head appears before a flickering, pixelated background. The cheap clinic is gone.

"So, you want to know about North America? I'll teach you about North America." (beat)

"You will mourn all of my dead."

I can feel the panic of the epic horror trip coming. I had had enough difficulty the first night.

"I am profoundly grateful for what you have done. But I have a body. I can't mourn all of your dead. It will kill me. I have a body, I must dwell on this embodied plane, remember?"

"Oh no, you will."

In short, I had learned to negotiate.

Back and forth we went, knick knack, paddywack

give the dog a bone

but in birdsong—

it sang, I listened, I sang back—it went like a whistle, thistle

blooming incandescent

Faschizzle

until:

"Ok, look, you want to integrate this knowledge into your life back in North America? In your work, will you testify that you have been healed by an ancient Indian technology?"

"I will."

Scene shift as I hear: "And now for some joy."

4

LSDNA

Creative Problem Solving, Consciousness Expansion, and the Emergence of Biotechnology

I had to struggle to speak intelligibly.
—Albert Hofmann on his self-experiment with LSD-25

Finding a place to start is of utmost importance. Natural DNA is a trackless coil, like an unwound and tangled audio tape on the floor of the car in the dark.
—Kary Mullis on the invention of Polymerase Chain Reaction

LIFE BEGINS WITH COILING-MOLECULES BE NEBULAE.
—Michael McClure, as quoted by Francis Crick

SO THERE IS A REFRAIN, ORDER AND NOISE, RHYTHMIC ENTRAINMENT, and symmetry breaking. In short, at perhaps its most compressed, laughter. Indeed, perhaps laughter is something like the bardo between ordered speech and its noisy, babbling, babeled other. In its endless search engine for eloquence in ego death, that nonrepresentable, ineffable, Thing That Happened to Me and What It Means, ecodelic discourse must of course asymptotically approach nothing but babbling, bubbling, laughter, dissolving the ordinary ego and making space for the extraordinary interconnections of collective laughter.

But what role could such laughter possibly play in *scientific practice*? In the first two books of this trilogy, *On Beyond Living* and *Wetwares*, I attempted to trace out the ways that the life sciences were molded by "information." Talking, representing, and writing about living systems in terms of information—such as the "genetic code" or "program"—yielded "rhetorical softwares" that enabled forgetting as well as insight, focusing human attention on the molecular scale of living systems and sometimes obscuring the larger scale embodied and ecosystemic levels of development and evolution. In a continuation of that rhetorical map of the life sciences, I want to map molecular biology's rhetorical and conceptual evolution and its debts to those forms of agency best indicated by laughter but available to many extraordinary feelings endemic to even the scientific will: laughter, terror, ecstasy. Such modes of response, I will argue, were crucial to a conceptual evolution in the life sciences whose feedback loop arrives at cloning and tends toward a nanotechnological impasse: DNA information, at first understood by molecular biology as a fundamentally stable semantic phenomenon or "secret," became a spectacularly mutable technology of replication and differentiation by the early 1980s. Considered spatially, and as an object of knowledge, DNA becomes distributed across a field of forces directed but not mastered by human attention.

While rumors of the so called "death of life" have perhaps been greatly exaggerated, something changed when human beings began to directly select organisms at the level of genotype, recombining them. A vector of evolution that had already been operative—the effect of human attention on the biosphere named by Vernadsky the *noösphere*—intensified when it became linked to the direct manipulation of nucleic acids and their expression in novel organismic and ecosystemic contexts and scales. This was among other things a *scalar shift* in the effect of human consciousness and actions on the biosphere, giving rise to novel organisms less "bred" than selected out of the space of all imaginable traits. This scalar shift is also a speeding up of evolutionary time, which goes nonlinear as improbable admixtures of traits—a flounder gene is expressed in a tomato in the service of putative human gustatory pleasure—outside the spatial, temporal, and species boundaries separating fish from fruit. Slowly, but unmistakably, a biotechnological earth hosts enormous extinction events and the emergence of species literally *conceived* in the human imagination.

This is a familiar story, but how did we get here? This veritable "undoing" of life, and the concomitant "loss" of integrity in organismic and taxonomic distinctions—wrought first by recombinant DNA and then Polymerase Chain Reaction (PCR)—seems to occur in response to an undoing and doing of identity sculpted by that most tabooed and double-entendred scientific enterprise: the self-experiment. In particular, the necessary role of the self-experiment in the scientific study of psychedelics—an inquiry not into life but into consciousness—provides the ecology for the emergence of these innovative and even ecstatic modes of interaction apparently necessary to this particular unfolding of biotechnological evolution. While in the last chapter I tried to reframe Jeremy Narby's provocative claim that the shamans of the upper Amazon shared the blind spots of reductionist molecular biology, this chapter will trace out the ways in which shamanic rhetorical practices are themselves at the heart of molecular biology's transformation as well as the genetic medicine associated with it.

My discussion will respond to the common self-experiments at play in the seemingly diverse ecologies of the ecodelic "expansion" of "consciousness" and the engineering of evolution and biotechnology. These ecologies will be treated as twin or replicated domains where an *informatic desire*—a desire to become information—distributes and disperses both consciousness and life into inhuman, inorganic, and extraterrestrial realms.

TAKING DNA, OR TRIPPING OVER THE ORGANISM

In my informatic genealogy of the life sciences thus far, I have claimed a shift from an emphasis on the interiority of organisms and the associated revelation of their essence to a harnessing of DNA's capacity to replicate and its subsequent "distribution" of life. No longer simply the attribute of a perceived sovereign organism, life now emerges out of the connections of a network, involving an essential impropriety—it is life's habit of refusing containment that gets biotech capitalism's attention. The capacity to be owned—"Hey, that's my pig!" or *Diamond v. Chakrabarty*!—requires rhetorical articulation that is continually undermined by the radical intertwingling of ecoystems.[1]

These dual aspects of living systems—life as an autopoietic system of inside and outside, and life as a meshwork of information systems—offer extraordinarily different models of "life," both of which take place under the sign of "DNA" and molecular biology. One, the cracking of the code, looked to reveal vitality as an attribute of a "aperiodic crystal," an orderly rather than mysterious enterprise best apprehended by physics and understood through an intensive albeit technical practice of crystal gazing.

James Watson and Francis Crick could behold a Rosalind Franklin crystallograph and regard it as encrypting the "secret" of life. The second model of living systems we associate with biotechnology, and its collateral capital and publicity markets, is less interested in what DNA might "mean" than with what it can *become*, and the relation between what it can do—replicate—and its production of value.[2] Indeed, an emphasis on the primordial importance of *copying* reminds us of the impossibility of keeping secrets, as replication allows genes and ideas to travel to multiple contexts, some intended and some not. Contemporary life science, despite all the (New Testament) chatter of God associated with the rough draft of the human genome, is interested less in clear cut "beginnings" and "ends" than in experimentation and mutability, capacity for deterritorialization that generates value in the economy.[3]

While molecular biology was busy on its "eighth day of creation" (Judson 1979, 1) discovering, decoding, and analyzing the "secret" of life, LSD-25 was also proffered in the labs, living rooms, and communes as the secret of consciousness. Contrary to its usual representation as a seamless technology of unveiling associated with the instant gratification allegedly sought by an entire decade—the 1960s—LSD was treated as a paradigm shattering and highly tuneable tool for the probing, revelation, and "expansion" of consciousness. Writers, researchers, and experimentalists such as Timothy Leary, Humphry Osmond, Richard Alpert, and Stanislav Grof all sought to study the function of "set and setting" in the instantiation of psychedelic states and their capacity to transform human consciousness. These writers *took DNA* and used it to frame and articulate ecodelic sessions as programmable but not controllable events, experiments with and on the self. As such, theirs was a fundamentally *pragmatic* rather than *semantic* relation to the "information" of DNA, more recipe than message.

How did these writers *take* DNA? In fact, there were experiments among researchers attempting to ingest DNA and RNA itself as a hallucinogen, sometimes in the hope of developing a "learning lozenge" that would inscribe the experience of LSD onto the brain (Stafford and Golightly, ch. IX). But nucleic acids were also crucial *rhetorical* vectors composing psychedelic discourse of the 1950s and '60s: the talk, thought experiments, manuals, images, sounds, and technical papers that resulted from variously intentional and unintentional ingestions. It was here that, time and time again, nucleic acids would play a crucial role as rhetorical adjuncts to the narration and programming of psychedelic experiences.

Ecodelic discourse, both of scientific and "recreational" nature, faced a similar rhetorical dilemma as the rest of the ecstatic traditions to which it responds: it must report on an event which is *in principle* impossible to communicate. Writers of mystic experience from St. Theresa to William James have treated the unrepresentable character of mystic events to be the very hallmark of ecstasis. In a scientific context, ecodelic discourse faced a similar struggle in the effort to report on the knowledge beyond what Aldous Huxley, sampling from William Blake, described as the "doors of perception." To deal with this rhetorical ordeal, a long running "cosmic game" (Grof) where maps of the ineffable are improbably rendered by participants across space/time, many psychedelic researchers had recourse to the rhetoric of nucleic acids—DNA and RNA became privileged networked characters in the stories and practices of psychedelic science. One could even say that DNA was a kind of psychedelic archetype, so frequent was its image and language in twentieth-century psychedelic science.

Nucleic acids were more, though, than "content providers" for the channeling of ecodelic knowledge into quasi-scientific protocols and the new modes of life enabled by them such as artificial life, transgenic organisms, and enormous amounts of science fiction. As carriers of the news of molecular biology's informatic vision, rhetorics of nucleic acids were also set and setting for psychedelic sessions themselves: more than reporting devices, the rhetoric of nucleic acids helped to suggest that these sessions were themselves, like DNA, endlessly programmable.

DOUBLE TAKE

Problems of reportage troubled the discourse of LSD almost from its very inception, and certainly from its very first ingestion. In his fundamental ergot studies in 1938, Albert Hofmann first synthesized Lysergic acid diethylamide, abbreviated LSD-25 (Lyserg-saure-diathylamid) for laboratory usage. This novel molecule was primarily noted for its strong effects on the uterus, but as Hofmann retroactively reports in his autobiography named for a molecule, *LSD: My Problem Child*:

The research report also noted, in passing, that the experimental animals became restless during the narcosis. The new substance, however, aroused no special interest in our pharmacologists and physicians; testing was therefore discontinued. (12)

That an ergot-derived compound would affect the uterus was not a surprise: Hofmann's agenda was "to isolate the active principles (i.e., the effective constituents) of known medicinal plants to produce pure specimens of these substances" (ibid., 36). And "ergot . . . had been used since olden times by midwives" (ibid., 40). Hofmann continued his work in the ergot field, but LSD-25 was thought to have little pharmacological value, so between 1938 and 1943 "nothing more was heard of the substance LSD-25."

And yet Hofmann still had ears for the crying of LSD-25. According to his own account, LSD wouldn't leave him alone:

> And yet I could not forget the relatively uninteresting LSD-25. A peculiar presentiment—the feeling that this substance could possess properties other than those established in the first investigations—induced me, five years after the first synthesis, to produce LSD-25 once again so that a sample could be given to the pharmacological department for further tests. This was quite unusual; experimental substances, as a rule, were definitely stricken from the research program if once found to be lacking in pharmacological interest. . . . Nevertheless, in the spring of 1943, I repeated the synthesis of LSD-25. As in the first synthesis, this involved the production of only a few centigrams of the compound. (14)

Although Hofmann's attribution of presentiment must be placed within its context as an autobiographical confession, it nonetheless well names a peculiar agency that often adheres to those self-experiments that are survived: the inability to forget. Note the exogamous agency of Hofmann's synthesis: he was "induced" to produce LSD-25 once again. The memory of LSD and, as we shall see, LSD-25 itself, seems to have little truck with the usual operations of will. Indeed, according to Hofmann, his response to the crying of LSD-25—synthesis—resulted paradoxically in an interruption or dissolution:

In the final step of the synthesis, during the purification and crystallization of lysergic acid diethylamide in the form of a tartrate (tartaric acid salt), I was interrupted in my work by unusual sensations. (15)

This interruption of the I, rather than ending an experiment, instigates one. In a report to a superior, Hofmann did his best to offer a description of the phenomenon that ensued after the interruption, but he only "surmised" a connection with the LSD-25. Intriguingly—given the origins of the ergot pharmacopeia—Hofmann's neighbor, supplying him with milk he had requested as an antidote to the "poisoning"—literally, nursing him—took on the appearance of a witch:

The lady next door, whom I scarcely recognized, brought me milk—in the course of the evening I drank more than two liters. She was no longer Mrs. R., but rather a malevolent, insidious witch with a colored mask. (49)

Hofmann's analysis of the cause of the interruption itself was interrupted by the question of ingestion. Given his method of synthesis, Hofmann reasoned that his accidental passage must have been through the skin, and that the substance must therefore be of extraordinary—indeed, unprecedented—potency.

Thus Hofmann's causal analysis of the unusual sensations seemed to offer two extraordinarily unlikely—i.e., unprecedented—alternatives. On the one hand, the cause could remain unknown, and the tasteless and odorless LSD-25 had merely been associated with the experience rather than causing it. In this instance the strange interruption retained its enigmatic status, a non sequitur of Hofmann's experience. On the other hand, if the minute quantity of LSD-25 were the causal agent of the interruption, then Hofmann was faced with the equally unlikely scenario that he had discovered the most potent compound known in history. To resolve the situation, Hofmann had recourse to an extraordinary non sequitur, itself seemingly emerging without cause: "There seemed to be only one way of getting to the bottom of this. I decided on a self-experiment" (20).

In what sense can one "decide" on a self-experiment? What warrants this decision? If it seems obvious that indeed Hofmann did decide on such a course of action, it must also be noted that such a decision is, of necessity, itself an experiment, one that emerges not from any deliberative logic but from an incalculable action, a breakage in the chain of reasoning: just do it. Hofmann's deliberation on his possible responses to the interruption was itself not subject to anything like a procedure, an algorithm shorter than repetition by which he could arrive at a resolution of the two equally enigmatic, if thoroughly differentiated, outcomes.[4] Thus this decision to self-experiment is a testing of a self as well as a hypothesis. The outcome of this experiment—Hofmann's implication of himself into the research—cannot be meaningfully differentiated from the experimental dosing of LSD-25. Only an additional experiment—the synthesis and ingestion of LSD-25—will, retroactively, provide this experiment in decision making with anything like a result. As if to mark the extreme danger that the implication of a self and body into an experiment with a compound of possibly unprecedented potency entails, Hofmann, with extraordinary courage, writes oxymoronically of a self-experiment embarked upon with "extreme caution":

> Exercising extreme caution, I began the planned series of experiments with the smallest quantity that could be expected to produce some effect, considering the activity of the ergot alkaloids known at the time: namely, 0.25 mg (mg = milligram = one thousandth of a gram) of lysergic acid diethylamide tartrate. (16)

The danger here, though, is not only the risk of an unknown compound of apparently extraordinary psychic potency. Rather, the instance or event of danger emerges coincident to the interruption of work itself: how to proceed? For as Hofmann notes after his now deliberate ingestion, LSD-25 is nothing if not the incessant and yet irregular arrival of the question: *How to go on?* This is a question of endurance for the experimental self: usual experimental protocol demands that everything is involved in an experiment *except* the self, and yet here it is precisely only the self and its responses that are the very assay of LSD-25. Only the variable examples of history provided anything like a protocol for self-experiment, calibration for the assay.[5]

This assay had great difficulty generating any readout:

> 4/19/43
> 16:20: 0.5 cc of ½ promil aqueous solution of diethylamide tartrate orally = 0.25 mg tartrate. Taken diluted with about 10 cc water.

Tasteless.

17:00: Beginning dizziness, feeling of anxiety, visual distortions, symptoms
of paralysis, desire to laugh. (16)

LSD-25 did little to present itself for inscription. Without flavor, within
forty minutes it produces predominantly anticipatory symptoms, events
which were about to make themselves more fully known. The "desire to
laugh," as such an experimental anticipatory symptom, was particularly difficult to assay. For by what means would any desire to laugh be registered
by an observing self, except by laughter itself and its subsequent attempted
blockage? What agency would interrupt said desire and, interrupted, in
what sense did one desire to laugh?

Obviously, anyone can wish for laughter: this is the unlikely hope
marked by the sitcom laugh track. But what is named by Hofmann is perhaps less a wish than a tantalizing inclination, a becoming laughter that is
neither a cackling nor its absence, a meanwhile in which the proximity of
the future—I am about to laugh—is unbearable to the self of the present. For
Hofmann, this desire indeed becomes unbearable, the doing of the experiment veritably undone:

Supplement of 4/21: Home by bicycle.

From 18:00–ca.20:00 most severe crisis. (See special report.)

Here the notes in my laboratory journal cease. I was able to write the last
words only with great effort. By now it was already clear to me that LSD
had been the cause of the remarkable experience of the previous Friday, for
the altered perceptions were of the same type as before, only much more
intense.

I had to struggle to speak intelligibly. (Ibid.)

Hofmann's certainty regarding the causal role of LSD-25 in the visions
and disturbances he experienced was equaled by his inability to communicate the character and nature of the struggle. While under the variable
influence of LSD-25, Hofmann periodically ceases to be capable of even an
attempt at communication and thus, an attempt at experimentation. In place
of the usual and invisible expectation of communicability that is the rhetorical arena of scientific observation, the struggle suggests Hofmann's reliance

on another mode of knowing altogether: the agon(y) of the ordeal, an epic event in which knowing is not separable from an irreducible participation. "I had to struggle to speak intelligibly."

This inability to communicate is not a deficit in the ecodelic experience, but a highly variable symptom of it. Recall that such a challenge, as in the Greek rhetor Demosthene's use of a pebble to increase his eloquence, can also paradoxically activate human rhetorical capacities. In this sense, the self-experiment is both a failure and a success: as an experiment *with* the self, the outcome is close to null. No meaningful empirical report can be generated, and therefore the knowledge of the psychedelic experience can in no way be gathered or repeated. As an assay, the self is found wanting. If the experiment is an occasion at which, strangely, the self is to be present as the very apparatus through which the inquiry is to be made visible and replicable, then the apparatus has faltered and the experiment is nothing but artifact.

And yet the need to communicate the ineffable persists. Later, Hofmann would nonetheless prepare a special report to his supervisor, Professor Stoll. Here Hofmann was struck by the *capacity to remember* the experience with great, even machinic precision:

> What seemed even more significant was that I could remember the experience of LSD inebriation in every detail. This could only mean that the conscious recording function was not interrupted, even in the climax of the LSD experience, despite the profound breakdown of the normal world view. For the entire duration of the experiment, I had even been aware of participating in an experiment, but despite this recognition of my condition, I could not, with every exertion of my will, shake off the LSD world. Everything was experienced as completely real, as alarming reality; alarming, because the picture of the other, familiar everyday reality was still fully preserved in the memory for comparison. (20)

As an experiment *on* the self, the ingestion of LSD-25 was indeed a resounding success—the experimental object was unmistakably transformed, alteration extending even to the agency of Hofmann himself and his rhetorical capacities as a veritable medium for the experience. As a recording, Hofmann could not avoid remembering as, for the purposes of LSD-25, he became what Merlin Donald describes as an "external symbolic storage" device for LSD. (Donald 1991, 17) Combined with Hofmann's own memory of LSD's agency—"And yet I could not forget the relatively uninteresting LSD-25"—Hofmann's transformation by the experiment is subtly, but nonetheless inescapably, inscribed in the confession: LSD-25 is not easily erased from the experience and memory of the experimental subject, a subject who, like it or not, is recording. Indeed, in some sense Hofmann is both recorder and recording here, as he must respond exegetically to the demands of memory. Hofmann would later describe this structural coupling between observation and reality in terms of "tuning," a rhetoric resonant with Leary's post-cybernetic "Turn on, Tune in, Drop out":

> If one continues with the conception of reality as a product of sender and receiver, then the entry of another reality under the influence of LSD may be explained by the fact that the brain, the seat of the receiver, becomes biochemically altered. The receiver is thereby tuned into another wavelength than that corresponding to normal, everyday reality. Since the endless variety and diversity of the universe correspond to infinitely many different wavelengths, depending on the adjustment of the receiver, many different realities, including the respective ego, can become conscious. These different realities, more correctly designated as different aspects of the reality, are not mutually exclusive but are complementary, and form together a portion of the all-encompassing, timeless, transcendental reality, in which even the unimpeachable core of self-consciousness, which has the power to record the different egos, is located. (196–97)

Hence while the conclusions to Hofmann's exercise of extreme caution remained to be determined, there could be no argument but that the risk had yielded interesting data, namely, the transformation of Hofmann himself into a being seemingly incapable of forgetting through recourse to "the unimpeachable core of self-consciousness." On this factor alone, Hofmann could determine that he had happened upon a substance of unprecedented potency. Its usefulness and character would, of course, call for further research, but Hofmann's understanding of the causal nature of LSD-25 in his experience was certainly an important result capable of representation to his colleagues.

But if Hofmann had great faith in the splitting capacities of LSD as a molecule that allows for an almost unique position as a retroactive observer and real-time participant, where "universe and self, receiver and sender, are one" (198), his supervisors, at least at first, did not:

> As expected, the first reaction was incredulous astonishment. Instantly a telephone call came from the management; Professor Stoll asked: "Are you certain you made no mistake in the weighing? Is the stated dose really correct?" Professor Rothlin also called, asking the same question. I was certain of this point, for I had executed the weighing and dosage with my own hands. Yet their doubts were justified to some extent, for until then no known substance had displayed even the slightest psychic effect in fraction-of-a-milligram doses. An active compound of such potency seemed almost unbelievable. (21)

While the incredulous questions focused repeatedly on issues of quantitative importance, clearly it was also the qualitative transformations wrought by LSD-25 that inspired disbelief. The missing special report, whatever it says, was evidently unable to translate the delirium of the self-experiment into the allegedly intersubjective space of scientific communication. So little convinced were Hofmann's colleagues of his report that they took the almost unbelievable step of repeating the self-experiment:

> Professor Rothlin himself and two of his colleagues were the first to repeat my experiment, with only one third of the dose I had utilized. But even at that level, the effects were still extremely impressive, and quite fantastic. All doubts about the statements in my report were eliminated. (22)

That is, LSD-25 was in this instance both the object of scientific inquiry and the medium for the communication of the results of that inquiry, the translation of a solo experiment into the general equivalent of scientific truth. While in the case of Maria Sabina and many others, ecodelics seem to induce an increased capacity for discourse, Hofmann's was an eloquence of silence in which the psychedelic *was* the discourse, the medium was the message. While Hofmann's wife returned from Lausanne upon hearing reports that he had had some sort of "breakdown," his fellow scientists willingly and immediately ingested LSD in order to eliminate any doubts fostered by Hofmann's report, a psychedelic republic of letters.

It is as an experiment *on* the self that Hofmann's discoveries are replicated by the community. Only by encountering a veritable undoing of self— a submission to the possible transformation one is in fact testing for—can interesting data from this novel pharmacological agent be gathered, evaluated, and transmitted.

INFORMATIC PRAYER

The combination of ineffability common to many mystic traditions and the necessity of communication proper to scientific practice continued to pose problems for the study of psychedelics as they migrated from Sandoz Pharmaceuticals, where Hofmann worked, to places like the pre-1962 psychology department of Harvard University, the working home of Timothy Leary, Ralph Metzner, Richard Alpert, and Michael Horowitz. Among a parade of intellectuals and artists that moved through Leary's burgeoning circle was writer William S. Burroughs, whose cut-up techniques had, a few years earlier, been used to strip a written text of its meaning—what Burroughs called the virus of the Word—and to allow such texts to interrupt normal consciousness, à la the earlier techniques of Dada. After a visit with Burroughs, Leary hit upon the idea of using Burroughs's cut-up technique as a framework for reporting the psychedelic experience. In this framework, both Burroughs's texts and LSD were experiences one less understood than underwent and, in undergoing them, recorded. These recordings—as with Hofmann's precise recall of his LSD trip—were not, however, communicative in the usual sense: they could be understood only retroactively, after the subject had encountered LSD-25. With the cut-up method, Leary hoped to interrupt the grip of the authorial ego that might interfere with the more direct recording of the LSD experience. In other words, the LSD experience was treated less as an event to be reported on than an experience to be assayed by writing—the cut-up was a symptom and not a description of the encounter of LSD. In this sense, the information gathered about LSD was understood less as a semantic production than a pragmatic one: all data concerned not an *understanding* of LSD but instead consisted of testing LSD for

what it could *do*. Writing, in this context, becomes less a struggle for intel-
ligibility than a prolix signature of the LSD experience.

Indeed the apparent need to write in response to LSD provoked a
graphomania of sorts among Leary's group. By 1962 they had all either quit
or been removed from their positions at Harvard University, and were now
leading itinerant seminars from Zihuatanejo, Mexico; several islands in the
Caribbean; and then finally the Hitchcock brothers' mansion in Millbrook,
New York. *The Psychedelic Review* regularly published the group's prolific
work and, in 1964, Leary, Metzner, and Alpert published a manual for use
in association with LSD entitled *The Psychedelic Experience*. The manual was
presented as a source of protocols for the management and study of psychedelic sessions, protocols "based on *The Tibetan Book of the Dead*." As such,
much of the writing was oriented less to "understanding" LSD experience
than with dealing with it, tuning it. These texts were intended to be part of
the very interface of psychedelia, algorithms for attaining and prolonging
particular states:

> One may want to pre-record selected passages and simply flick on the
> recorder when desired. The aim of these instruction texts is always to lead
> the voyager back to the original First Bardo transcendence and to help
> maintain that as long as possible. (Leary 1995, 97)

These recipes and techniques for ecstasy were repeatedly and explicitly
linked by Leary to the writing and execution of a sequence of steps in a computer environment, programming:

> A third use would be to construct a "program" for a session using passages
> from the text. The aim would be to lead the voyager to one of the visions
> deliberately, or through a sequence of visions. . . . One can envision a high
> art of programming psychedelic sessions, in which symbolic manipulations
> and presentations would lead the voyager through ecstatic visionary Bead
> Games. (Leary 1995, 98)

In this framework, psychedelic subjects become both authors of, and
platforms for, "symbolic manipulations and presentations," interactive wetware of infinite experiment and transformation. In the preface to a later

work, Leary would write that *"The Psychedelic Experience* was our first attempt at session programming" (Leary 1966b, 37).

Crucial to this vision was that the function of information here was to "program." With this articulation, psychedelic experience becomes not only a recording by the subject, but an intentional one: one becomes a self who writes the self. Less an activity of understanding or even communication than of repetition and transformation, these programs had to be actualized, altered, and remixed. Indeed, the writings "based" on *The Tibetan Book of the Dead* were often not understandable or recognizable by readers until, like Hofmann's colleagues, one had undergone an encounter with LSD. "The most important use of this manual is for preparatory reading. Having read the Tibetan Manual, one can immediately recognize symptoms and experiences which might otherwise be terrifying, only because of lack of understanding as to what was happening. Recognition is the key word" (Leary 1997, 97). Thus *The Tibetan Book of the Dead* (a translation commissioned by W. Y. Evans Wentz, a student of William James and William Butler Yeats, and translated by Tibetan scholar Kazi Dawa-Sandup) was treated as a source book not to be decoded as much as shared and deployed in divergent contexts, wielded more than understood. Leary made this pragmatic understanding of information even clearer in *Psychedelic Prayers*, his 1966 "translation" of the *Tao Te Ching*; what Leary called a "time tested psychedelic manual":

> Like all great biblical texts, the Tao has been rewritten and re-interpreted
> in every century and this is how it should be. The terms for Tao change in
> each century. In our times, Einstein rephrases it, quantum theory revises it,
> the geneticists *translate* it in terms of DNA and RNA, but the message is the
> same. (1966b, 37)

Translation, for Leary, is both universally available—"the message is the same"—and utterly variable and reliant upon context, what Leary will refer to as "set and setting" for the LSD experience but that can be extended to his own work of translation as well:

> These translations from English to psychedelese were made while sitting
> under a bamboo tree on a grassy slope of the Kumoan Hills overlooking the
> snow peaks of the Himalayas. (Ibid., 38)

Like the drugs and plants with which they are to be used, psychedelic manuals are catalysts for transcendent experiences—"or can be, given the appropriate preparation, attitude, and context" (the "set and setting" in Leary's felicitous phrase) (ibid., 10). Crucial to this facility for novel settings is not, strictly speaking, its universal message, but rather its capacity to be translated and travel into novel contexts: "The advice given by the smiling philosophers of China to their emperor can be applied to how to run your home, your office, and how to conduct a psychedelic session" (ibid., 38). This "deterritorialization" is explicitly treated by Leary as an attribute of nucleic acids and their processes of transcription and translation. Writing of the aforementioned Evans-Wentz, Leary writes:

> W. Y. Evans-Wentz is a great scholar who devoted his mature years to the role of bridge and shuttle between Tibet and the west: like an RNA molecule activating the latter with the coded message of the former. No greater tribute could be paid to the work of this academic liberator than to base our psychedelic manual upon his insights and to quote directly his comments on "the message of this book." (Leary et al., 17)

Note here that despite the emphasis on the "message of this book," the psychedelic manual is not concerned with the communication or elucidation of a meaning, although it also does so. Rather, the text is seen as a how-to tool for managing and transforming diverse contexts with the help of exotic and yet thoroughly debugged techniques for "activation" or transduction. Its translation is less the production of an equivalent meaning than the "porting" of code to a different platform: "It became apparent that, in order to run exploratory sessions, manuals and programs were necessary to guide subjects through transcendental experiences with a minimum of fear and confusion" (Leary 1966b, 36). This lack of interest in the semantic operation of such manuals is itself pragmatic: reportage continually fails, even while context is itself a powerful constituent of the psychedelic experience. Hence over time, repetitive failure would tend to be avoided as a negative feedback loop, and other models of abstraction would be explored. The incommunicability of the psychedelic experience was taken to be a measure of the complexity of human consciousness as well as the insufficiency of most concepts to it:

No current philosophic or scientific theory was broad enough to handle the potential of a 13 billion-cell computer. (Ibid.)

As "prayers," Leary's translations also remind the reader of the actively sacred and rhetorical register involved in the LSD context: prayers, above all, demand prayer. For Leary, their effects emerge out of their very utterance, a whispered utterance that needs to be serenely said more than heard: "they should be read *very* slowly and in a serene voice. They should be considered prayers to be whispered" (ibid., 33). Ideally, then, these rhetorics should approach pure distributed action. Only a trace of the utterance will persist, a whispered scar of language's ecstatic embodiment.

The prayer program was divided by Leary into six parts, and the very center of the sequence is occupied by a twelve-part sutra (literally "thread") entitled "Homage to DNA." Here Leary, "translating" and transforming *The Tibetan Book of the Dead* into epideictic praise of the book of life, instructs the reader/tripper to "contact cellular consciousness" via the utterance of these meditations. The first, "The Serpent Coil of DNA," invokes the old and now familiar image of the Ouroboros, a figure also used by biologist C. H. Waddington in his discussion of feedback:

We meet it everywhere, but we do not see its front. . . .
When we embrace this ancient serpent coil
We are masters of the moment, and feel no break in the curling
Back to primeval beginnings.

This may be called
unraveling the clue of the life process. (Ibid., 67)

Rather than containing a "message," DNA is hailed as a molecule of ceaseless activity whose very embrace leads to an unraveling, an unfolding perhaps of the implicate order. But even as a "master of the moment," this "unraveling" reveals no central wisdom greater than this: our implication in an ancient coil of transformation without beginning or end. Thus rather than triumph, Leary's poetry suggests an affirmation of primeval complicity, a complicity owed to the very self-replicating Ouroboros of DNA and, perhaps, with LSD and the prayer itself. *The universal message alluded to by Leary*

Behold, the Ouroboros.

is none other than transformation itself, the effect of consciousness on itself for which the Ouroboros is a compressed figure.

Contrasted with the molecular biology of the same period—the transmitter of the rhetorics of nucleic acids that were the set and setting of psychedelic knowledge—research into psychedelics, both in the lab and the commune, were strikingly pragmatic in their understanding of information. Both Hofmann and Leary wrote of the need for endurance, a practice that would enable repetition within and beyond the psychedelic experience even while the very assay of the experience, the self, was becoming variable and even breaking down and transforming. Indeed, it was the practice of repetition itself that seemed to emerge as crucial to the acquisition and transmission of ecodelic knowledge. For Leary and the programmers of psychedelic practice, sessions were sequences of information to be actualized differentially—no two sessions could be enacted in precisely the same fashion, precisely because of the transformative effects of LSD and its teachings, repetitive teachings of undoing the self—"unraveling the clue of the life process." While molecular biology continued to write and talk of decoding the book of life and practicing a hermeneutics of DNA, psychedelic discourse suggested that information was less a phenomenon to be understood than it was a potent mutagen of human experience, mutagens that could only be understood retroactively.

HIGHWAY 128, REVISITED

If I had not taken LSD ever would I have still been in PCR?
I don't know, I doubt it, I seriously doubt it.
  —Kary Mullis, BBC Interview

THIS IS THE POWERFUL KNOWLEDGE. We smile with it.
  —Michael McClure, as quoted by Francis Crick[6]

In his 1998 autobiography *Dancing Naked in the Mind Field*, Nobel Prize–winner Kary Mullis details his invention of Polymerase Chain Reaction or PCR. While both the topic—the invention of a veritable Xerox machine for nucleic acids—and the genre—scientific autobiography with a hint of scandal and the promise of a forbidden, "naked" truth—encourage the telling of a heroic tale of innovation, Mullis's narrative continually highlights the thoroughly contingent and ungovernable arrival of an idea. Far from simply inflating the role of a lone scientist struggling to know life and the cosmos, *Dancing Naked in the Mind Field* offers a testimony to the thoroughly other, even alien, character of a scientific vision.

By every account, Mullis is himself an unpredictable creature. A biochemist by training, Mullis's most important publication in graduate school treated time reversal and its cosmological implications. In love with the craft of producing new compounds in a more and more efficient fashion, Mullis was every bit as much a tinkerer as a theorist.

This tinkering extended to Mullis's own consciousness. Like many of his UC–Berkeley chemistry colleagues, Mullis considered the potent new, and not so new, psychedelics to be experimental objects as real and interesting as any other in their practice (163). These molecules, and their differences, though, could only be studied through participatory interaction, "playing the molecules." Indeed, after one particularly difficult episode with the incorrect dosage, Mullis underwent an entire personality change: overtaken by amnesia, Mullis had the sense that indeed he had become someone else as a result of the encounter (169).

Nor was Berkeley Mullis's sole connection to psychedelics. During the period of PCR's emergence as a technique, Mullis visited with Albert Hofmann at the house of his friend Ron Cook. Indeed, Mullis later compares Hofmann's three-lettered discovery with his own:

> The famous chemist Albert Hofmann was at Ron's that night. He had invented LSD in 1943. At the time he didn't realize what he had done. It only slowly dawned on him. And then things worked their way over the years as no one would have predicted, or could have controlled by forethought and reason. Kind of like PCR. (11)

This conjunction of LSD and DNA is, of course, strikingly resonant with Leary's treatment of both above: for Mullis, the main connection between the two was contingency itself, a practice "no one would of predicted, or could have controlled by forethought and reason." Only the unfolding of events allows for a retroactive ascription of narrative and reason—no description of the events could have been compressed into anything shorter than the events themselves. In short, the practice of translating LSD and PCR into technoscientific objects involved nothing less than the transformation of knowledge without precedent, a non sequitur not available to prophecy. There appeared to be no shortcut to either, with only the narrative of accident and coincidence making retroactive sense of the event. Unforeseeable, the effect of both could only be endured, responded to like the surfer Mullis is, where the "sender and receiver are one" (Hofmann, 196).

But Mullis himself goes further in his account of the connection between PCR and LSD. Retaining the contingency of the concept, Mullis nonetheless credits his experience with LSD with a certain capacity for the sheer strangeness of the idea of PCR. In a BBC interview Mullis hesitantly ascribed a paradoxical agency to LSD in the invention of PCR:

> PCR's another place where I was down there with the molecules when I discovered it and I wasn't stoned on LSD, but my mind by then had learned how to get down there. I could sit on a DNA molecule and watch it go by and I didn't feel dumb about that, I felt I could, I mean that's just the way I think is I put myself in all different kind of spots and I've learned that partially I would think, and this is again my opinion, through psychedelic drugs. If you have to think of bizarre things PCR was a bizarre thing. It changed an entire generation of molecular biologists in terms of how they thought about DNA.[7]

Unlike Mullis, who appears to have been inspired, but not dosed when he worked out his idea for PCR, Francis Crick appears to have envisioned the double-helical structure of DNA while using LSD.

> FRANCIS CRICK, the Nobel Prize–winning father of modern genetics, was under the influence of LSD when he first deduced the double-helix structure of DNA nearly 50 years ago. (Rees)[8]

Was this a feedback loop between evolution and ecodelic states as exemplified by Crick and Mullis, enabling a new scale, speed and precision with which consciousness could alter evolution? Astronomer Carl Sagan echoed Julian Huxley when he declared that "We are a way for the cosmos to know itself," and this knowing is sometimes steeped in awe (Sagan, 1980). By 1975, Crick would admire poet Michael McClure's capacity—with his iterative explosion of word—to convey "the effects of the hallucinogen" (Crick, 1975). Yet Mullis's idea was a remarkably simple one. Indeed, in his cabin off of Highway 128 in Mendocino County, California, the idea struck Mullis as far too simple to be workable:

> If the cyclic reactions that by now were symbolized in various ways all over the cabin really worked, why had I never heard of them being used? Why wouldn't these reactions work? (8–9)

Key to Mullis's understanding was the cyclic or iterative character of DNA itself. Unlike earlier molecular biologists who had sought the "meaning" of the genetic code, Mullis sought only iteration: he wanted to replicate enormous quantities of any given sequence. Mullis writes that because of his knowledge of computer programming, "I understood the power of a reiterative mathematical procedure" (5).

It is precisely as an iteration machine that Mullis treats DNA. While H. G. Khorana and his team had provided the foundation for PCR as early as 1965, they did so within a cryptographic paradigm of discovering the nature and meaning of the genetic code. Francis Crick, writing the lead article for a volume devoted to the triumph of the newly decrypted code, declared that the "historic occasion" was the understanding of the four-letter alphabet of ATCG and its triplet combinations that produce the twenty amino acids that in turn fold up into proteins: "From all of this we were able to work out the *meaning* of several of the remaining doubtful triplets" (Crick 1966, 3; emphasis mine). Khorana's article appears thirty-six pages later in the volume, and it is clear from its frame and its content that Khorana's group sought understanding and even proof: "It was therefore clearly desirable to try to *prove* the total structure of the genetic code by this method" (39). As Paul Rabinow points out in *Making PCR*, the problem of generating DNA sequences was solved by cloning methods, so Khorana's group left the techniques they had discovered for another decade and perspective.

While Rabinow writes that Mullis's approach was the "opposite" of Khorana's, it is perhaps more precise to notice that each had a fundamentally different understanding of information. For Khorana, the genetic code was a space of revelation—the production of a gene was in the service of an understanding of the entire genetic code, an understanding attentive to the rhetorical and epistemological genres of a proof. By contrast, Mullis sought less to understand DNA—for that, after all, was accomplished—than to transform it. It is this pragmatic understanding of genetic information through which PCR emerged.

Consider, for example, Mullis's account of the PCR process:

> If I could arrange for a short synthetic piece of DNA to find a particular sequence and then start a process whereby that sequence would reproduce itself over and over, then I would be close to solving the problem. . . . The concept was not out of the question because in fact one of the natural functions of DNA molecules is to reproduce themselves. (6)

This treatment of DNA as an Ouroboros—not unlike that of Leary's— looked to the pragmatic capacities of nucleic acids: the capacity to replicate. And while Rabinow, in his anxiety to avoid a heroic vision of scientific innovation, seeks to distance his own account of PCR from Mullis's, he perhaps overlooks the fact that Mullis was not hero but host, as his instantiation of an iterative, pragmatic understanding of information replicated not just DNA but psychedelic culture itself, as the notion of expansion and decontextualization that dislocated consciousness in the 1960s now deterritorialized life, allowing it to become just another sample. Where to begin on the effects of such a deterritorialization or expansion of life? Perhaps one must begin uncoiling the implications by pondering the replication of a conjunction of molecules, hosted by human consciousness: LSDNA.

When first I drank ayahuasca, the ayahuasca drinker was overwhelmed by the sheer multitude of entities both visual and vocal. Faces both human and otherwise swam before him competing for his now splintering attention. I had not yet learned the method of the "blank screen," taught to me in a tutorial-like session with what I was now calling "mamahuasca" two years later in Peru. Before the tutorial from mamahuasca, I had again faced the prospect of drinking ayahuasca with more than a modicum of dread. Perhaps I always shall. Despite my best intentions and preparations, the ayahuasca drinker again arrived in Iquitos exhausted, sleepless, and anxious. The mototaxi buzzed me to an empty conference hotel with a lone soldier who watched as I tried to divine the function of a payphone. Everyone had already left and were headed for the jungle. Yes, I could meet them if I took a mototaxi to milemarker 31.

The moto arrives to a compound rumored to be the property of an oil executive. Oil was the other thick viscous liquid transforming and mobilizing the denizens of Iquitos. Years later, Peruvian troops will open fire on an indigenous protest seeking to block oil expansion in the upper Amazon.[9]

All the mellow psychonauts are ready to head for the jungle, and they lounge in hammocks and talk in clusters as I emerge from the moto twitching, nervous, lugging my bag. I stop and begin slathering myself in Deet, wondering about the anti-malarial Lariam that I took that had been providing me with most unpleasant hallucinations and lucid dreams for the previous week while the Iquitos Dharma Bums look at me with loving amusement. Didn't I know that Alfred Wallace had conceptualized natural selection in the delirium of a malarial fever? And U.S. troops in Iraq were balking at their prescribed Lariam, preferring the surreal horror of preemptive, fabulated war and prescription amphetamines to the dark visions induced by the horse pills.

My friends had already signed on to drink with a young mestizo curandero named Percy. Would I drink without my teacher, Norma? I crammed into the van with fifteen others—psychologists, graduate students, doctors, system administrators, painters, philosophers, and school teachers and we made our way into the jungle from Iquitos. Then after an hour's hike into the muddy jungle, we arrived at Percy's rotting platform. "Platform" and "rotting" are, I suppose, redundant in the rain forest, but I wonder where I am going to sleep when it comes.

We gathered flowers and herbs, trying to learn the names and pronouncing them incorrectly, over and over. One by one we forced ourselves into the uncannily cold waters of a stream and emerged to be covered in blossoms, leaves, and tobacco smoke. I waited with no small measure of cynicism, almost sneering at the baptismal aspect of the ablution. The alluvial cold focused my attention, and a great wind of energy came over me direct from Percy's mouth. I think I remember weeping.

5

HYPERBOLIC

Divining Ayahuasca

This specially designed tour is for those serious seekers who long to have
a direct experience of other dimensions of themselves and the universe.
    —Soluna Tours

HAPTIC GAZE SMACK:
AYAHUASCA PILGRIMS IN THE RAIN FOREST

IQUITOS, PERU'S LARGEST JUNGLE CITY OF FIVE HUNDRED THOUSAND, cannot be reached by road. The pilgrim's route toward "mother ayahuasca," the ecodelic brew prepared for millennia by upper Amazonian shamans, usually begins with a boarding pass and a security check. With great quotidian efficiency, guards sort through twined boxes covered in tape, crates of wool, and soft drinks toted by an indigenous family behind whom we wait, and wave my radio producer through. Fears that his recording equipment would be the occasion for a border hassle were as unfounded as my decision to wear shorts on a flight to Lima in August. The other passengers had the decency not to laugh, but I envied them curled and coiled under tiny airline blankets shielding them from the aggressive air conditioning, and they were ready for the chilled coastal morning when we arrived.

An hour and a half after shuffling through customs bleary-eyed and sniffling, our plane soars over the Andes. Iquitos announces itself through a green labyrinth—the Amazon and its tributaries begin their fractal "S" fifteen minutes before we are due to land. The pilgrim is greeted by handlers, guides to help navigate the impoverished, but quite friendly, streets. The ubiquitous mototaxis of Iquitos sketch a lattice work with their routes through the dirt grid, and the two stroke engines play a layered, rhythmic soundtrack over broken—very, very broken—Spanish as the pilgrim makes his way to a canoe, prepared for a practiced, motorized drift toward the camp of an ayahuascera. There, he will drink a noxious and viscous liquid and begin his interdimensional journey, which often begins with vomiting.

Finding ayahuasca in Iquitos is not difficult. One does not need a sense for occult locales to locate it—it is, according to anthropologist Marlene Dobkin Del Rios, an integral part of the medicine of the region. But the pilgrim/tourist who seeks the enlightenment of the yage way of knowledge has probably begun his training well before the departure gate, or should have. For by all accounts, ayahuasca (a potent admixture of various DMT and Monoamine Oxidase Inhibitor–containing plants found in the region) is hardly a recreational drink. Like other ecodelics, ayahuasca can yield very different kinds of journeys, depending on the "set and setting" of the tea drinker, including programming offered by curanderos in the form of icaros—rhythmic and often whistled songs that accompany and guide the tea drinker on their journey. Anxious, even terrifying trips are not uncommon, and unlike the legendary brown acid of Woodstock, it is usually not the psychedelic agent that is the ultimate or even proximate cause of the distress. The problem, the drinker discovers, is *the self*, which must give up its attachments if it is to abide the massively parallel consciousness induced by ayahuasca. This parallel consciousness is often presented as a multitude of entities and forms for whom death is a transition but not a destination—"ayahuasca" can be translated as "vine of the spirit" in Quechua, and is sought out for its ability, among other things, to erode the very distinction between the living and the dead. But to abide this parallel presentation, an enormous flow of information not verifiable in the serial time of the body, the pilgrim prepares the self for its momentary disappearance through a culling of the self and its wants: each pilgrim begins with a regime of selective self-negation or denial. The would-be interdimensional traveler must fast prior to the ayahuasca ceremony, or face the wrath of a possible inadvertent serotonin crisis provoked by a piece of cheese or chocolate and their MAOI ingredients.

Some will, of course, rely on chance to be their shaman: if the mood is right, they will drink whatever disgusting beverage is put before them. But most pilgrims seeking out the ayahuasca experience in the upper Amazon will have been drawn as much for the context as the content. The plants that make up classic ayahuasca are legally and readily available on the Internet. Recipes abound on the Web that enable would be ayahuasceros to concoct brews out of plants that grow readily and widely around North America, Europe, and Australia.[1] Some contemporary ayahuasca drinkers even create the admixture entirely through pharmaceuticals—so called pharmahuasca, which combines MAO inhibitors (such as the prescription antidepressant Nardil) and a source of DMT, the reputed active ingredient of most brews and an endogenous product of the human brain.

The proliferating and extensive literature on ayahuasca seldom fails to inform the reader that the brew is essentially a mixture. A plant like *Banisteriopsis caapi* (yage, "force") is needed to block the stomach's destruction of DMT, and a source of DMT (such as *Psychotria viridis*, a relative of coffee) needs to be potentiated. But given the well-established importance of programming to ecodelics, there is at least a third, crucial component to the experience, one tuned to the legendary accounts of modern psychonauts such as William S. Burroughs, Terence McKenna, and the innumerable anonymous posters to the Internet: the ecosystem, knowledge, and spirituality of the rain forest.[2] What draws many of the tourists seeking the ayahuasca experience in South America is a unique, albeit globalized context of an erudite and healing shaman guiding visions in a vanishing rain forest, the very habitat of the alkaloid exuding plant allies. This is most certainly what is being sold in ayahuasca tourism, as a glimpse at Web sites offering these trips make clear. In this sense, a look at ayahuasca tourism helps us to understand the nature of contemporary tourism in general: Why do globalized citizens bother to go anywhere in an increasingly interconnected universe? With my bare legs, surrounded by bundled passengers, perhaps I sought a capacity to sustain all of the interconnection, the formation of a commons.

Pilgrims to an apocalypse, ayahuasca tourists are, nonetheless, tourists. And they bring with them the penumbra of Western consumerism even as they seek enlightenment through an older and perhaps more sustainable mode of being. Already ecodelic, the ayahuasca tourist has become sensitive to all of his inputs and outputs: the diet was begun two weeks earlier, a cautious tourist wondering about the body's chemistry as cheese begins to appear as waxen lard, a repulsive leaden block throwing images of a future serotonin crisis right into my gourd.

And then there were the shots and preemptive malarial dreams induced by the Lariam, and deliberations over the proper concentration of Deet, the military grade mosquito repellent. Reading through the Web sites and browsing the travel section in a Barnes and Noble megastore, it sometimes feels as though I am surrounded by contradictory instructions and imminent insect assault. Back in my house, I stop, smoke herbs, and reflect. I get inside my mosquito netting bivouac and just breathe, wondering if the green of the jungle will smack my vision with haptic force. I can still recall the stunning, overwrought precision of the world revealed to me in fifth grade, when I was first fitted for glasses. Snapping into focus, the world suddenly became all angled, distinguished, a sharp and angry crystal. Everything was about to slice me open, so fine and well honed were these new lines bent out of myopia. Would the Amazon hit me with this haptic gaze smack? At seventeen I visited Ireland *en familie*, and learning to drink Guinness with my brother, an athlete of the bar stool, playground, and bong, my eyes seemed to get just a little wider as they were stretched out by the force of the almost bioluminescent green of the land, a wind tunnel for color vision.

AYAHUASCA HOSPITALITY, OR DISTRIBUTED DRUG ACTION

If ecodelics manifest a capacity to be affected by an environment, then what kind of effects would the upper Amazon have on an ayahuasca drinker? The British explorer Richard Spruce, writing about his 1852 visit to the "Amazon Valley and Orinoco," responded to "accounts given by travelers of the festivities of the South American Indians, and of the incantations of their medicine men," sought out a "feast," and was sufficiently overwhelmed by the hospitality of his hosts that he was unable to complete the ceremony. As a gesture of welcome, Spruce was presented with an enormous cigar, "2 feet long and as thick as the wrist," and "etiquette demanded that I should take a few whiffs of it—I, who had never in my life smoked a cigar or a pipe of tobacco" (Luna and White, 84). After puffing and, one can only imagine, fighting off a massive coughing jag, Spruce writes that he still wanted a full dose of ayahuasca, but was not up to the sheer simultaneity of this hospitality protocol, which was of the parallel rather than serial sort:

> I had gone with the full intention of experimenting the caapi on myself, but
> I had scarcely one cup of the nauseous beverage, which is but half a dose,
> when the ruler of the feast—*desirous, apparently, that I should taste all of his*
> *delicacies at once*—came up with a woman bearing a large calabash of caxiri
> (mandiocca beer) of which I must needs take a conspicuous draught, and as
> I knew the mode of its preparation, it was gulped down with secret loathing.
> (Ibid., 8; emphasis mine)

After the beer and the secret loathing, paradoxically secreted through an
opening—the obligatory production of hospitality—came the enormous cigar,
which was anything but just a cigar, which was then followed by a large cup of
palm wine. Spruce goes on to suggest that the dose of ayahuasca was crowded
out by "a complex dose," through which he approached something like group
mind in the form of events that might be, by the reader, "readily understood."

> It will be readily understood that the effect of such a complex dose was a
> strong inclination to vomit, which was only overcome by lying down in a
> hammock and drinking a cup of coffee which the friend who accompanied
> me had taken the precaution to prepare before hand. (Ibid., 84)

Spruce, overwhelmed by an excess of hospitality, recovers with the help
of more: "drinking a cup of coffee which the friend who accompanied me
had taken the precaution to prepare before hand." What is the nature of such
a "complex dose?" First, note that while Spruce is the model guest, accepting
the good things put before him, he nonetheless refuses an exchange: he over-
comes the inclination to vomit. This vomit is, though, part of the ayahuasca's
"dose"—ayahuasca's very action is implicated in its capacity to purge. This is
what ayahuasca "gives" you—a capacity to open. Indeed, "dose," we find in
the *Oxford English Dictionary*, is given to us by "dosis," which is quite simply
a "giving," and to be dosed with ayahuasca means to be given the capacity
to give (up) the contents of an interior. German toxicologist Louis Lewin,
whose invention of the discipline occasioned his inquiry into various "phantastica," includes vomiting as part of the very action of ayahuasca:

> If the stomach and small intestines are empty the passage of the drug into
> the lymph-tracts takes place much more rapidly and with greater force.

These conditions are realized when caapi is consumed in the usual manner, because certain doses of the substance give rise to vomiting, which is desirable and to a certain degree necessary, as a preparation for the final action on the brain. (Lewin, 117)

Indeed, so connected is ayahuasca's action to purging that it is sometimes difficult for drinkers to keep the brew down long enough to have any action other than nausea. This gift that spawns further giving, then, is not simply an effect of the allegedly nauseating *flavor* of the (extremely variable) ayahuasca brew, but is instead the very *action* of the mixture on a human drinker.

This infinitive "to purge," often presents itself as an imperative to ayahuasca drinkers. Among the crowd of visual conventions found repeated in the Peruvian painter Pablo Amaringo's work are jeweled cities of numinosity, converging rainbows of twisting triple helical anaconda assemblies, green feathered bird-men in rapt discussion and wearing belts, numerous ladies of the reptilian rainbow feather with ceramic pots balanced upon their noggins, flying saucers, DNA, and men gathered together in a common, and thoroughly violet, projectile vomit. Spruce failed, though, to let go of his interior, to form such a commons, to actively relinquish his absolute difference from the outside. To do so, he would have had to submit to the inclination to vomit, an inclination which must itself be part of any scientific or otherwise useful model of ayahuasca drug action. It is in this sense that the self is an aspect of the "apparatus" of any inclination to know ayahuasca: the assemblage of observation entails massive attention, in a more than cognitive way, to the stomach and small intestines. Such attention paradoxically requires the self to recede from the field of observation, an involution analogous to the involutions and invaginations of embryological development. Here, rather than extracting and representing information about the external world, the exploration focuses on the interior of the explorer himself. Galactic in magnitude, this interior presents as much complexity and diversity as the external world. Perhaps most crucially, this involution—the transformation of the interior into the site of exploration, an interior whose itinerary leads to a vanishing point, spiking toward infinity—renders difficult if not impossible any simple delimitation between the explorer and the realm explored, creating a feedback loop that spikes toward infinity.

Psychonauts who *have* formed a commons with ayahuasca, and thus opened themselves toward an investigation of themselves with it, often treat this spike toward infinity with the language of the sacred, as in the naming of ayahuasca as an "entheogen," an agent that awakens the divinity within.

And yet even such a coinage perhaps unnecessarily preserves the very distinction between inside and outside at stake in ayahuasca drug action. And to simplify matters further as we explore the complexity of divining ayahuasca's action, contemporary treatments of the irreducibility of even the simplest algorithms offer a map for this spike toward the infinite that does not yet require the resonance with a monotheist culture enthralled with a fundamentalist understanding of *theos*.

In accord with the continual re-discovery of the role of programming in psychedelic experience, Spruce discovers that the action of the brew is not reducible to itself. To actualize Spruce's "full intention of experimenting the caapi on myself," he would have had to actively give up control, give up his minding of the borders between inside and outside. Perhaps it was Spruce's sense of propriety, an understandable desire to be a polite, non-vomiting guest, that preventing him from truly being a guest to the host's hospitality.

Hence the *complexity* of the dose downed by Spruce does not decrease through the intake of a cup of coffee proleptically prepared, but instead undergoes a qualitative change: subject to a sudden dissolution, the difference between Spruce and the upper Amazon was reestablished through the (shamanic?) application of another plant, a relative of the chacruna with which the ayahuasca was likely prepared: coffee. Spruce, whose very name suggests the agency of plants in his own life, did not receive the gift offered to him by this plant assembly because he blocked it with another plant tonic. This opening requires an action—releasing the self from the throttling embrace of the self by minding the borders of the body in a non-Euclidean fashion, forming a commons. Spruce forms a commons, but it is with the familiar one: his friend. The biologist Robert Rosen offered an appealingly simple definition of complexity that is in accord with Spruce's non-encounter with ayahuasca.

> A system is *simple* if all its models are simulable. A system that is not simple, and that accordingly must have a nonsimulable model, is *complex*. (Rosen 1999, 292)

And yet though it looks simple, Rosen's own definition here is, of course, complex, since as a model we might deploy to differentiate complex from simple systems, it must be tested, used, fathomed, broken, mended, tinkered back into existence. Indeed, this is no doubt how Rosen fashioned his model—by trying it out. Complexity, the ongoing accumulation of novel dissipative structures, is an evolutionary phenomenon, and hence is constantly being hacked. One such hack is stigmergy, the use of already existing models from evolution.

Spruce samples Manuel Villavincencio, a geographer whose 1858 description of ayahuasca oscillates between the first person and third person:

> This beverage is narcotic, as one might suppose, and in a few moments it begins to produce the rarest phenomena. Its action appears to excite the nervous system; all the senses liven up and all faculties awaken; they feel vertigo and spinning in the head, then a sensation of being lifted into the air and beginning an aerial journey; the possessed begins in the first moments to see the most delicious apparitions, in conformity with his ideas and knowledge: the savages (apparently the Zaparo of eastern Ecuador) say that they see gorgeous lakes, forests covered with fruit, the prettiest birds who communicate to them the nicest and the most favorable things they want to hear, and other beautiful things relating to their savage life. When this instant passes they begin to see terrible horrors out to devour them, their first flight ceases and they descend to earth to combat the terrors who communicate to them all adversities and misfortunes awaiting them. (Harner 1973, 155–56)

Yet if one "might suppose" that the beverage is narcotic, one does so only under the weight of an enormous linguistic tradition. For immediately, without a rest, Villavincencio, "awakens" to ayahuasca's capacity to "excite" and "liven up." This awakening is followed by an enormous compression: in an "instant," "they see gorgeous lakes, forests covered with fruit, the prettiest birds who communicate to them the nicest and the most favorable things they want to hear." Even an investigation of what one "wants to hear" would fill a lengthy narrative, yet all of this, for Villavincencio, transpires in an instant that becomes suddenly horrific, if equally saturated with "communication."

About a century later, William S. Burroughs followed Spruce and another explorer of plant habitats, ethnobotanist Richard Evans Schultes, to the upper Amazon in search of yage or ayahuasca. Burroughs sought a clue to the killer of his wife Joan, shot dead in a game of "William Tell" in Mexico City. Burroughs, of course, held the gun, but what he sought was the "final fix," an experience of telepathy wherein he might comprehend and "write his way out" from the "bad spirit" that pulled the trigger, the Third Mind. Or noösphere?

DNA COMMUNICATION

> *In the mental dimension, in contrast to the physical, the all-pervasive*
> *experience of absolute certainty does not require further verification*
> *and will be structured according to current mythology or the belief*
> *system of a St. Francis, Pascal, or Ramakrishna. What is one man's*
> *loss of freedom, therefore, may be another's gain in creativity.*
> —Roland Fischer, "A Cartography of Ecstatic and Meditative States"[3]

It is difficult to discern the genre of *Invisible Landscape*; heterodox in its insistence on an organismic and holistic view of living systems, this 1975 text cowritten by Dennis and Terence McKenna nonetheless treats living systems as dimensions accessible to an informatic vision of life. Part systems science, part explorer narrative, and part trip report, this admixture nonetheless echoes classical scientific style with its use of the passive voice and its treatment of ayahuasca as a visual prosthesis for a scientific gaze:

> During the course of our investigation of the shamanic dimension, our attention was drawn to a report of ayahuasca usage among the Jivaro (Harner 1968); the shamans . . . are said to produce a fluorescent violet substance by means of which they accomplish all of their magic. Though invisible to ordinary perception, this fluid is said to be visible to anyone who has ingested the infusion . . . this suggests that ayahuasca may enable one to see at ultraviolet wavelengths. (McKenna and McKenna, 95)

Compare this passive-voiced observation, where the McKenna brothers' attention is, like Spruce's, "drawn" by something outside of themselves, to the famous opening of *On the Origin of Species*, where Charles Darwin's gaze, "much struck," can hardly avoid the "mystery of mysteries":

> When on board HMS *Beagle* as a naturalist, I was much struck with certain facts in the distribution of the inhabitants of South America, and in the geological relations of the present to the past inhabitants of that continent. These facts seemed to me to throw some light on the origin of species—that mystery of mysteries, as it has been called by one of our greatest philosophers." (1909b, 21)

Struck with the facts, Darwin spends several hundred pages listing and diagramming, in exquisite detail, traits and their variation over massive and impossible to conceive geological time. While Darwin's "gradualism"—both rhetorical and evolutionary—is legion, it is worth noting that his first paragraph immediately alights on his (recursive) quarry: the "mystery of mysteries." By comparison, the McKennas are rhetorically restrained: the "magic" of the shamans, while not marked by incessant scare quotes in the McKennas' text, is quite simply data from a report: "they are said" to produce a violet substance with which they accomplish their magic, just as Spruce (whose papers were edited after his death by none other than Alfred Wallace) responded to "accounts given by travelers of the festivities of the South American Indians, and of the incantations of their medicine men." While the notion of a shamanic "dimension" may provoke a double take in some readers, even a lay reader familiar with themes in twentieth-century physics and mathematics—from such texts as *Dancing Wu Li Masters* or the *Elegant Universe*—would not be alarmed by the notion that reality consists of more than three or even four dimensions. Recall, too, Albert Hofmann's model of diverse, tuneable realities manifested as distinct aspects of the One. And it is not only the McKennas that invoke such an "implication" or involution: Spruce's text, in responding to "accounts" of "incantations" becomes, upon inspection, itself something like an incantation, summoning the other evolutionist, Alfred Wallace, to assemble Spruce's papers after his death.

So too do the McKennas share contemporary technoscientific culture's fascination with the pragmatic capacities of science: as a molecular prosthesis for visualization, ayahuasca may in fact enable, the McKennas suggest, new modes of perception in segments of the electromagnetic spectrum not perceived by the human eye. If numerous earlier researchers compared LSD to an "electron microscope" or a "non specific amplifier of human consciousness," ayahuasca is, among other things, for the McKennas, a device for querying and exploring perception with mind.

And this ultra vision is paired with a similar capacity to tune human audio perception:

> We also had occasion to ingest synthetic tryptamines and had observed as a regular feature of the tryptamine intoxication a peculiar audible phenomenon. This is a very faint, but definitely perceivable, harmonic overtone of varying pitch and frequency that seems to emanate from inside the skull while one is under the influence of tryptamines; the exact nature of this harmonic tone eludes precise verbal description, as it varies in quality and amplitude during the course of the tryptamine experience, first manifesting itself as an extremely faint sound on the very edge of audibility, rather akin to the sound that might proceed from distant wind chimes. (McKenna and McKenna, 95)

While Darwin goes on to describe a sumptuous and multifaceted external world upon which natural and sexual selection have been inscribed, the McKennas ask the reader to investigate, if only phantasmatically, the inside of one's own skull. The McKennas solicit investigation, while Darwin summarizes, reports, and argues. Of course, such an "internalist" investigation—an inquiry in which, as in phenomenology or Zen, introspection of different states of mind and the fathoming of interiority are not simply understood as *devoid* of objectivity—presents difficulties for the modest scientific reporter, such as Spruce, who would seek credence in a community based on the distinction between the subject of investigation and its object of analysis, not their unpredictable admixture.

As ethnobotanist and "shamanologist" respectively, Dennis and Terence sought to understand these plants and their usage by registering the effects upon their own psyches, and they rhetorically modeled this investigation for readers and, even, for each other. As with the work of contemporary scientists Alexander and Ann Shulgin, psychonautical research recognizes as a first principle that these plants and compounds must be studied by collectives. In short, to investigate psychedelics, one must form a commons. Here, the resistance of ecodelic experience to language becomes the occasion for a prolix gathering seeking a commons, a paradoxical search engine for eloquence: "Dennis wouldn't stop talking, and it was really no longer possible to communicate with him" (T. McKenna 1993, 117).

And the reader, a virtual psychonaut, is very much solicited into the collective. Try, now, with me, to summon a sound that, ringing out from the interior of your skull, defies verbal description and is on the edge of audibility, tending toward the procession of distant wind chimes. . . . Here description morphs into recipe, an invitation to investigate ayahuasca from the "inside"—not (necessarily) by ingesting a mixture of *Banisteriopsis caapi* and *Psychotria viridis*, but by working deliberately on sequences of imaginative practices to fulfill the implicit bond, if not habitual contract, between authors and readers. Trying to understand this sound on the edge of reference, following along with the imbricating description of a barely audible sound from within, the reader must try to become still in an effort to resonate with the sound toward which the McKennas gesture, a sound of incipient resonance, a cognitive synesthesia. Psychologist Roland Fischer's hallucination-perception continuum maps this effort as the ergotrophic inclination toward ego death, "the rebound of ecstasy toward samadhi," an active inclination that seeks the zero or mobius point where the experience of a self folds inward, and takes itself to be an object of perception and, in perception, perceives the fundamentally recursive nature of an "I" beholding itself, beholding, but only on condition of a transformation in the observer, a befolding: *Tat Tvam Asi*. Such a tuning of the mind toward minding apparently yields a hyperbolic space, one spiking toward infinity. Darwin turns the reader's gaze toward an infinite (an empire in which the sun never sets, covered with life) without; the McKennas' focus observation on the (hyperbolic) infinite "within."

Mathematician Daina Taimina describes hyperbolic spaces in an interview with science writer Margaret Wertheim:

> There are many ways of describing the hyperbolic plane. In formal geometric terms it is a simply connected Riemannian manifold with negative Gaussian curvature. In higher-level mathematics courses it is often defined as the geometry that is described by the "upper half-plane model." One way

of understanding it is that it's the geometric opposite of the sphere. On a sphere, the surface curves in on itself and is closed. A hyperbolic plane is a surface in which the space curves away from itself at every point. Like a Euclidean plane it is open and infinite, but it has a more complex and counterintuitive geometry. (Wertheim)

Wertheim notes that to understand this spike toward the infinite characteristic of the hyperbolic space, you must make a model of such a space and inhabit it in time:

> I have crocheted a number of these models and what I find so interesting is that when you make them you get a very concrete sense of the space expanding exponentially. The first rows take no time but the later rows can take literally hours, they have so many stitches. You get a visceral sense of what "hyperbolic" really means. (Ibid.)

So too does the conceptualization of a hyperbolic space require the labor of model building. Mathematician and semiotician Brian Rotman's model of mathematics argues for a thermodynamic understanding of the mathematical process, discovering at the level of mathematical practice Landauer's 1961 analysis of the physics of information.[4] In Rotman's treatment of "counting on non Euclidean fingers," even the act of iteration necessary to any counting entails thresholds of quantitative increase in difficulty. In this dissipative semiotic model, any (serial) counting to "infinity" is haunted by sudden incapacity as the energy (or information) required to compute (count) the infinite eventually becomes greater than the energy in the universe. Rotman is here in accord with the observation of Edward Fredkin that the universe is computing the future as fast as it can.[5]

And it was not only thermodynamics that made the conceptualization of hyperbolic spaces difficult. While nature, it turns out, abounds in hyperbolic forms such as kelp, cabbage, and sea slugs, mathematicians had their attention focused elsewhere:

> There are also not that many mathematicians sitting around looking at sea slugs. It's only been fairly recently that mathematicians have been looking at things like leaves and trying to understand them. A lot of these can't be

understood in terms of analytic equations—neither can the physical model of the hyperbolic plane—so you have to use other kinds of methods, like geometric methods. It's not until recently that we've had the tools to do that. (Ibid.)

And yet if such conceptualization was difficult, it was not without its admirers and functioned as a kind of infinitely curvy femme fatale of nineteenth-century mathematicians. Wolfgang Bolyai (1775–1856) urged his son Janos Bolyai to avoid the study of hyperbolic geometry as one would avoid the fleshpots of Egypt: "For God's sake, please give it up. Fear it no less than the sensual passions, because it, too, may take up all your time and deprive you of your health, peace of mind and happiness in life" (ibid.).

And the patterns themselves are no less powerful attractors of attention:

Some of the models had great aesthetic appeal, especially given the enormous variety of repeating patterns that are possible in the hyperbolic plane. After the geometer Donald Coxeter explained these conceptual models to Escher, he used patterns based on these models in several of his prints. (Ibid.)

While Darwin's formulation of the "mystery of mysteries" involves the reader in a recursive loop of thought arrested only by a being much struck by the facts of the external world, the McKennas summon an inquiry into the "unstruck": the conditions of being an observer, that always mobile attention sink, are the recursive quarry here. The McKennas offer a *hyperbolic* map of subjectivity—one which queries the infinite regress of observers observing their observations, together. On this model, as with Wertheim's need to crochet the hyperbolic space in order to grok it, so too must the reader model the hyperbolic model of a self with a self, listening toward distant wind chimes.

Mathematician David Henderson notes that the hyperbolic space can be understood as a continuous involution of a sphere: "On a sphere, the surface curves in on itself and is closed. A hyperbolic plane is a surface in which the space curves away from itself at every point" (ibid.).

Opening toward infinity, the self must in this sense divine the infinite, engage in an ongoing search. Contemporary cognitive science—and the thermodynamics of information—teaches us that such a perception involves a catastrophic increase in the sensory-motor ratio and a radical increase in the processing of entropy or information:

> When we imagine objects and places, this imagining creates mental space that is constrained in many of the ways real space is constrained. Although you can imagine impossible movements like your feet lifting up and your body rotating until your head floats inches above the floor, these movements take time to imagine and the amount of time is affected by how large they are. (Stafford and Webb, 101)

While the McKennas do not yet ask us to imagine an object or a place, but instead invite us to participate in an infinite "regress," another, shamanic, dimension is nonetheless invoked, a relation with a transhuman and alternative ontology mapped hyperbolically. For following Wertheim's analysis, what does hyperbolic really mean? *The Oxford English Dictionary* notes that this is originally a term of rhetoric:

> A figure of speech consisting in exaggerated or extravagant statement, used to express strong feeling or produce a strong impression, and not intended to be understood literally.[6]

The Invisible Landscape is indeed in the tradition of the utopian explorer narrative, a "Kubla Khan" (itself composed "in a sort of Reverie brought on by two grains of Opium taken to check a dysentery") of plant tryptamines narrating another dimension, a world revealed by an irreducible participation with ayahuasca and psilocybin.[7] It is therefore more map than referent, a guide to hyperbolic spaces that, if they are to be understood properly, must not be taken literally even as the transformation provoked by the (guided) interaction is actual. In responding to a "report" to be tested empirically, the brothers sought to report on it, and in reporting, involve attentive readers in it. They model an interpretation for us that seeks a figure-ground reversal between "ordinary" late-late capitalist consumer reality, and one that persists in a dimension accessed through the molecular bit flip of ayahuasca, an ecstatic rebound toward samadhi that operates as programming of the "I" by folding it in, on, and through itself in an origami of subjectivity. Architect and topologist Haresh Lalvani characterizes the study of all such imaginable topological forms as morphogenomics:

> In biological terms, this is an epigenetic code that exists in parallel with the biological genetic code captured in the DNA sequences. The morph-code leads to the possibility of *morphogenomics* mentioned earlier. It provides a formal tool in the design and manufacturing processes. (Lalvani qtd. in Gans and Kuz, 117)

And their report is itself psychedelic and hence interdimensional—the psyche of the reader must be manifested to access the report, turned in on itself in involution; we must listen attentively and ergotrophically toward distant wind chimes.[8]

HYPERBOLIC REST

Even becoming still long enough to engage the possibility of such a sound requires "physical work," exergy expended even in the erasure of information. In *True Hallucinations*, Terence McKenna's lengthier and more markedly literary account of "The Experiment at La Chorrera," McKenna suggests that whatever protocols the brothers worked up to both induce and manage an "experimental schizophrenia" appeared to enable the compression of an enormous amount of information. Excerpts from Dennis McKenna's journal are woven into the pages of *True Hallucinations* as citations of "The Experiment." On March 4, 1971, what the group titles "March Forth!," Dennis describes psilocybin's function as a molecular antenna for human consciousness as it tunes into an enormous quantity of information:

> The psilocybin that occurs in the mushroom acts as an antenna for picking up and amplifying the harmonic ESR tones of all tryptophan-derived compounds of all living organisms within its range. Since the psilocybin undergoing metabolism is superconductive, this means that its range of reception is theoretically infinite. The antenna does, to some degree, pick up a signal whose ultimate origin is the *totality of living creatures*. (T. McKenna 1993, 99)

Turning hyperbolically toward the "theoretically infinite" reception of "the totality of all living creatures," this molecular antenna's transmission of "May Day" was an "experimental schizophrenia," though the experiment at La Chorerra involved more than Dennis. Terence narrates it thus, in a vision:

> I see this notion as an effort to explain the very real sense of informational interconnectedness that pervaded our experience, which occurred in one of the densest tropical rain forests on the planet. (Ibid., 98)

True, Terence, in the improbably titled "I Understand Philip K. Dick," wrote that he had been involved in a *folie a deux* with his brother Dennis, and suggested that Dick—whose 1974 *Valis* experience bathed him in pink light and too much information—made it *trois*. But the locution here is precise and heuristic at the same time: what Terence "sees," has a vision of, is an "effort," a trial, an attempt to articulate the informatic interconnection in the context of that poster ecosystem for a living system, the Amazonian rain forest.

Rain forests, recent research tells us, are the best producers of entropy on the planet—they are the most efficient dissipaters of the energy "between cold earth and hot sun." Nineteenth-century physicist Ludwig Boltzmann's phrasing for the gradient of energy that, his investigations concluded, must be dissipated as entropy. Contemporary ecologists James J. Kay and Eric Schneider link this capacity to degrade energy to the increasing complexity of ecosystems:

> In other words ecosystems develop in a way which increases the amount of exergy that they capture and utilize. As a consequence, as ecosystems develop, the exergy of outgoing energy decreases. It is in this sense that ecosystems develop the most power, that is, they make the most effective use of the exergy in the incoming energy while at the same time increasing the amount of energy they capture. (Schneider and Kay 1995, 168–69)

"Exergy" is thus a restated description of entropy, a measure of the overall disorder of a system that increases as "low hanging fruit" of energy gradients are consumed and others, with the help of consciousness, discovered. Entropy, of course, is also a measure of the information content of a system, thanks to Claude Shannon's formulation in *The Mathematical Theory of Communication* (Shannon, 60). Kay and Schneider note that an decrease in exergy—a greater dissipation and hence expenditure of energy—often emerges through changes in kind, forms of order emerging out of and enabling developmental and evolutionary transformation. These transformations yield "information databases about self organization strategies that work," enabling ever greater dissipation of energy.[9]

> Interestingly enough, the tropical rain forests with their coupled cloud system, with the sun directly overhead, have the same surface temperature as Canada in the winter. The low tropical rain forest OLR temperatures are due to the cold temperatures of the convective cloud tops which are generated by the underlying cooler forests. (Schneider and Kay 1994, 26)

So while the prospect of an informational interconnectivity struck the McKenna brothers as surreal in the context of a rain forest far from an "information technology" of any usual kind, they were in fact surrounded by, among other things, a massive system of informational connectedness, "information databases about self-organization strategies that work" (Schneider and Kay 1995, 171). Rain forests are apparently the most effective producers of entropy on the planet, and they seem to achieve this dissipation through the interconnection of massive biodiversity. Schneider and Kay wrote: "Preserving information about what works and what does not is crucial to the continuance of life. This is the role of the gene and, at a larger scale, biodiversity" (1994, 171).

It would not be the first time that schizophrenia was associated with the analogies and metaphors of a contemporary information technology and provided insight into them. Avital Ronell traces out the debts of early telephony to schizophrenia and its spiritualist double, the aptly named "medium."[10] But I want to suggest that the McKennas less engaged in an experimental schizophrenia than discovered, and rediscovered, the order to be found in resonance. Indeed, if "ecodelic" is to be dropped into the mix of interfaces whereby we interact with mind, I would suggest that it function as an informational compression device, one that reminds us of the "echo" built in to the use of these plants, as we resonate stigmergically with those models produced by evolution. Returning to my earlier discussion of eloquence and Henry Munn's notion of "ecstatic signification," this would indeed appear to be the very telos of psychedelics—the compression of information through the breakdown in symmetry that enables one to become resonant with evolution while mapping the hyperbolic space of human subjectivity. That's a pretty remarkable discovery for a couple of schizophrenics.

MANIFESTING PSYCHE, OR MODELING RESONANCE WITH AUM

> *From the midst of that radiance, the natural sound of Reality, reverberating like*
> *a thousand thunders simultaneously sounding, will come. That is the natural*
> *sound of thine own real self. Be not daunted thereby, nor terrified, nor awed.*
> —*The Tibetan Book of the Dead*[11]

While this use of a "schizophrenia" that would function as a veritable scientific imagining and "imaginating" technology resonates with the history of the use of ecodelics as simulation machines for psychoses, this rhetorical history itself must be tuned at least as much as an ayahuasca experience. While "experimental schizophrenia" in "psychotomimesis" provided the initial paradigm for researchers across the planet after Albert Hofmann's 1943 re-discovery of LSD, the term was determined to be a non sequitur to researcher Humphry Osmond in his use of LSD to treat alcoholics, and "psychedelic" was created as a new scientific instrument for the investigation of consciousness. Such a mode of innovation and knowledge production emerges not through the continual expansion of what is known, but emerges in evolutionary fashion through the capturing of attention—neologism. This very real work of "imaginating"—summoning, (and therefore at some level experiencing) an aspect of consciousness that was previously unknown or not noticed is fundamental to the production of new forms of knowledge or novel facets of extant knowledge.

With his brother Terence, Dennis suggests that schemes for compressing the massive volume of information work could be modeled and imagined, as a nested hierarchy:

> If we imagine the harmine-DNA complex as a radio-cybernetic matrix, then
> we can suppose that this matrix stores information in a regressing hierarchy

of interiorized reflections of itself, in a form similar to the familiar ivory balls carved one inside the other, each level free to rate independently. In response to the vibration of tryptamine-RNA charge—transfer exchanges, modulated by mind into a usable signal, information searches of any sort might be conducted through a process that we suggest might be much like the principle of retrieval of information from volume holograms. Such a process of information retrieval and image projection would never lag behind human thought. Indeed, conscious thought may be precisely this process, but occurring on a more limited scale. (McKenna and McKenna, 106)

The reader, already solicited toward distant wind chimes, becomes aware of reading about the very mechanism of conscious thought, consciously, coming to its limit, a doubling of the present, as an "image projection would never lag behind human thought." Such an infinite regress of self-reference is "static" only when the observer ceases to alter in recursive play in and through the investigation of consciousness. In this case, that trip manual, *The Tibetan Book of the Dead,* offers a prescription for simple practices of awareness: "It is quite sufficient for thee to know that these apparitions are thine own thought-forms. Recognize this to be the Bardo"(104). "Schizophrenia" is, to paraphrase Robert Anton Wilson, "a word." In this context, Tibetan advice to those who would manage and therefore alter their consciousness is: Remember the hyperbole!

During the interval between death and birth, Bardo entities seek to capture the attention of the soul otherwise concentrated on the Clear Light, and the "Bardo Thodol" suggests that awareness itself is ample protection from the catastrophe (from the Buddhist perspective) of such attention capture and its metonymy, rebirth. An awareness of one's own recursive linkage to the forms of thought we are beholding need not imply the existence of a revealed, true plane of reality, but instead insists on an essential emptiness of reality, a becoming devoid of form or content. "The Heart Sutra," a first-century Mahayana Buddhist text, treats it this way:

Form is emptiness; emptiness also is form. Emptiness is no other than form; form is no other than emptiness. In the same way, feeling, perception, formation, and consciousness are emptiness. Thus, Shariputra, all dharmas are emptiness. There are no characteristics. There is no birth and no cessation. There is no impurity and no purity. There is no decrease and no increase. Therefore, Shariputra, in emptiness, there is no form, no feeling, no perception, no formation, no consciousness; no eye, no ear, no nose, no tongue, no body, no mind; no appearance, no sound, no smell, no taste, no touch, no dharmas, no eye dhatu up to no mind dhatu, no dhatu of dharmas, no mind consciousness dhatu; no ignorance, no end of ignorance up to no old age and death, no end of old age and death; no suffering, no origin of suffering, no cessation of suffering, no path, no wisdom, no attainment, and no non-attainment. Therefore, Shariputra, since the bodhisattvas have no attainment, they abide by means of Prajnaparamita.[12]

While a section named more explicitly as a mantra follows this treatment of the emptiness of consciousness besides its own beholding, the *repetitio* of NO that composes this passage should remind us that even here, in perhaps the most straightforward description of Buddhist ontology and epistemology, rhythm has its say, marking the temporal and hence transient passing of even this sutra. There is "no attainment and non-attainment," and they abide precisely by chanting these rhythmic patterns of continual reversal: "Form is emptiness; emptiness also is form. Emptiness is no other than form; form is no other than emptiness."

On this reading that is a chanting, "The Heart Sutra's" insight into mind consists of a repeatable recipe for reminding readers of their recursive role in their apprehension of worlding. This can entail working *at* and *through* the sutra (literally, "thread" or "path"), as the reader notices the form of the content describing emptiness, and the content of the form that seems to empty itself through (formal) repetition. In the first two lines above, one chiasmus follows upon another: "Form is emptiness; emptiness is also form. Emptiness is no other than form; form is no other than emptiness." In such a context, the awareness that one sees through a glass, darkly, must not recoil onto an epistemological certainty regarding a true, non-bardo consciousness. Instead, one must meditate on and repeat practices of emptiness to avoid filling in one bardo with another, blocking the Clear Light with an egoic attention ensnared by the prospect of the truth. In this sense the Clear Light of the "Bardo Thodol" resists Eliade's description of mystic light as an inseminating vision that gives life, and instead must be seen as a metaphysical Temporary Autonomous Zone that allows for openness to immanence that does not seek fulfillment.

And the McKennas seem to know they are in the bardo: *True Hallucinations*, even as a title, rhetorically induces, à la the repetition of "The Heart Sutra," an essential oscillation at play in this admixture text. Holograms, for example, are extremely effective compression techniques, and yet the very nature of the information compressed by them, or by a holographic model, is itself uncertain, available less for "belief" than for "processing." Consider, for example, Terence's vision of a flying saucer in the rain forest:

> As I watched, the clouds re-coalesced in the same way that they had divided apart, taking another few minutes. The symmetry of this dividing and rejoining, and the fact that the smaller clouds were all the same size, lent the performance an eerie air, as if nature herself were suddenly the tool of some unseen organizing agency. (T. McKenna 1993, 157–58)

This uncanny scene becomes even eerier when we consider that ecosystems are, of course, *precisely the tool of some unseen organizing agency*: the symmetry breaking and resonant evolution and development of information ecologies tending toward the increased dissipation of exergy for which the rain forest is exemplary. And Terence McKenna, with his language of "symmetry" and its organizing force, seems almost to echo it: an extensive scientific literature describes the emergence of order from otherwise random oscillators. John Collier and Mark Burch summarize their arguments in the abstract of their 1998 article, "Order from Rhythmic Entrainment and the Origin of Levels through Dissipation":

> Rhythmic entrainment is the formation of regular, predictable patterns in time and/or space through interactions within or between systems that manifest potential symmetries. We contend that this process is a major source of symmetries in specific systems, whether passive physical systems or active adaptive and/or voluntary/intentional systems, except that active systems have more control over accepting or avoiding rhythmic entrainment. The result of rhythmic entrainment is a simplification of the entrained system, *in the sense that the information required to describe it is reduced*. (165; emphasis mine)

Aldous Huxley's refrain—the continual return to the flower—could be described precisely as a rhythm, and it is a rhythm in the service of a compression of information. What perhaps oscillates, in Terence McKenna's account, is truth itself.

> Entrainment can be communicated, passing information from one system to another. The paradigm is a group of jazz percussionists agreeing on a complex musical progression. The process of rhythmic entrainment is complementary to that of symmetry breaking, which produces information. *The two processes account for much, if not all, of the complexity and organization in the universe.* Rhythmic entrainment can be more or less spontaneous, with the completely spontaneous form being uncontrollable. A balance between the two forms can produce a more robust system, requiring less energy to maintain, whether in physical, biological or social systems. (Ibid.; emphasis mine)

Collier and Burch contribute to a maturing scientific literature that has described the self-organizing order that is apparently ubiquitous in our cosmos and intensified on biotic planets. Here no central organizer, such as, say, DNA, serves as the source of the orderly unfolding of organisms and their ecosystems. Instead, through interaction, otherwise random (devoid of symmetry) oscillators begin to resonate, resulting in more order, compressing the information necessary to describe or even dance to it until yet another symmetry is broken, and the song or sentence is over. And *True Hallucinations*, in oscillating between a declaration of truth and its erosion, begins to resonate:

> The siren sound was rapidly gaining pitch, and, in fact, everything seemed to be speeding up. . . . In the last moment before it was lost, I completely threw open my senses to it and saw it very clearly. It was a saucer-shaped machine rotating slowly, with unobtrusive, soft, blue and orange lights . . . as it passed over me I could see symmetrical indentations on the underside. It was making the whee, whee, whee, sound of science fiction flying saucers. (T. McKenna 1993, 158)

Everything in Terence's prose here is caught up in the first part of the rhetorical algorithm, *True Hallucinations*, until it is not, "before it is lost."

No sooner does McKenna attract the reader's attention to the runic "symmetrical indentations" inscribed in symmetrical indentations than the infinite regress is interrupted in slapstick fashion, and the "sighting" becomes a "citing": "it was making the whee, whee, whee, sound of science fiction flying saucers." What are we to do when a credible witness (perhaps hyperbolically) compares their own vision to something marked as false, a triplet of onomatopoeia followed by fakery? Note, for example, that not only is the sound itself emerging from a craft that seems to parody itself, but that the onomatopoeia of "whee, whee, whee" remains unmarked as a citation. Devoid of quotes and yet, obviously, quoting, it both marks and fails to mark its falsehood in a single gesture.

Nor is this "psychedelic sophist" content with this, perhaps, stridulating oscillation between true and false. For as the disk whirls something recognizable comes into focus, an archetypical hoax:

> The inevitable incongruous detail that seems to render the whole incident absurd . . . as the saucer passed overhead, I saw it clearly enough to judge that it was identical with the UFO, with three half-spheres on its underside, that appears in an infamous photo by George Adamski widely assumed to be a hoax. I had not closely followed the matter, but I accepted the expert opinion that what Adamski had photographed was a rigged up end-cap of a Hoover vacuum cleaner. (T. McKenna 1993, 159)

So scrambled are the categories of fiction and reality here that the mind cannot come to rest on a conclusion concerning the veracity of the text. In the McKennas' admixture text, the induced oscillation between "true" and "hallucination" becomes less a hypothesis about reality than a rhythm for browsing the hallucination-perception continuum. Such a (fundamentally Popperian) model tunes our thought toward experiments, a technology or "rhetorical software" for our imaginations, prospecting for evolution in morphogenomic space, topologist Haresh Lalvani's map of the combination of all forms. By continually oscillating between true, false, as if true, and true, as if false, Terence McKenna helps teach readers and psychonauts to avoid the resolution of the inquiry, and foregrounds the status of his report *as a model*. In Mahayana Buddhist theories of knowledge, it is not the failure of language that poses the greatest obstacle to enlightenment, but the failure to apprehend and inhabit this gap between language and actuality in a way that is mindful of language's "suchness"—its nature as a model. By fluctuating between an affirmation of the truth and its uncanny double, *True Hallucinations* helps readers welcome the instabilities of a psychedelic experience and forestalls simple conclusions concerning the ontology, meaning, and worth of ecodelic apprehensions.

In short, the self must remain at stake and in process in the observations. Terence's constant and subtle interruption of any simple "agreement" between author and reader (or between Terence and Dennis) more or less requires the attentive reader to continue their inquiry. Rather than "converting" readers to ecodelics, Terence solicits them into continual investigation. As with Popper's negative model of truth discussed in chapter 2, "true" in this context means "not yet disproved," and its conjunction with "hallucination" demands a dance between the affirmation of a particular account of psychedelic experience and a memory of the fate of all other "explanations" or descriptions of reality, scientific or otherwise: they are always about to be falsified.

If we apply this model to our ecodelic perception, human consciousness *acts* as or *models* a transcendental expression of an immanent universe, a figure-ground involution of Atman for Brahman that apparently gives our species, and perhaps the cosmos, pleasure. Human perception *acts* transcendental, in the image of a transcendental god, in so far as it is rendered as hierarchically distinct from the universe. Yet this transcendental expression is precisely an expression, even involution, of an immanent ecology in which the very distinction between self and cosmos emerges from a nested hierarchy with as much interior as exterior connectivity, as recursive thought becomes hyperbolic in both the mathematical and rhetorical sense.

The will to narrate, explain, recount, and remix is more than an attribute of psychedelic culture—it is the human tendency to form a commons with narratives, gestures, solicitations, icons, and supplications. Look! You are completely enmeshed with the cosmos! The "fourth drive" of the universe, its very *telos*, appears to include the imaginary investigation of unprecedented states for which only a continuum of true and false is sufficient. By some accounts, such as The Anthropic Principle, the universe is finely "tuned" toward conditions giving rise to a human consciousness that would behold it. If so, this tuning also tunes an experience of wonder that certainly qualifies as an alteration of consciousness putatively linked to the VMAT2 gene, where set and setting again proves crucial (Silveira). The Psychedelic Anthropic Principle (PAP) suggests that the universe is tuned toward conditions for the emergence of a consciousness that would behold the cosmos, and itself, in awe, a complex admixture of surprise and recognition in which the dissipation of information spikes toward infinity.

During my own second ayahuasca ceremony, the ayahuasca drinker was given some time in which to simply query the plant intelligence presenting itself first as birdsong, and then fashioned after an internal Socratic voice, searching for bandwidth between the buzz of the Amazonian insects, croaking amphibians, and birds in full courtship display. Without thinking, I asked it what the Universe was.

"The universe is a way of tricking itself. Next question." This trickster ontology might remind us that the cosmos is capable of transforming itself in suddenly novel ways, forgetting its own premises, breaking symmetry, and suddenly experimenting with an increased capacity to compress information, again, hyperbolic. By 1973, Roland Fischer would write of his investigations of the effect of psilocybin on the visual field, "Thus far, our studies suggest that certain hallucinogenic drug induced transformations in visual space may be regarded as an optimization of information" (1973, 258). This optimization involves both a compression—holograph, fractal, and geometric forms serve to reduce the *amount* of information processed in the experience by in-forming them with templates from the implicate order—and an exorbitant sacrifice, as each compression serves to enable the processing of yet more information. In this context, the capacity of human consciousness to explore and actualize the morphogenomic field of the imagination in search of entropy production is no more astonishing an invention by evolution than *Canis lupus familiarus* learning to hang its head out the window at thirty-five miles per hour for the sheer haptic gaze smack of it, and no less so.

THE EXERGY OF THE DOG HANGING ITS HEAD OUT THE CAR WINDOW: TERENCE MCKENNA'S LITERARY PRECURSORS

Throughout the six realms of samsaric existence this teaching is known
as "Liberation through Hearing and Reflecting." It is not necessary to

understand what this teaching means, but rather in simply listening to the
words and letting one's own interpretations do the rest. The title means
just what it says, particularly the use of the word "through."[13]

By 1971, it had become inescapably clear that psychedelics were not going to be investigated though the usual avenues of technoscience. The "scheduling" of compounds and plants by the U.S. government, and the subsequent adoption of international protocols governing inebriants illegal in the United States, meant that psychedelics were no longer an attention attractor for scientists getting grants and publishing results. Research, of course, continued, and it is a delicious irony of this prohibition on science that researchers such as Alexander Shulgin and David Nichols have continued to browse the hallucination-perception continuum for novel bioactive molecules and have expanded the molecular universe of ecodelics by an order of magnitude since the prohibition began. But the programming of psychedelic experience is a crucial avenue of practice and research that emerges out of the intensive investigation of psychedelics in the 1950s and 1960s, and the possibility of organizing above ground psychonautic programming collectives—sacred or otherwise—seemed rather slim when Dennis and Terence McKenna formed their research group at La Chorrera, so far from the NSF and so close to Gaia. And the rain forest, as noted above, abounds in methods for the production of entropy, among them the remarkable blue glow of the Blue Morpho butterfly and the eyespots of the Owl butterfly:

> The yellow eyespots remind most observers of owl eyes, but some behavioral lepidopterists believe the butterflies mimic a toxic tree frog that resides in the same ecosystem. (Microscopy Resource Center)

Through the production of an eyespot, this species survives either by impersonating a predator or by pretending to be in the possession of chemical weapons. The ecology, in either case, is one of information and fiction, and the rhetorical acts of the butterfly's wing entail not high-fidelity reproduction, but the scrambling of the codes between the fictional and the real. "Is that an owl I just saw? A frog?" And it was with this "Batesian" mimicry that Terence McKenna, along with his literary precursor, Vladimir Nabokov, became so fascinated.

Hence evolution, the epitome of "nature" even as a construct, has been operating on the fictional dimension of ecosystems for quite some time. In order to have responses to the world that do not simply repeat themselves, evolution browses for solutions that "virtualize," and therefore render mobile, entire aspects of itself. Having developed a chemical weapons factory for the Red-eyed Tree frog, evolution either cites it or, with a visual form apparently indistinguishable in the imagination of humans from the eyes of an owl, impersonates a predator and hence engages in a kind of ecosystemic mirroring. From this perspective, the information necessary for the butterfly to survive is compressed—rather than going to all of the cost of either being megafauna or developing chemical weapons capabilities, evolution selects for the "citing" of these capacities. But what is cited can itself be compressed, as the eyespot need only evoke and not reproduce the ocellus.

The induction of a (difficult to resolve) fluctuation between true and false is the very essence of the uncanny, and is apparently sufficiently disruptive to the perception of predators to provide butterflies with the spots with a selective advantage over those without. The compression of representation into the induction of a fluctuation is an evolutionary experiment; perhaps predators need only be subject to the double take, itself an uncanny doubling of the reindeer head twitch, for these gorgeous butterflies to survive, reproduce, and seduce humans into seeking them out, comparing them to fiction.

Reporting on the experiment at La Chorrera, Terence transmits something essential to the scientific reportage of these states. Usually confined to the serial first person narrative testimony, psychonautic investigation acquires a new form of enunciation and visualization with the McKenna brothers' "folie a deux."[14] Within this already scrambled rhetorical domain, Terence continually unravels or (to use a word he seemed to like) deconstructs the position of "knowing what psychedelics are." By miming the butterflies which induced his initial travel to South America, Terence's prose in *True Hallucinations* oscillates in iridescence between markers of the real and the false. McKenna works a gradient on the hallucination-perception continuum between the imaginary and the real, and does so in a way that encourages a recursive "undecideability" about his reports. One can neither completely dismiss the account nor fully embrace it, deferring any possible capture of ecodelics by an ontology that would know what they "are." This ensures that continued research will have the best possible chance of not foreclosing any of the practically infinite knowledge and visions apparently made available by properly guided ecodelic investigation. The reader oscillates between thinking "this is true" and "this is an hallucination," propelling her through the text in search of more evidence one way or the other, and the fluctuation continues. Ronald Siegel, whose work on the animal use of psychedelics is fundamental to any understanding of their action, repeated Klüver's search for a form constant with human subjects:

> In the controlled set and setting of the experiments . . . the psychonauts held on to the experimental demands and kept up their detached reporting. The untrained subjects, however, frequently described themselves as having become part of the imagery. At such times *they stopped using similes* in their reports and asserted that the images *were real.* (Siegel 2005, 235; first emphasis mine)

If there remains, for the reader, any question that the experience of gnosis attested to by psychonauts is real, Siegel's research suggests that very real learning takes place in repeated psychonautic sessions: *they learn to manage the information of the experience without ascribing an ontology to it.* Indeed, the deployment of rhetorical devices (similes) useful to blurring but not disavowing the distinction between a hallucination and a perception becomes the very marker of a trained psychonaut. In this sense, the psychonauts have been trained with the use of psychedelics to become better rhetors. The trained, i.e., experienced, psychonauts enter into Fischer's realm of the "as-if-true," and the explicitly rhetorical nature of their formulations allows work on detaching from any ontology entailed by their descriptions. The McKennas take psychonautic reportage to an even further level of detachment: neither real nor unreal, *True Hallucinations* transmits the "indescribeableness" of tuning into Gaia by systematically resisting a resolution of the kōan title. This oscillation appears to be essential to the warp and woof of ecodelic experience. Siegel weaves the metaphor of the loom into his text:

> Altered perceptions of real and imaginary events were woven together as they experienced the full glory of what was going on inside and outside their brain—"that enchanted loom," as the neurophysiologist Charles Sherrington called it. (Ibid., 232)

EXCURSUS:
TO THE ALIENS

> *In brief, just a like a feather drifting in the wind, one roams*
> *among things without becoming attached to them.*
>   —The Life and Teaching of Naropa

> *The thought came to me with certainty that they were*
> *manipulating my DNA, changing its structure.*
>   —DMT volunteer, University of New Mexico, 1991[15]

While Siegel's psychonautic adepts seemed able to remember the heuristic character of their descriptions, researcher Richard Strassman reports a denial of rhetoricity among many of his psychonautically experienced patients in human subject trials of DMT. DMT is a potent psychedelic compound endogenous to the human brain. Smoked, injected, or used in combination with a MAO inhibitor such as Syrian rue or *Banisteriopsis caapi*, DMT consistently produces visions that are among the most compelling in the human archive. Psychonauts regularly describe the experience of smoking DMT in ballistic terms, a kind of psychedelic "shock and awe":

> My soul was propelled forward as if from a cannon. An electric neon clown
> face, like one would see at a carnival funhouse, came screaming toward me.
> It opened its mouth and swallowed my soul and I flew down a long track
> which kept bending downward, ever more downward, even when I must
> have been freefalling vertically, I was pulled more and more 'downward'
> along these two parallel infinitely downward-bending lines, like a rider on a
> psychedelic rollercoaster.[16]

Perhaps the truly remarkable thing about Strassman's study was that he was able to carry it out. In the midst of an intensifying War on Drugs, Strassman legally administered an extremely potent psychedelic to human volunteers. Compared with this extraordinary act of knowledge production, the reports of the DMT volunteers perhaps risk banality—a kind of tripper's *Penthouse Forum*. Yet it is worth noting that some of the most persistent and consistent reports produced by Strassman's volunteers concerned alien contact, as in the alien biotechnologists narrated in this section's first sample from an (apparently) transgenic tripper.[17]

Ubiquitous indeed, those iconic alien "greys" show up even in a realm persistently described as an alternate dimension. Familiars of deterritorialization, these aliens are primarily informational beings:

> Many volunteers' encounters with life-forms in these non material worlds involved the powerful sense of an exchange of information. The type of information varied widely. Sometimes it concerned the biology of these beings. (Strassman, 190)

At once biological and technological, this "powerful sense" is indeed sensory, and packs a paradoxically informatic wallop. San Francisco Bay area psychonauts Gracie and Zarkov write that for a one to five minute interval after smoking DMT, "For all practical purposes, you will no longer be embodied. You will be part of the intergalactic information network."

But if amateur psychonauts and Strassman's volunteers experience DMT as a veritable password into an encrypted network of information, they nonetheless insist that the beings are no mere simulacra: "this is not at all a metaphor. It is an independent constant reality" (ibid., 95).[18] And while the volunteers repeatedly parried attempts to highlight the rhetorical nature of their encounters—definitively differentiating the DMT experience, from, for example, a dream—they nonetheless narrated their experiences in the terms and characters of a space topologically, if not ontologically, other than ordinary reality:

> A high-tech nursery with a single Gumby, three feet tall, attending me. I felt like an infant. Not a human infant, but an infant relative to the intelligences represented by the Gumby. (Ibid., 193)

What exactly are "the intelligences represented by the Gumby"? Gumby, as many readers are no doubt aware, was the claymation character created by Art Clokey who could "jump into any book with his pony pal Pokey too." Gumby was thus a three-dimensional interdimensional traveler whose very vehicle was intertextuality—a capacity to be spliced into any, other, text. Gumby, ecodelic: familiar of a world enmeshed.[19]

LEARNING FROM GUMBY: CLOSE
ENCOUNTERS OF THE NTH KIND

This ecodelic mesh yields an experience of reality that is both deeply participatory and hardly subject to individual control. On the newsgroup alt. drugs.psychedelics, the appropriately handled Twitchin sums it up after an experience smoking DMT's legal (and much more potent) chemical cousin, 5-MEO-DMT:

> DMT Aliens are there . . . and they are not purpleheaded beasts! (unless you want them to be!) What is reality? It is what WE as humans make it . . . its impossible for ALIENS to see it the same . . . I mean, our senses . . . we make it . . . thats what makes OUR reality . . . THEY will be different . . . information highway . . . hyperspace . . . that's where things MERGE . . . that's when ALL are ONE . . . live through it . . . experience it . . . its temporary . . . 30 Mins later . . . I'm back (?)—I love this space where you try to 'recover.'[20]

While this posting would seem to lean toward a solipsistic recuperation of human agency—"What is reality? It is what WE as humans make it."—Twitchin negotiates the ontological conundrum of the alien by simultaneously affirming human participation in the emergence of our reality and entangling us with a thoroughly differential and densely interconnected informatic other. The encounter with the alien becomes fundamentally unknowable—"THEY will be different"—and palpably involved with human experience, the experience of experiment and ordeal characteristic of the "MERGE" into information: "that's when ALL are ONE . . . live through it . . . experience it."

Crucial to these encounters is that they must be experienced to be communicated or understood; as with the simple algorithms of physicist Stephen Wolfram's *New Kind of Science*, no shortcut exists between the informatic encounter and the patterns they yield. These are encounters of the nth kind, unknowable in advance even as they are thoroughly informational.[21]

How to navigate such an enmeshed ecology? How can we, with Twitchin, "try to 'recover'"? For if allopoiesis—the dynamics of becoming other, even perhaps alien—must be folded into our experience of being in an ecosystem in crisis, so too do we continue to dwell fitfully within the egoic and all too human ecologies of heredity, labor, and value *extraction*—the autopoiesis of self-propagating capitalism. These practices of compression (more product, less time) and deterritorialization—information shrinks and kinks space, enabling the sudden proximity of the alien—tell a very truncated story of the future, a telos whose vanishing point into nanotechnology augurs what one Web site calls the "big scrunch," a reductionist apocalypse that forgets the Whole. The McKennas, fascinated by the *I Ching*'s capacity for complexity from simple rules, located an apocalyptic and informational mutation associated with the end of the Mayan calendar in 2012. How can a highly distributed and increasingly allopoietic open-source noösphere continue to sustain beings so addicted to encrypted and privatized forms of life? I would suggest that psychonauts and abductees are early adopters of a transitional, transhuman identity precipitated by our intensified and amplified ecologies of information in the context of an ecosystem in distress. These forms of identity are, of course, pathologized, even criminalized. But can the identity practices of the last century—citizen, soldier, human—survive the info-quake that abduction records?

In short, no. If they are to do so, these identity practices must avoid the algorithmic bottleneck presented by a regime of verification—it is a computationally intense labor, for example, to sort out an experience of DMT or abduction into categories of "true" and "false," categories that may be non sequiturs in a world of information constantly subject to change. Indeed, as Strassman and Terence McKenna both found out, it is sometimes even difficult to differentiate between DMT experiences and abduction events themselves. In place of verification, we need tools that help us "live through it," even as it transforms us and "it" remains an unknown.

Philip K. Dick, a veritable psychonaut of Gaia's newly informatic mesh, wrote incessant exegesis of his experience of becoming information in the early 1970s. Dick wrote until his death in 1982, finding an adequate response to his psychedelic, informatic contact only in graphomaniacal autobiographical fiction that claimed to be telling the truth in at least four novels and an eight thousand page, mostly handwritten, unnumbered manuscript. "Nailed by information," Dick rendered a name and an outburst of information— VALIS, the Vast Active Living Intelligent System. VALIS, or, as he sometimes called it Sophia, Firebright, Zebra, the AI voice, was symptomatic of Dick's perception that the universe was composed entirely of information. Fittingly for a novelist who felt that he had "fallen" into one of his own novels, VALIS told Dick the same thing it told one of his characters, Nicholas Brady: he had "tuned into an intergalactic network."

Dick's response is instructive. It was only by jettisoning his sense of an autonomous self—he became a character in one of his books, his books were fiction—that he was able to manage the ontological shock of nonlinear time and alien contact.[22] But his turn toward fiction was not a retreat into illusion but a disciplined treatment of the world (not unlike both Gumby's and Terence McKenna's) as both "intertwingled" with metaphor and irreducible to it, a practice that experiences fiction as differential to reality and not opposed to it.

This fluctuating continuum of truth and falsehood resonates with Fischer's hallucination-perception continuum, and frequently provokes a search for a "multi-valued logic" for navigating the differences and distinctions between "ordinary" states of consciousness and shamanic ones, what Michael Harner dubs the SSC (Shamanic States of Consciousness). Fischer sought to move beyond an oppositional understanding of hallucination (or fiction) and reality, but in so doing he crafted a deeply participatory ontology wherein states not subject to 3-D verification palpably exist for the subject:

> Hallucination and dream experiences may not only be "true" or "false" but also 2) "as-if true" and "as-if false," while the subject is playing an actor's role and is, at the same time, a captive audience of his own drama. It is like the experience of alternating figure and ground while being included in the change. (1972, 166)[23]

This fluctuation of figure and ground is entangled with the subjective experience of the investigator: an attempt, and failure, of verification. "What needs to be specified is the inability of the hallucinating subject to *verify* through *voluntary motor* performance in *Euclidean three dimensional physical space* the phenomena experienced in the *conceptual and sensory* dimension" (ibid.; emphasis in original).

This disjunction between motor performance and sensory stimuli is, for Fischer, the very crux of hallucinatory phenomena, as his research suggests over and over that a high sensory-to-motor ratio provokes hallucinatory experience in human subjects. Perhaps today we are witnessing the manifestation of a similar mismatch, as human consciousness undergoes a figure-ground fluctuation as it becomes transhuman. The two-fold expansion of information—both a quantitative increase and its technical expansion into new domains—is not matched by any significant increase in voluntary motor performance, e.g., judgment or deliberation, but it can perhaps be modeled by the rush of information in a DMT session and the avatars of greys that apparently enable the replication of DMT reports.[24] If so, these citations of the alien are most certainly true hallucinations, not (yet) subject to verification in *"Euclidean three dimensional physical space."* Yet these hallucinations announce the difficult truth of their own existence, signatures of a transhuman condition enmeshed with the dynamics of information, the reception of an uncanny, proliferating, and uncertain other: close encounters of the nth kind.

ECODELIC PERCEPTION

The periodic table of chemical elements and the tables of elementary particles in physics are finite. In contrast, the periodic table of form is infinite and open-ended, the way biological genomes are, and the way the biological "tree of life" is. This enables new forms to be continually discovered within the ever-expansive morphological hyperuniverse.
—Haresh Lalvani, "Genomic Architecture"

The imagination is thus a vehicle for traversing the morphogenomic field, and music can act as a memory device for its navigation. As pattern-breaking tools that resonate with any template of music, sound, vision, smell, speech, touch, or idea, ecodelics are imaginary fuel, meta-archetypes whose de-patterning effect of symmetry breaking is a most compressed algorithm for "Find an Archetype and explore it! Now!" It is in this sense that we can understand the McKennas' notion of a "lens" that,

would be the practical equivalent of a transdimensional vehicle in that it would, in common with holographic imagery, actualize the Hermetic axiom that "What is here is everywhere, what is not here is nowhere." It would be comparable to a flying saucer that moved through time and space, not in any conventional sense but one that *is* all time and space, warped through a higher topology into the boundaries of conventional space-time. This is similar to Jung's understanding of the psyche and is the effect that this holo-cybernetic matrix would tend to create. (McKenna and McKenna, 106–7)

What kind of space is this invisible landscape, and how do sound and vision work to render it for the McKennas' readers? Psychologists note that the imagination requires energy and is subject to constraint, and as we saw with Aldous Huxley's flowers, one way of mitigating the expenditure is to model the language and gestures of others stigmergically. The motor theory of language suggests that this bootstrapping through the mimesis of evolutionary and developmental patterns is fundamental to the very origins of human speech, as well as its development:

> The fundamental idea is that language was constructed on the basis of a previously existing complex system, the neural motor system. The programs and procedures which evolved for the construction and execution of simple and sequential motor movements formed the basis of the programs and procedures going to form language. At every level of language, from the elementary speech sounds, through the word-forms on to the syntactic rules and structures, language was isomorphic with the neural systems which already existed for the control of movement. (Allot)

Roland Fischer provides another model:

> During the "Self"-state of highest levels of hyper or hypoarousal, this meaning can no longer be expressed in dualistic terms, since the experience of unity is born from the integration of interpretive (cortical) and interpreted (subcortical) structures. Since this intense meaning is devoid of specificities, the only way to communicate its intensity is the metaphor; hence, only through the transformation of objective sign into subjective symbol in art, literature, and religion can the increasing integration of cortical and subcortical activity be communicated. (1971, 902)

For the McKennas, this non-dual apprehension was something like a figure-ground dissolution: "At that moment the world turns lazily inside out and what was hidden is revealed." (T. McKenna 1993, 112). A more recent text treats this classic figure ground inversion in terms of a "mind hack":

> If you flash a figure up to people for just long enough for them to see it and make an interpretation—to see a duck or a rabbit—then they can't flip their mental image in their mind's eye to see the other interpretation. (Stafford and Webb, 104)

Rabbit Season!
Duck Season!

I am an infant, learning to lie on the earth. I am repeatedly told to "listen to my mother." I again protest that this is too much; I can't handle seeing my mother again. My mother tells me that she won't appear, silly, just lie with her a bit. "A mother needs some love too." I lie with her and as I lie, I realize I am lying with her in death—I felt that she was indeed dead, but that death was not an absence of being, but a state of being among others, and that I was sharing it. She told me that, of course, she was my mother, and that the earth was my mother too. And that ayahuasca was my mother.

Gnosis:
Unfolding allegories of consciousness as the capacity to manipulate genomes. Sexual selection seems the likely ecology for the emergence of consciousness, which is not the same as an ecology of information. Why: how expression is the leading edge of information; Wolfram automata sprout against a backdrop of peacock mathematics, where peahens grok the symmetries and divergences of tail-feather patterns. Evolution of breeding systems as involved in other systems of cognition—avian, cetacean, plant, machine.

6

THE TRANSGENIC INVOLUTION

A consideration of a non-semiotic notion of communication as the
sharing of genetic material across traditional species barriers.
　—Eduardo Kac, "GFP Bunny"

*Is not cultivated wheat (*Triticum sativum*) a plant brought by man to the condition*
in which we see it today? Can anyone tell me in what country a similar plant lives
naturally, that is, without being the result of its cultivation in some place nearby?
　—Jean Baptiste Lamarck, *The Zoological Philosophy*

This glowing description of the effects of the haschisch, though
given by one who had often used it, is on that very account,
like the pictures of the opium-eater, open to suspicion.
　—James Johnston, *The Chemistry of Common Life*

FALLING INTO A WHOLE WITH A RABBIT

IT MAY SEEM PERVERSE TO SUGGEST THAT IF YOU WANT TO UNDER-
stand the telos of contemporary biotechnology, then you ought to sample
some hydroponic White Widow and look at artist Eduardo Kac's transgenic
bunny Alba. And indeed I would suggest no such thing.

WAITING TO INHALE, OR BUGGING OUT

Inhaling White Widow, a potent and recently evolved hybrid of *Cannabis indica*
and *Cannabis sativa*, titrates a taboo, a legal infraction, an act of cognitive liberty, and a relationship to a newly biotechnological plant.[1] Since the massive

236

surveillance and law enforcement campaigns of the drug wars have forced its cultivation to become both domestic and indoor, the quality, variety, and potency of cannabis has, er, *grown* at seemingly fantastic rates. In response to an intensified prohibition and subsequent deterritorialization—"I hear choppers, let's get these plants inside, quick"—cannabis undergoes a technical transformation that should be the envy of more mainstream biotechnological enterprises. And although it is the THC levels that get the most attention— some estimate that levels of this psychoactive and even psychedelic compound in high-end cannabis have increased nearly fourfold since the 1980s—it is the new genetic diversity of cannabis that is truly dizzying, a diversity that can itself only be encountered through the smoke or vapor of inhaled cannabis.

True, one can sample the cannabinoid porn to be found in such publications as *The Big Book of Buds* and see that hybrids like White Widow differ not only in biochemistry but in phenotypic presentation: the low, shrubby Afghani asks only to be grown in a closet and subjected to a high-pressure sodium lamp, while the fractal and filigreed crystals of Haze arrest the gaze as expertly and incessantly as an orchid summons a wasp. The massive proliferation of these spectacular images of cannabis on the Internet and in magazines such as *High Times*, *Cannabis Culture*, or *Heads* hardly supports the claim made by bestselling horticultural writer Michael Pollan that:

No one would ever claim marijuana is a great beauty . . . no one is going to grow cannabis for the prettiness of its flowers, those hairy, sweaty-smelling, dandruffed clumps. . . . The buds are homely, turdlike things, spangly with resin. (Pollan, 122, 137, 138)

Spangly?[2] While I can agree with Pollan that the flowering tops of a female cannabis plant do not charm the eye with the classic beauty of an orchid, it is difficult to explain the rampant dissemination of these new images of glistening green colas without grappling with the effects of cannabis "porn" on viewers. It becomes tautological, but vital to recognize that these images act as attractors on viewers, that many cannabis users spend time looking at, and not just inhaling, cannabis.

While the enthusiasm that growers and users might have for their favorite intoxicant might seem to need little explanation, it does bear noticing that the Internet has not yet become home to photo galleries of home-brewed liquor, beer, or inhalants. There is a function to images of cannabis in the community of growers and users that is simply absent in many other demographics of intoxication. Googling and oogling images of green buds such as those at www.marijuana-picture.com teaches one immediately that, of course, the rhetorics of cannaboid porn are as diverse as cannabis itself. Here are images of an accursed share of buds, heaped harvests that contest the creeping sense of scarcity that always haunts a criminalized habit. Then there are spectacles of health perhaps most appreciated by growers, medium range shots of plants in full bloom. Their exuberant vitality images regimes of water, light, phosphorus, potassium, nitrogen, trace elements, and carbon dioxide that are absolutely precise and thoroughly supple in relation to the always shifting needs of the plant. But easily the most ubiquitous image of sinsemilla (literally, "without seeds") is the close-up on a ripe green cola enmeshed with, and refracted by, shimmering crystals of THC. Here one looks not at a crystal ball but into a crystalled bud for the glistening evidence of cannabinoid production, as if the future effects of the plant were made achingly and vertiginously visible. But the crystals, in arresting the eye, also solicit it further, leading the viewer *inside* the flowering female. "Some of the pictures almost take you inside the bud" (Rosenthal, xv).

Given the likely demographics of the viewers of such images, these veritable entries into the flower do resonate with the sexual gaze of a dominator culture, a gaze that transforms the enveloping imbricated surfaces of calyxes and trichromes into yet another (seedless) receptacle for male passion. And yet it must also be recalled that if this is a pornography drawing on tropes found most frequently in *Penthouse* rather than in eighteenth-century botanical prints, then it is a pornography of *plants*. Is it not striking that this familiar but fantastic entry should map not only a metonymy of the eye and the phallus, but a veritable becoming insect? Catching a buzz indeed. The perspective hailed by the spangled buds is less inseminator than pollinator, more enflowering than deflowering.

Hence another, parallel and bugged out, reading of cannabinoid porn presents itself here. Spectacular close-ups of flowering cannabis, spangled with resin, work to blur the very boundary between human and plant. In soliciting a more or less sexualized gaze at a plant, such "bud shots" articulate an assemblage of plant, machine, and human that has driven THC levels through the roof. The flower-pollinator relation mapped by a *High Times* centerfold marks

Exploring the visual contact high.

the tangled but hardly dominating relation of users and growers of these plants. Here viewers are as charmed by the plants as they are instrumental to the radical differentiation of the plant genomes. Some writers have indeed looked at the (un)canny differentiation of cannabis and suggested that the relation between cultivator and cultivated has itself become blurred, that it is cannabis that is perhaps an agent of its own proliferation. As the appropriately named Pollan puts it, "So who is really domesticating whom?"

Traveling Stoner Problem

Still, if the ubiquitous glossy shots of cannabis teach us that for users and growers of cannabis, appearance matters, it seems equally clear that appearance itself is but one trait selected for by contemporary cannabis breeders. Flavor, aroma, ease of cultivation, and a remarkable variety of qualitatively different highs are all the object of selective pressure, and it is here that the biodiversity of cannabis becomes at once obvious and inaccessible. The diversity is obvious as the Sensei Seed Bank Catalog, *Cannabis Culture*, or just plain Google teems with hundreds of different strains (and possibly three different species) of psychoactive plants. The spectrum of names—Big Freeze, Sour Diesel, Northern Widow, California Orange, K2, Millenium, Flo, Master Kush—speaks to the diaspora, the sudden differentiation of the cannabis genome transmitted via a sprawling list that would defeat any rhetorical urge toward taxonomy, or at least excessively divert it. For if one looks to the naming practices of the cannabis ecology as a way of assaying its diversity, as I am, then the researcher buzzes with a veritable contact high. As mnemonic devices, the crowd of cannabis names primarily testify to a joyful and often synesthetic disarray: Purple High, Mazar, Oasis, Shaman, Nebula, Voodoo, Free Tibet. . . . One looks hilariously, but in vain, for a structuralist algorithm that would reveal a secretly referential character to cannabis nominalizations, what we might call "ganjanyms."

But the diversity is also, essentially, inaccessible. Surely this rhetorical but nonetheless dizzying disarray is of a different kind than that famously and yet cryptically induced by *Cannabis sativa and Cannabis indica*? Only one way to find out. . . . Like other psychedelic allies, cannabis requires a human assay for its diversity as an organism to be evaluated.

And yet where to begin? Marc Emery's Seed Bank, an online Canadian vendor of high-end cannabis seeds, offers 611 different strains. How is the would-be cannabis biotechnologist to proceed? The evaluation of each individual strain, not to mention the combination of strains that is the province of contemporary cannabis breeders, presents an unfathomable and incalculable enterprise, available only partially to those willing to self-experiment. Mapping the diversity of cannabis requires not only a quantitative and/or molecular genetic description—its lineage, preference, and habitat—but demands an active and paradoxically stoned deliberation.

An example from contemporary mathematics helps to situate just how confused the (necessarily, intermittently, stoned) cannabis biotechnologist must be. It is a cause of much fascination and embarrassment to mathematics that the seemingly simple computation known as the "traveling salesman problem" presents so much difficulty to modern day Pythagoreans. The problem is as follows: imagine you are a traveling salesman with responsibilities for fifty different towns in northern California, Oregon, Washington, and British Columbia. Given the knowledge of the distances (and/or costs) between the towns, what is the shortest (or cheapest) route for you to take as you make your rounds, distributing, showing, and selling your wares?

Despite the simplicity and ubiquity of this type of problem, its solution is nontrivial. It turns out that there is no general procedure for determining the shortest route other than the measurement and comparison of the different routes. And it gets worse: at around fifty different towns, the number of different possible routes approaches the estimated number of particles in the universe.

By analogy, even an algorithm to determine a sequence of assays—first smoke Haze, then Widow, then—becomes a highly dubious enterprise when dealing with combinations of 611 strains. Less a question of "distance" than "difference," the combinatorial practice of cannabis genetics, if it is to proceed from a deliberative logic, finds itself faced with an enormous calculation. And more strains are being developed all the time.

Dude, Where's My Car?

Yet, à la my ability to arrive (by bike) at work this morning despite the prediction of Zeno's paradox to the contrary, cannabis biotechnologists all over the globe do precisely calibrate, combine, and integrate the differences between different strains. We might be tempted, therefore, to suppose that cannabis breeding involves a slovenly departure from deliberation, less a practice than a passing out: the so called "couch lock" associated with certain strains of Indica-influenced cannabis. And yet if contemporary cannabis genomics cannot, *a priori*, operate through a careful calculation or deliberation of the usual, algorithmic sort, it nonetheless involves a set of heuristics that bring the fundamentally interactive nature of cannabis breeding into relief.

DJ Short is one of a number of emerging cannabis breeders who have achieved a measure of (paradoxically anonymous) celebrity through their innovations in cannabis breeding. Blueberry—a Sativa/Indica mix with, yes, the flavor of blueberries—exemplifies the innovative effects of DJ Short's breeding methodology. Aptly named, this DJ treats the cannabis genome as an immense mix to be sampled, recombined, and *scratched*. Like the conceptual artist and sonic shaman DJ Spooky, DJ Short highlights the fundamentally interactive and entangling processes of creative production.

First and foremost, of course, is the sampling of the plant itself to determine which plants to breed together. But DJ Short's sampling procedure involves more than the casual twist of bud into a bowl or joint. Instead, DJ Short carries out a veritable dance with the plant under consideration, even pausing briefly to rub up against it:

> A sort of scratch and sniff technique is first employed. With clean, odor free fingers gently rub one plant at a time, on the stem where it is well developed and pliable. . . . The newer leaves at their halfway point of development may also be rubbed and sniffed. (2003, 92)

These strange antics give a topological, and biological, spin to the boundary blurring introduced by Pollan above. The transformation and combination of cannabis genetic information—that is, cannabis sex[3]—takes place here through a veritable mixing of bodily fluids, as DJ Short and a cannabis plant momentarily but undeniably share a territory. The question of where the plant ends and DJ Short begins momentarily, but unmistakably, means nothing.

This human-plant alliance suggests that in DJ Short's methodology, selection favors those plants that *excel at dissolving boundaries*. In this case, the incredible array of flavonoids coaxed out of the plant must be present, but also mobile: the gentle strokes of the breeder, over multiple generations, renders an amplified flow of flavor.

And, of course, it is not only physical boundaries that must become fluid in this selection. As with Pollan's question—"Who is growing whom?"—cannabis seems almost uniquely capable of inducing the collapse of figure and ground that questions the agency of grower, grown. Indeed, in *On Being Stoned*, a quantitative and qualitative study of the effects of marijuana on human subjects, psychologist Charles Tart notes that "figure-ground shifts become more frequent and easier to control when stoned."

Alba and Biotechnological Enlightenment

In a role reversal for a sometime model organism, Alba too, requires a human assay. Kac's biotechnological rabbit glows with the Green Fluorescent Protein when, like so many 1970s psychonautic basements, it is bathed in the proper spectrum of light. In glowing, it too involves the passage of an increasingly plastic taboo structure—*What is an animal? How ought we to treat them?* Hence Alba's glow provokes questioning and debate, as if discourse were the real output for which bioluminescence is a catalyst. In contact with a human audience, Alba becomes an imaging device for the solicitation and registration of a rhetoric of transgenic genomics.[4]

But if Alba, who was quite white, and White Widow, which is not at all, are linked through their need for a human hosting, the entanglement speaks to their status as recently evolved familiars, border creatures who both extend and hack strangely into our agency as humans. How do a rabbit and a plant hack human agency in the context of biotechnology? For surely biotechnology is nothing if not the intensified application of human consciousness to evolution and its ecosystems. Homo sapiens' recently amplified capacity to manipulate genomes would be, in this light, a qualitative shift as well as a quantitative increase in human control over the living environment. Cloning technology, for example, promised to end the alleged nightmare of human reproductive difference as early as 2003, as humans become asexual as well as sexual reproducers.

But both the cannabis hybrid and the transgenic rabbit expose us to a more liminal agency than the conjunction of consciousness and genomes might suggest. If the promise of genomics was a "triumph over death" (Jaçob) or a revelation of "what life is" (Watson) then its delivery has been rendered more in anxiety than gnosis. While Alba glows, her light does not signal an epistemological enlightenment but the sudden arrival of an affect. In bioluminescence, Alba lights up a habitat whose fundamental output is interconnectivity. Alba is, er, *living proof* that machines, signs, and organisms, in their newest promiscuities, no longer dwell in definable taxonomical domains, but are instead differentials of intensity: networks. Alba's glow indicates that organisms are now indeed online, logged into the evolutionary network and transforming Darwin's "tree" of life into a fabulous mesh of interconnection. An interaction with Alba solicits not merely novelty and surprise, but a sudden sense of implication, a linkage between humans and Alba no less actual than her relation to the *Aequorea victoria* jellyfish that is the source of the GFP gene. Hence, if discourse is Alba's output, so too does she solicit a practice of affective connection. As an icon, Alba functioned as a sort of neon sign for transgenesis, but she was a sign who did much more than signify. In a less replicated, but no less revelatory image, Kac the artist is seen to be practically entwined with Alba. Selected, cropped, and zoomed, the image reveals a hospitable but entangled grapple.

Kac writes of his first encounter with Alba:

> As I cradled her, she playfully tucked her head between my body and my left arm, finding at last a comfortable position to rest and enjoy my gentle strokes. She immediately awoke in me a strong and urgent sense of responsibility for her well-being.

While one may hear prolepsis in Kac's testimony—a preemptive response to the objection that he somehow abuses Alba by making her glow with the status of "art"—Kac's account also highlights an essential effect of the bunny. If the "big blue marble" shots of Earth from space provoked a sense of global unity and interconnection among many otherwise isolated viewers, Alba seems to provoke an outburst of hospitality, an urge to loosen the boundaries that otherwise divide any particular human and animal. "Between my body," Alba provokes a multitude.

DARWINIAN COMPLICATION

*It is interesting to contemplate a tangled bank, clothed with many plants of many
kinds, with birds singing on the bushes, with various insects flitting about, and
with worms crawling through the damp earth, and to reflect that these elaborately
constructed forms, so different from each other, and dependent on each other
in so complex a manner, have all been produced by laws acting around us.*
    —Charles Darwin, *On the Origin of Species*[5]

As recently evolved familiars, White Widow and Alba are Darwinian to the
core. There can be no question but that selection is responsible for the emergence of these novel life-forms. But when it comes to responding to the glow
of Alba or the buzz of White Widow, it is not natural or even artificial selection that is constitutive of these organisms and their peculiar traits. It is
perhaps obvious that it is not fitness in any usual sense that is the metier of
either the rabbit or the plant. Glowing under 488 nm light does nothing to
help the rabbit in its ongoing struggle for survival, and the sheer surplus of
cannabinoids produced by White Widow goes beyond any utilitarian project
of chemical, albeit natural, warfare. And yet surely both are poster creatures for Darwin's analysis of variation under domestication, as wrought by
human deliberation as the bulldog.

Of course it is the case that humans have cultivated these organisms,
and that cannabis evolution has hinged on human preferences and choices.
Yet in what sense are the selection procedures of DJ Short or Eduardo Kac
choices? DJ Short, besides the scratch and sniff *implication* with his plants,
also enters into states in which the distinct categories proper to deliberation
and choice are themselves incessantly scrambled and recombined.

So, for example, when DJ Short teaches us how to select males for breeding, he asks us to cease scratching and sniffing and begin smoking:

It is possible to test males by smoking or otherwise consuming them. This
practice may be somewhat beneficial to beginners as it does involve a sort
of obvious discretion . . . make sure this test smoke is the first smoke one
consumes in a day in order to best discern its qualities, or lack thereof.
(2003, 94)

It may seem obvious that in order to test the quality of a spliff, one must smoke it. And yet male cannabis plants present difficulties to the would-be biotechnologist because of their relative lack of THC vis-a-vis the females. Besides learning the timely and accurate sexing of plants—the ability to read the signs of maleness from a seedling so that they might be removed from the environs, a skill necessary to the production of buds—cannabis biotechnologists must also learn to read the signs of quality from male plants. In other words, *one must learn to "get high" from low-potency marijuana.* The would-be grower must therefore learn to become less an athlete of the bong than a sensitive to her plants, capable of being affected by even traces of THC.

Thus stoned, however, the grower is taught by the plant. Even a beginner can learn to select a good plant for breeding in this fashion. While cannabis is notorious for its (prized) ability to *impair* judgment, in this context it becomes the very agency of judgment itself. The practice of becoming a sensitive involves an increased capacity to respond, less an act of agency than the arrival of a feeling: *I'm stoned.* Like an attempt at sleep (or enlightenment), one must learn less what to do than how to let go.

Still, even if the choice to become affected by another is a paradoxical one, an act of agency that undoes or stones identity, a sense of choice remains. The experience of this recipe of selection may itself not *feel* deliberate or even precise—"uh oh, smoked too much"—but its procedure nonetheless constitutes an algorithm. While the successful assay necessarily scrambles the human-plant boundary, smoking a male demands an absolute sobriety that would bring its difference into relief: "Make sure this test smoke is the first smoke one consumes in a day."[6]

Yet if this assay depends upon the integrity of the category sober/stoned, DJ Short seems to argue that one knows the plant only as a mixture. Indeed, DJ Short ends his discussion of selection with less sobriety than ecstasy. While the smoking of males seeks to assay the difference cannabis might make to ordinary, baseline consciousness, whatever that is, DJ Short's final test for selection involves a modulation of extraordinary consciousness. He suggests that the best test for the character of a plant is to use it in conjunction with another psychedelic such as psilocybin or LSD-25:

Ideally, the psychedelic substance will further the range of noticeable subtleties by one's psyche . . . if the herb is truly blissful it will become readily apparent under such psychedelic examination." (Ibid., 96)[7]

Again, a choice is made, but *its mechanism is the departure of agency itself.* Psychedelics are sought out precisely because they put the control of an ego into disarray, a manifestation of mind or psyche not amenable to the usual strategies of control.

Darwin noticed that sexual selection—that abjected, other vector of evolution that has spurred everything from peacock feathers to bioluminescence—seemed to rely on a similar sort of breakdown. Writing of the effect of male birdsong and plumage on females during courtship (aka selection), Darwin writes:

> Are we not justified in believing that the female exerts a choice, and that she receives the addresses of the male who pleases her most? It is not probable that she consciously deliberates; but she is most excited or attracted by the most beautiful, or melodious, or gallant males. Nor need it be supposed that the female studies each stripe or spot of colour; that the peahen, for instance, admires each detail in the gorgeous train of the peacock—she is probably struck only by the general effect. (1909a, 430)

While feminist researchers have rightly noted that Darwin was both attracted to, and troubled by, female choice precisely because it was *female*, this and other passages also emphasize the problematic nature of what Darwin called "charm" in evolution. "Struck by the general effect," females are both agents of selection and the charmed subjects of the feather, sufficiently seduced that the very boundaries of male and female breakdown into indiscernible zones necessary to reproduction.[8]

It was sickeningly obvious to Darwin that the feather train of a peacock was hardly the result of any struggle for fitness of the usual sort:

> Nor can we doubt that the long train of the peacock and the long tail and wing-feathers of the Argus pheasant must render them an easier prey to any prowling tiger-cat than would otherwise be the case. (Ibid., 411)[9]

A peacock struts and frets his hour upon the stage. Photo by Kathleen Pike Jones.

Yet strategies suited to courtship rather than survival abound in nature: birdsong, colorful plumage, insect stridulation, perhaps even language itself.[10] Darwin's intense and exquisite study of the mechanisms of sexual selection—studies barely noted by most contemporary researchers on the subject—continually focused on signaling tactics for inducing the dissolution of boundaries, a sudden fluctuation of figure and ground.

Consider, for example, Darwin's analysis of ocelli, eyespots that adorn the feathers of said peacock. If Darwin is convinced that indeed an array of said eyespots charm the peahen, how exactly do they do so? Darwin puzzled over the effect, and was at first disappointed by the peacock's charm, little appreciating the perspective of the peahen:

> When I looked at the specimen in the British Museum, which is mounted with the wings expanded and trailing downwards, I was however greatly disappointed, for the ocelli appeared flat, or even concave. But Mr. Gould soon made the case clear to me, for he held the feathers erect, in the position

in which they would naturally be displayed, and now, from the light shining
on them from above, each ocellus at once resembled the ornament called a
ball and socket. (Ibid., 46–47)

What so impresses Darwin in this remarkable event of sexual selection—he himself plays the role of peahen, subject to the charms of a certain
bespangled Mr. Gould—is not only the precision of the ocelli,[11] but their
capacity to render three dimensions in a two-dimensional medium. The
flat eyespots are practically an ornithological cinema, throwing images of
depth and clarity through the deployment of an iterated but flat surface.
This remarkable performance of *tromp d'oeil* captures not the peahen but her
attention. Much "struck" by the display, both Darwin and the peahen are
persuaded by the charms of the peacock.

What images were thrown? These "ball and socket" images achieve not
only three but n dimensions: uncanny, they present nothing other to the eye
than an eyeball itself, instilling a momentary but actual dissolution of the
boundary between viewer and viewed.

Hence the importance both of a language of choice, and the experience of
seduction, when attending to the mechanisms of sexual selection. Provoking not fitness but entanglement, sexual selection excels at the momentary
breakdown of inside/outside topologies.

BACK TO THE WHOLE, OR THE EARTH AGLOW

The firefly is a poster creature for both bioluminescence and sexual selection, and recent research has even sought to study the differential reproductive success of transgenic zebra fish in competition with their less spangled
competitors. Alba veritably glows with sexual selection.[12] And if a bunny is
considered as a sign of a most semiotic sort, it is, of course, a sign of reproduction. If Alba, too, in her bioluminescence, charms us, she does so by
revealing and even inducing our mutual entanglement in practices of evolution. So too does cannabis seem to remind us of our co-implication, producing the effect of a strangely distributed and sexually articulated agency.
These familiars are thus exemplars of contemporary biotechnology whose
methods are citational and recombinant: biotech vectors us towards distri-

bution and involution, a weaving together of life-forms whose name is best understood as a verb: Gaia. Such a technology is essentially ecodelic, as we leave the world of reference whereby the narrating "I" can maintain any reliable differentiation from its object of knowledge or love as "consciousness" and "life" once again converge.[13]

As with certain peacock feathers, if we look properly at Alba and White Widow an additional dimension of experience and even consciousness suddenly unfolds, as the earth becomes less a globe than a mesh. Fade to bioluminescence: the earth, aglow with a clear light.

Gaia Mandala

7

FROM ZERO TO ONE

Metaprogramming Noise, with Special Reference to Plant Intelligence

VALIS (acronym of Vast Active Living Intelligence System, from an American film): A perturbation in the reality field in which a spontaneous self-monitoring negentropic vortex is formed, tending progressively to subsume and incorporate its environment into arrangements of information. Characterized by quasi-consciousness, purpose, intelligence, growth and an armillary coherence.
    —*Great Soviet Dictionary,* Sixth Edition, 1992

People have three spiritual states. In the first they have no thought of God at all, but worship and pay service to everything else: friends and lovers, wealth and children, stones and clods. Once they gain a little knowledge and awareness, then they serve nothing but God. Yet, after learning and seeing more they enter a state of silence.
    —*Discourses of Rumi*

And SPAZZ is a letter I use to spell Spazzim
A beast who belongs to the Nazzim of Bazzim.
Handy for traveling. That's why he has 'im.
More easy to pack than a suitcase or grip,
Those horns carry all that he needs on a trip.
    —Dr. Seuss, *On Beyond Zebra*

"The youth that was I," you may come to speak of him in the third person, indeed the protagonist of the novel you are reading is probably nearer to your heart, certainly more intensely alive and better known to you.
    —Erwin Schrödinger, *What Is Life?*

THE REST OF MY SECOND AYAHUASCA JOURNEY CONCERNED THE plant intelligence that rather matter-of-factly runs the planet. My journeys since have often repeated this message: human consciousness, from this perspective, is just a way for plants to move plant and bacterial genes around, but we have apparently forgotten this and take ourselves to be the center of planetary telos. True, we are integral to the planetary telos. The near term futures of this planet are very much linked to the consciousness, or lack thereof, of human beings. And the new thermodynamics of Margulis, Sagan, Salthe, Schneider, and Swenson et al. suggest that we indeed have a purpose: the ongoing elaboration of the void through the dissipation of ever greater amounts of information through rhythmic entrainment and symmetry breaking, an expression of the cosmos' need to creatively destroy information to defer what systems scientist Stanley Salthe has characterized as the "senescence" of the information overload of thermodynamic systems trolling for increased entropy production. But the plants seemed to remind me that we are by no means at the center of planetary telos because there is no center, only the distributed emergence of complexity and its dissolution (cosmos). Our consciousness is an aspect, even a manifestation of, a distributed consciousness accumulating since the evolution of bacteria three-and-a-half billion years ago, and perhaps before. Consciousness, I came to believe with a fervor rivaling the McKennas', is the distributed capacity to manipulate and transform living systems, and this manipulation intensified with the human capacity to imagine the future of different organisms, selected and mated, genomics hacked, whacked.

Here the imagination becomes crucial as a tool for comparing competing thermodynamic gradients—humans heuristically project the effect of different combinations of foods, drugs, or alleles on the future. Artificial selection—Charles Darwin's name for the selective effects, conscious and unconscious, initiated by humans in selected breeding of domesticated plants and animals—was, then hardly artificial: according to this ayahuasca-inflected consciousness of a distributed plant intelligence, it is the very bootstrapping of (human) awareness. By continually selecting for different plants that, these selecting primates predicted, would affect their health, hunger, or attention, human beings deployed and honed a gradient of organized attention in ecosystems. While ecosystems do not become, *in toto*, aware through this process, the process does lead recursively to an organism becoming aware that it is embedded in an ecosystem with differential outputs and inputs that can be toggled on and off in a practically infinite, and certainly galactic in magnitude, set of combinations. Bacteria, Sorin Sonea tells us, have been sharing DNA through transduction over the entirety of the planetary surface, the biosphere, for at least three-and-a-half billion years, so perhaps we ought not be surprised to find that humans have tuned to an analogous commons where DNA and other molecules are shared and remixed through the organized attention of humans and their practices of conscious and unconscious selection over time.

Timothy Johns, for example, notes that humans indigenous to the high-altitude Andes have learned to combine edible clays with the ingestion of potatoes in order to make the former palatable and nontoxic. While for most contemporary readers the introduction of *the earth* into their diet would seem to be a non sequitur, an arbitrary introduction of an element foreign to it, this is perhaps due more to our system of taxonomy—animal, vegetable, or mineral?—than anything else. To an organism seeking to survive and leave behind progeny, it helps to have a sense of one's environment *as a system* in order to visualize and actualize its possible combinations and re-combinations of inputs and outputs, including inputs and outputs aphrodisiac in effect. While every other element of this ecosystem either seeks invisibility or separation from human beings over time, ecodelics remind them that even the reductionist aspect of any ecosystem—the way it can be cut up for thermodynamic profit—depends upon its aspect as a whole. In becoming aware of its ecosystemic implication, perhaps this organism can ensure the sustained livelihood of the ecosystem itself.

This kind of systemically recursive thinking requires narratives with which to orient each subsequent recursion, and my own emerging narrative of ayahuasca experience became my orienting "program," a rhetorical practice for learning and self-transformation, such as that necessary to the writing of a book devoted to inquiry into minded systems. Such a narrative only exists when it is put deliberately into language and continually revised, grappled with and shared. Here's what I wrote down in some notes I compiled after my second investigation of ayahuasca with Norma Panduro, the mestiza healer with whom I had now begun to collaborate. I posted it on a Web forum where ayahuasca drinkers, shamans, and just curious browsers can discuss the brew and its effects, and share insights, guidance, testimony, and recipes of preparation:

> The plant intelligence that joyfully runs the planet sees human consciousness as a tool for shuffling genes around. The plant complex's aesthetic sensibilities and desires drove early hominid evolution toward its current form and vice versa—we have coevolved.

I opened an online bank account that allowed me to send money to Norma to help run a clinic for addicts and children, sometimes both. I traveled to Peru thinking I would find a drug that defeats language, and instead found a medicine, and this medicine has proven very useful for treating addictions. A UN-sponsored clinic in Pucallpa, Takiwasi, regularly treats addicts with ayahuasca for addictions to cocaine, opiates, and alcohol, and Norma sought to bring a more shamanic perspective and practice to a clinic of her own outside Iquitos. In sending the money to support the clinic, I felt no sense of righteousness or even justice—I was continually responding to the insights of the journey toward the galactic dimensions of my "own" brain as I reflected on and worked through what I very much appear to have been taught. The narrative had become planetary in scale:

> Now consciousness, forgetful of its plant host, finds itself involved in a runaway spiral of self-selection most precisely discussed by R. A. Fisher in his discussion of sexual selection. This runaway spiral has consciousness amplifying or "expanding" absurd aspects of its nature, interesting mostly to itself. A very particular version of itself, I might add: the all powerful ego. Hence the forgetting of our plant host. . . . These selective forces, to say the least, ain't helping us survive, but are as unwieldy as a peacock's tail and far less interesting. We are fleshbots in love with an inflatable ego doll: ourselves.

Here I was, writing to a network that seems to replicate the mycelial form—the Internet—and discussing myself as if I were an extended phenotype—of plants. I posted this rhetoric about plant intelligence in the hope, yes, of propagating it. Jünger's notion of plants as an "autonomous power" had ceased to be "merely" a thought experiment for me, reminding me of the fundamental importance of photosynthesis and plant respiration to our ecosystem: it became an experience of responding continually to an attribute of the world I had forgotten or, rather, failed to perceive.

I found remarkable resonance in the writings of other drinkers of ayahuasca. I laughed when I read Dennis McKenna's experience of the "green teacher" who suggested a Copernican shift of consciousness toward plants:

> You monkeys only think you are running things. . . . You don't think we would really allow that to happen, do you? (Qtd. in Luna and White, 157)

The ecodelic insight I experienced in the upper Amazon seemed to enable remarkable health and personal, er, growth. My asthma was markedly improved, and continued to improve asymptotically as I approached a complete healing, moving my baseline level of health up several notches. This increased health fed back onto more rigorous swimming workouts, which increased my health baseline even further, which prompted me to replace all of my car trips with a bicycle. Soon, I stopped using an inhaler altogether, a liberation that allowed me to feel in control of my asthma and, yes, become healed. Able to breathe, I then learned to focus my awareness on my whole body atopic dermatitis (eczema), and can now report with grateful astonishment that of this, too, I have been healed. I rediscovered music and began to play ontological space drones with friends on a Moog synthesizer that just fell into my lap from a former student. I hadn't played music since the age of twelve, yet I found myself able to let go of my ego sufficiently to just play, and in just playing I found myself playing stride piano riffs I recognized from hearing my older brother jam on our old black upright in my South Jersey childhood home.

This attention to sound folded into my scholarly investigation of ayahuasca, and reminded me of the importance of the sonic dimension of the ayahuasca experience as discussed in the last chapter. All of this growth enabled other growth, and I found myself meditating more, panicking less, and navigating the increasing complexity of my life with more grace and health. The distributed drug action of ayahuasca was a reminder, too, that *ayahuasca was not the cause* of these transformations. This helped me avoid, most of the time, reifying the plant intelligence and treating her as an entity somehow separate from my encounters with her. This in turn was good practice for my meditation. And so on.

Suffice to say that all of this recursion was starting to crowd out older versions of myself. I came to recognize that all of my former self-attributes that couldn't survive the ayahuasca experience, didn't, and that I was grateful for that. I had nothing against my former iterations, but the lives they were leading didn't even approach being sustainable. What intent or sense of purpose did the others bring to my life? I loved my family, yet I did not previously know why I was going into the future with them except that I was afraid of dying, and of losing them. After the ayahuasca and the works it helped induce, I felt that my home was a cluster of practice oriented toward a purpose—new and frequently inarticulable understandings now seemed to solicit me into the future. Music, hope, health, and love began to fill my life. We welcomed another child into our lives, a girl, Violet. In short, I had a very clear sense of meaning, narrative, and an inclination to succeed. No, I had by no means Figured it Out, but I had been graced with an interruption of my former broadcasts that seemed to enable unprecedented gratitude and a desire to share the joy. Without such an ergotropic narrative, I didn't think any of my former selves would have survived for long, and indeed they haven't.

Now, years later, it is only through the concept of stigmergy—self-organization that uses the evolutionary past as a template—that I was able to accept the fact that I appear very much to have learned something from my ayahuasca experiences. It seems to have increased my capacity for interconnection *because I have become more comfortable with interconnection*. This learning brought me up against a taboo of which I was only vaguely aware: my own rejection of the possibility of acquiring knowledge "directly" from the planet, gnosis. Knowledge, I believed, had to be gathered and incorporated. One does not acquire anything but sensory information directly from an environment. Writer Philip K. Dick's gnosis of pink light was of an informational kind:

> The external information or gnosis, then, consists of disinhibiting instructions, with the core content actually intrinsic to us—that is, already there (first observed by Plato; viz: that learning is a form of remembering). (Dick, 1991, 239)

And, indeed, I did seem to be remembering: how to play music, how to detach, how to dwell simultaneously in the Euclidean-samsaric world of everyday life and bureaucratic reason and what I came to call the "jewel"—

the sparkling, fluctuating, living cosmos I was beginning to affirm and interact with. And I knew that fundamental to my response-ability to ayahuasca had been my writing practice, where I was already comfortable with entities other than my conscious mind getting into the act. Burroughs called it the Third Mind; Freud, the unconscious; and Darwin, aesthetic selection. I understood that creativity often involved sending my ego in to investigate areas not available to it, and that I learned to calmly and curiously accept the insights and solutions to problems that would occur when, in the ecstasis of writing, I seemed only to host texts and not control them.

And yet, I found myself telling myself while writing, this was *your* situation. Others might not be in a position or have (painfully) gathered the tools to integrate an ontologically shattering experience like ayahuasca into their lives. What if I was writing a book that would somehow mislead someone into drinking ayahuasca or taking another ecodelic without preparation or guidance, or even the time to integrate the practice and its effects? At minimum, they could have a most difficult time, and at worst, they could be presented with material not digestible by any of the hegemonic realities—school, work, family, state—they would find themselves living in. Although strictly speaking the plants composing it are legal, drinkers of ayahuasca are even subject to criminal prosecution unless done in a country, like Brazil or Peru, where it is legal in a sacramental context. A possible landmark U.S. Supreme Court victory for the UDV, a branch of a legal Brazilian ayahuasca church in New Mexico, might help change things (since I am now a member of a similar church in Peru), but for now my obligation to my family meant that I could not, within the context of harm reduction, counsel the use of these plant allies in the United States while we, my country, wages war against what I increasingly, to my great surprise, was calling my sacrament.

It is not lightly or for legalistic reasons that I have found myself invoking this word, "sacrament." More investigation suggested that "sacrament" was involved in what Darwin called the "mystery of mysteries." By going to our code book, the Oxford English Dictionary, I learned that in Christian Latin from the third century the word was the accepted rendering of the Greek for "mystery." "In ancient Greek it occurs chiefly in the plural, denoting certain secret religious ceremonies (the most famous being those of Demeter at Eleusis) which were allowed to be witnessed only by the initiated, who were sworn never to disclose their nature."

I was writing this book to make sense of my own investigations into the following question: How did humans practice a science about states that could not be written down? As a rhetorician of science, it was a fascinating puzzle to me, no more and no less, to study the rhetorical protocols of a science whose very "data" could not be captured by conventional means. As my research continued, I learned that most pre-prohibition scientific researchers considered it unethical to research a compound or plant they had themselves never assayed, and so I found myself replicating some of the experiments of some researchers. An apprentice psychonaut, my goals in the project now grew to include harm reduction and a reevaluation of the human relationship to the plant adjuncts with whom they have, it became obvious to me, coevolved.

My own brother had been a casualty of the drug war, a fearless and immensely talented human being killed by a speed ball[1] in 1987 at the age of 27. I knew that the government's policy of propaganda (DARE) misinformation ("This is your brain; this is your brain on drugs.") and incarceration hadn't been the answer for him and was indeed part of the problem. I also knew that his investigations were more than "recreational" and that he had evolved a worldview gleaned from the writings of Castaneda, the early music of Pink Floyd and Syd Barrett, the large numbers of Kurt Vonnegut texts he consumed one after the other, and the sublime power of the ocean on the South Jersey shore where we lived. I wanted to make sure to provide all of the information, let readers know the source of this testimony to the remarkable efficacy of properly deployed ecodelics.

I searched around the various online forums for evidence that others were having a tougher time with the integration of what I had found so "impossible" and yet actual, my ayahuasca gnosis. Instead, I found remarkable self-organizing communities scattered across the planet, drinking ayahuasca in modern and post-modern cities, pursuing diverse spiritual and scientific goals but almost all bringing, as far as I could tell, a sense of serious intent to their drinking of ayahuasca. Ayahuasca tea has been seized in the United States, and as the case of the UDV made its way through the legal system, I was resolved that I would only drink the brew where it was understood and supported by the legal system. Set and setting is too important to allow it to be programmed by the state, and there was no question but that any ecodelic experience in the United States was imprinted by the criminalization and prohibition of altered states of consciousness. Millions smoke cannabis in this context, and my investigations suggested that the programming of psychedelic states was effectively narrowing the potential of the spiritual use of that plant, which many report to be a remarkable meditation adjunct as well as guide to its own cultivation. Besides, at this point I was frankly very much afraid of drinking ayahuasca alone, legal or not. Its shamanic footprint was such that I did not wish to summon the plant intelligence in my backyard and on my own, without the help of an experienced guide. And so I continued to meditate, but did not journey with ayahuasca brew in the United States.

Everyday family life and job realities affirmed themselves with their usual consistency and power. I wrote other sections, including one on Philip K. Dick. Then, improbably, I received an unsolicited e-mail from one of the moderators of the Ayahuasca Forums, one Lil Merlin. I clicked on the message with great interest, for I had come to greatly respect Lil Merlin's knowledge and experience. Maybe the rain forest was a dense informational network after all.

ON BEYOND VALIS

> *"A hologram," Fat echoed.*
> —Philip K. Dick, *VALIS*

> *While I was writing this work, l was a bit too fearful to express candidly in writing the direct experience, uninterpreted. I felt that a group of thirty persons' salaries, a large research budget, a whole Institute's life depended on me and what I wrote. If I wrote the data up straight, I would have rocked the boats of several lives (colleagues and family) beyond my own stabilizer effectiveness threshold, I hypothesized.*
> —John C. Lilly *Programming and Metaprogramming in the Human Biocomputer*[2]

My research and meditation kept me focused on the peculiar mantric power of language and the remarkable role of rhetorical repetition and programming in the composition of psychedelic experience. Laura Huxley repeated the "Bardo Thodol" for Aldous as he lay dying, and I found myself spontaneously repeating the Prajnaparamita mantra during meditation and other psychedelic practices—such as when George W. Bush visited central Pennsylvania and the Future Farmers of America. Here, with friends, I greeted the war criminal with sonic weapons powered by mantra. No, we didn't "believe" that we were firing actual projectiles at the president, nor would we want to do so, sharing as we do a commitment to ahimsa, a concept of thoroughgoing nonviolence from the Upanishads deployed by Mahatma Gandhi against the last English-speaking empire. Ours was a laughter cannon, a rhetorical practice designed not to impact but to transform. It introduced an uncanny and comic fluctuation between high seriousness (this regime must be stopped) and high comedy (its power is meager in the context of the history of the planet and the enormity of our ecosystem). And, of course, we were aiming the fluctuation directly at our selves, such that we might be capable of sufficient involution to move beyond anger, critique, contestation, and violence and begin to build a community and life other than that exemplified by Bush the Second's visit to the Future Farmers of America. Out at the Center for Sustainability, where I had begun to volunteer, we called ourselves the Future Farmers of America because what we were growing was indeed the future itself. Yes, we thought the mantric power of the words were powerful indeed, and not simply because of their content. Barack Obama's remarkable shift in tone and his foundation in the practices of eloquence seem to confirm that politics, too, are susceptible to more than spin in the rhetorical tradition.

Our confidence in these nonlethal defense technologies was well founded. Leary's work determines over and over, in its own kind of mantra, that *if* LSD is the most potent psychic compound yet discovered (Hofmann), *then* the rhetorical practices that program psychedelic experience—such as the criminalization of cannabis—are the proximate and highly tuneable cause of any given experience:

> Here again we stand on the threshold of the scientific re-discovery of
> the relationship of sound to consciousness. Psychedelic voyagers have to
> develop their own mantras in a pragmatic, experimental way, using sounds,
> rhythms, words or music that are meaningful and that work. (Metzner and
> Leary, 10)

Leary calls for an experimental program actualized by the McKennas and others who would tinker with the sounds, words, images, and contexts of psychedelic experience all through the criminalization of these states of mind, itself a powerful programming of psychedelic experience. Practices such as Stanislav Grof's Holotropic Breathwork, drumming, brain machines, and meditation itself have thrived, but the collective sharing of the rhetorical programming of the self modulated by plants and compounds has assumed a more distributed, "underground," and mycelial structure not available for the most part in official university research practices. Hence the prohibition has ironically regulated and restructured crucial research on the development and capacity of human ecodelic programming practices more than it has restricted investigation of the chemistry of these compounds. (Shulgin, Nichols, Dennis McKenna, Holloway). And transformations in technoscience such as genomics promise to further wither the effectiveness of any prohibition on psychedelic research, but the "problem" of involutional practice of a human observer in contact with psychedelic experience remains for psychedelic science and requires a commons in which to share in investigations.

Leary's rhetoric of "program," of course, links psychedelic practice directly to the computer culture that would emerge from it, as well as the culture of information that would form the ecology of Mullis's innovations with PCR. And the rhetoric of "program," like Aldous Huxley's "reducing-valve," summons its own amnesias and mantras. The history of molecular biology suggests that the rhetorical software of "genetic code" and "program" were successful but partial representations of living systems that often occluded the complexities of ecosystems and embodiment. Scientist and involutionary John C. Lilly's 1966 samizdat science text, *Programming and Metaprogramming in the Human Biocomputer*, sums up a framework that recognizes the role of such programs and metaprograms only within the dynamic wetware assemblage of developmental systems embedded in highly interconnected ecosystems over time:

> Until we have thoroughly explored this genetic code, until we can specify the organism and the conditions under which it can reach maturity, and become an integral individual, we will not have the data necessary for specifying all of the characteristics of the human computer which are brought to the adult from the sperm-egg combination. (Lilly 1972, xxiv)

Lilly's treatment of such "programming" provides useful and very straightforward descriptions of rhetorical "software" with which our minded embodiments are enmeshed with DNA, one that remembers the freedoms and constraints of any given genetic sequence in its developmental and eco-systemic context:

> This genic determinism of thinking can turn out to be a will-o-the-wisp. It may be that in the subsequent development of the computer it has become so general purpose that the original genetic factors and the genes are no longer of importance. Even as one can construct a very very large computer of solid state parts or of vacuum tube parts or of biological parts, it makes little difference as long as the total size, the excellence of the connections and the kinds of connections are such that one can obtain a general purpose net result from the particular machine. So may we possibly cancel out genic differences. So may each one of us, as it were, attain the same kinds of learning and the same kind of thinking machine little modified by genic differences. (Ibid., xxvi)

Indeed, it was precisely Lilly's desire to understand differences other than genetic ones that he deployed the isolation tank in pursuit of what he called his "scientific problem":

> Given a single body and a single mind physically isolated and confined in a completely physically controlled environment in true solitude, by our present sciences can we satisfactorily account for all inputs and all outputs to and from this mind-biocomputer (i.e., can we truly isolate and confine it)? (Ibid., xiii)

In other words, the evidence that genetic variation was causally related to human experience was less assumed by Lilly than investigated by him in total solitude: it was the occasion for an investigation of "internal" states of mind that may or may not correlate with genotype, an investigation of DNA's capacities through the systematic isolation and transformation of every other input and output. Lilly suspects that the attribution of individuating causality to one's genotype may be nothing but a program, in particular a "wishful phantasy":

Some kinds of material evoked from storage seem to have the property of passing back in time beyond the beginning of this brain to previous brains at their same stage of development; there seems to be a passing of specific information from past organisms through the genetic code to the present organism; but, again, this idea may be a convenient evasion, avoiding deeper analysis of self. One cannot make this assumption that storage in memory goes back beyond the sperm egg combination or even to the sperm egg combination until a wishful phantasy constructed to avoid analyzing one's self ruthlessly and objectively is eliminated. (Ibid., 12)

So too, of course, can the sheer plasticity of embodied DNA function as an equally egoically charged fantasy blocking actual inquiry, as in the repetitious cry of "culture" in response to any equally rhetorically charged "nature!" that serves as a rhetorical shelter for different political and epistemological "positions" on that obviously spiraling coevolutionary unfolding we might dub "transhuman naturing." Lilly's "ruthless" program required a suspension or isolation of the investigator in attempt to *calibrate the observer such that it might be transformed within the constraints of its genome.* Recognizing with quantum mechanics, as well as the psychoanalysis in which he was trained, that an observer was an unavoidable component of any scientific system, Lilly sought an adjustment of the very observing apparatus of science that would correct, if only a bit, errors due to the fantasies the self has about itself. If Leary emphasized the programmability of psychedelic experience, its capacity to be "played" by the collective assemblage of set and setting, Lilly deployed the tank as a kind of debugging tool for set. For example, being unaware of the recursive effect of consciousness on itself, one might phantasmically inhabit a world that posits a degree of separation from the environment that is false in degree and therefore false in kind. Just as the beginning of any musical practice involves the tuning of instruments and ears together, it is only by detaching from and observing the effect of the self on the self that this involutional science of consciousness even begins.

It is in this involutionary investigation into his own DNA that Lilly's work was its most heterodox. Anticipating some response to the scientific taboo against introspection, Lilly did some rhetorical programming of his own.[3] In his introduction to the second edition of his book, Lilly admits that he deployed his own encryption programs in the text, seeking to shelter the knowledge from "the usual reader":

> I had written the report in such a way that its basic messages were hidden behind a heavy long introduction designed to stop the usual reader. Apparently once word got out, this device no longer stalled the interested readers. Somehow the basic messages were important enough to enough readers so that the work acquired an unexpected viability. Thus it seems appropriate to reprint it in full. (Ibid., vi)

Lilly had encrypted *Programming and Metaprogramming in the Human Bio-computer* as a summary of his National Institute of Mental Health–funded studies from 1953–1958, seeking to avoid the incredulity of his colleagues. Despite his encryption strategies, the content of the report was enough to create doubts about Lilly's sanity. As with Albert Hofmann's experience in the discovery of LSD, Lilly's report was written for an audience who did not yet exist—one who had experienced the ontological and transpersonal bliss and disarray of Hofmann's problem child programmed with the "sensory deprivation" tank:

> I heard several negative stories regarding my brain and mind, altered by LSD. At this point I closed the Institute and went to the Maryland Psychiatric Research Center to resume LSD research under government auspices. I introduced the ideas in work to the MPRC researchers and I left for the Esalen Institute in 1969. (Ibid.)

It is difficult to reconstruct the exact segments of Lilly's text that may have provoked these narratives and their judgments, if only because there are so many. Lilly's glossary, for example contains an entry on mind:

> Mind: the computer brain detectable portion of a supraphysical entity tied to the physical biological apparatus. The remainder of this entity is in the soul spirit God region and is detectable only under special conditions. (Ibid., 130–31)

Perhaps it was Dr. Lilly's report that

thus the subject found whole new *universes* containing great varieties of *beings,* some greater than himself, some equal to himself, and some lesser than himself. (Ibid., 44)

Or maybe it was Lilly's history of physical self-experiments that brought him to the brink of death that led to the gossip about this interdimensional investigator. Yet Lilly exudes psychonautic caution. He is careful to bracket the ontological status of the states he accessed in the tank and with the use of LSD-25 as an adjunct, referring to the projection of divinities, aliens, etc. "external" to the mind as a "mistake." Lilly notes that "Instead of *seeing or hearing* the projected data, one feels and *thinks* it," and in this cognitive synesthesia, a substitution for an activity of mind for a sensory one, the human biocomputer turns itself not on but inside/out, becoming an embodied category error.

> This is one basis of the mistake by certain persons of assuming that the projected thoughts come from outside one's own mind, i.e., *oneness with the universe,* the thoughts of *God in one,* extraterrestrial beings sending thoughts into one, etc. Because of the lack of sensory stimuli, and lack of normal inputs into the computer (lack of energy in the reality program), the space in the computer usually used for the projection of data from the senses (and hence the external world) is available substitutively for the display of thinking and feeling. (Ibid., 78)[4]

In this sense, the isolation tank is, for Lilly, a device for substituting the non-Euclidean, synesthetic, perceptually cognitive "internal" world for the Euclidean outside, a sensory-motor slow down of the "human biocomputer" asymptotic to zero that enables an unfolding of mind, releasing "the throttling embrace of the self" and allowing for the perception, *real or imagined,* of *n* dimensions. For while human bodies dwell and survive in the perceptually Euclidean space of insides and outsides, where a distinction can nearly always be neatly drawn between one moment and another, the human mind is capable of dwelling in n-dimensional arrays ordinarily *crowded out* by the programmed attention orientation devices of the external world, the flowers of perception with which our ordinary mind is often understandably enthralled, "the space in the computer usually used for the projection of data from the senses (and hence the external world) is available substitutively for the display of thinking and feeling." In this framework, the tank becomes a way of not only blocking but dissipating information. Interrupting the incessant broadcast of self with the white noise of LSD, Lilly learned to metaprogram by tuning his attention away from who he believed he was and toward the phenomenologically infinite "invisible landscape" of subjectivity. Later, Lilly will characterize some of his work as the very "science of belief," a moniker worthy of rhetoric that practices to seek to instill and dispel belief.

The "sensory isolation tank" is hence an allosteric mechanism for testing the limits of any given genotype/phenotype dyad, a mechanism analogous to the operation of the Lac operon.

> When lactose is present, it acts as an inducer of the operon. It enters the cell and binds to the Lac repressor, inducing a conformational change that allows the repressor to fall off the DNA. Now the RNA polymerase is free to move along the DNA and RNA can be made from the three genes. Lactose can now be metabolized.[5]

In blocking one sequence of information, the operon enables the expression of another, no less actual, sequence of information. So too does the REST tank, for Lilly, block the ordinary sensory broadcast in order to enable another. Here human minding screens sensory states less than becomes a lens through which the self appears in its expanded, non-actualized and non-Euclidean capacities. As the dissipative structure through which these visions are screened, *the observer becomes subject to radical and nonlinear transformation.* Here the particularities of any given self can disappear, and the floater becomes subject not only to the vagaries and gravities of any particular ecosystem, but to the entire morphogenomic field of the cosmos. In place of the attention attractors found in the external world, the tank tunes one's attention toward an identity of self and consciousness itself by still reflection on the void.

As a visualization technology for interrupting the usual "monocast" perception wherein a self "laminates" or stitches together the diverse channels of perception, a tank practice loosens the grip of the ordinary self on consciousness and creates the possibility of focusing attention on a "multicast" of n-perceptual dimensions and recombinations. This recombinant and hence synesthetic perception visualized by the tank with the mind and the space of its metaprogramming functions isomorphically to the putative visualization technologies of Eleusis, where the focused attention on a filigreed stone seems to have enabled bouts of involution similar to Lilly's investigation of what he called the "cyclone" of self:

> Initiates were led to behold the Organic Light by gazing softly at the pillars in the telesterion hall, but seasoned initiates made a special rite of gazing at the omphalos stone. A peculiar braiding of knotted and bisected chromosome-like structures adorned these stones. (Variations occurred: e.g., the small omphali at the feet of the Ephesian Artemis.) Mystae who has attained the autopsia, unassisted beholding of the Light, were able to enter deeply into what they perceived. By holding their bodily concentration steady, they not only saw the Light interpenetrating the omphalos, they entered it, visually. Gazing into the depths of the substantial shadowless white luminosity, they saw the molecular structure of living matter, not just the form of DNA, suggested by the cruciform seed-like nodules engraved on the omphalos, but also the peculiar kinking and twisting of nucleic threads. This was the epiphany of the divine serpentine force to be encountered within the ornamental marble dome. Thus the omphalos, navel, was the entry point for perception of the cosmic umbilicus, navel-string, DNA-RNA.[6]

As in my discussion of Narby's meditations on the shamanic knowledge of DNA in the upper Amazon, it is worth going slowly and unpacking the rhetorical practices indicated by this Eleusian scene cited from the Internet, their analogies with Lilly's tank experiments, and my emerging arguments about the ecodelic involution of mind.

Recorded here are instructions in the science of attention. Initiates begin their learning through a subtle letting go, "gazing softly." A kind of bardo or between agency is opened up here, as the initiate is asked to both practice agency—the gaze is a deliberate, intentional action—and let go of agency, a softening of the gaze undoes any perspective that understands itself as occupying a "point" of view acting "on" a world. Gazing softly, the world, and then the self, becomes a field of perception, perhaps a "message field of communication" analogous to the ecstatic signification reported by Henry Munn among the Mazatecs. Acquiring "soft eyes" in the manner of an Eleusian initiate is a tuning of visual perception away from its ordinarily sharp distinctions between self and other, allowing for a blending of viewer and viewed. What is "softened" is precisely the borders between observer and observed, with pillars functioning to provide azimuth orientation to the initiate, easing the transition from egoic to ecodelic consciousness. In short, the first practice in the Eleusian scene chronicled here in the Telesterion hall involves the release of attention in the visual field, a literal blurring of borders between self and other that is not drastic enough to induce a panic or vertigo in subjects not practiced in the non-Euclidean space of the Mysteries.

The next Eleusian rhetorical algorithm induces a further refinement in this focusing of attention:

> Seasoned initiates made a special rite of gazing at the omphalos stone.
> A peculiar braiding of knotted and bisected chromosome-like structures
> adorned these stones. (Variations occurred: e.g., the small omphali at the
> feet of the Ephesian Artemis.) (Lash 2007)

Readers familiar with the Buddhist tradition will recognize this visual rhetorical practice as yantra, the focusing of attention on a highly imbricated and symmetrical visual form as a method for stilling the mind. Through the deliberate attention captured by the lotus or the jewel, a dissolution or dissipation of the perceptual self is accomplished, again enabling an apprehension of an order only implicated in ordinary consciousness: Sunyata, the void, all-that-is, *istigkeit*. Next in the involutionary spiral, itself a mnemonic for this practice of subjective investigation and visualization (a flowering of the "perennial" philosophy indeed), this stilling of the mind enables the (relatively) undistorted reflection of mind on itself, a self-beholding, *autopsia*. Again this is a practice possible only through the stilling or steadying of thoroughly embodied attention:

> Mystae who have attained the autopsia, unassisted beholding of the Light,
> were able to enter deeply into what they perceived. By holding their bodily
> concentration steady, they not only saw the Light interpenetrating the
> omphalos, they entered it, visually. Gazing into the depths of the substantial
> shadowless white luminosity, they saw the molecular structure of living
> matter, not just the form of DNA, suggested by the cruciform seed-like nod-

ules engraved on the omphalos, but also the peculiar kinking and twisting
of nucleic threads. This was the epiphany of the divine serpentine force to
be encountered within the ornamental marble dome. (Lash, 2007)

As with Narby, the first implicit claim here is: this use of embodied mind
as a technoscientific visualization device enabled the revelation of the reductionist vision of a molecular cosmos. Yet the initiate's attention is focused
still further on the specificity and even singularity of the dynamic topological forms, "the peculiar kinking and twisting of nucleic threads." This
"epiphany" hence revealed something more complex than the molecular
order of DNA, informed as it is, like Mullis's PCR vision, by the *form* of the
DNA delivery system as well as its informational content. Indeed, if there
is a revelation here, it is a revelation of transduction, a consciousness of an
organism's role in genetic expression, or an initiate's role in the focusing and
tuning of their attention and consequent altering of consciousness in involution. No act of belief takes place here, only an opening into the complexity
of another.

As with Aldous Huxley's continual return to the flowers, it appears that
one cannot go slowly enough when it comes to sharing these investigations
with others, who are perhaps subject precisely to belief, the cessation of said
opening:

> As one opens up the depths, it is wise not to privately or publicly espouse as
> *ultimate* any *truths* one *finds* in the following areas: the universe in general,
> beings not human, thought transference, life after death, transmigration of
> souls, racial memories, species-jumping-thinking, nonphysical action at a
> distance, and so forth. Such ideas may merely be a reflection of one's needs
> in terms of one's own survival. (Lilly 1972, 38)

Such (proleptic) concern over the proper reception of the description of
ineffable states available through properly focused attention is, as we have
seen, well founded. Yet who, exactly, is this "one" about whom Lilly speaks
from within this description of a swarm of metaprograms? If it is precisely
the self that must become a field of information to enable the cosmic consciousness often induced by the tank operon, who is this "one" who would
preside over its disappearance? If LSD and other ecodelics resonate with

"sacramental" attributes, their sacramental effects are linked to a capacity to be noisy, to allow for the noise and avoid the certainty of control and belief:

> In the analysis of the effects of LSD-25 on the human mind, a reasonable hypothesis states that the effect of these substances on the human computer is to introduce *white noise* (in the sense of randomly varying energy containing no signals of itself) in specific systems in the computer. These systems and the partition of the noise among them vary with concentration of substance and with the substance used. (Ibid., 76)

Lilly's notion that it is by "introducing" noise into the system that one can paradoxically but practically cultivate further order and transformation is resonant with researcher Craig Reynolds's practice of "jiggling" his robots to induce the more rapid evolution of wall-following behavior as well as the neurology of birdsong discussed in chapter 3. Lilly offers a vertiginous description littered with semicolons listing the effects of such noising, itself a kind of rapid fire cut up of the reader's consciousness. Each symptom is interrupted with a successive, yet apparently unrelated, effect moving from temporal to visual to synesthestic to somatic to cognitive registers in a seemingly ordered, staccato sequence or program only pretending to be an explanation:

> One can thus "explain" the apparent speedup of subjective time; the enhancement of colors and detail in perceptions of the real world; the production of illusions; the freedom to make new programs; the appearance of visual projections onto mirror images of the real face and body; the projections and apparent depth in colored and in black and white photos; the projection of emotional expression onto other real persons; the synesthesia of music to visual projections; the feeling of "oneness with the universe"; apparent ESP effects; communications from "beings other than humans." (Ibid., 76)

This introduction of noise parallels methods such as Burroughs's and Gysin's cut-up method, wherein the deployment of chance elements break the hold of authorial control and generate surprising order.

The increase in *white noise* energy allows quick and random access to
memory and lowers the threshold to unconscious memories *(expansion of
consciousness)*. In such noise one can project almost anything at almost any
cognitive level in almost any allowable mode: one dramatic example is the
conviction of some subjects of hearing-seeing-feeling God, when "way out."
One projects one's expectations of God onto the white noise *as if the noise
were signals*; one *bears the voice of God in the Noise*. (Ibid., 76–77)

In my own programming, I was finding this heuristic describing the
action of LSD (and, perhaps by extension, ayahuasca) to be useful as I tried
to navigate all the interconnection, and I thought hard about how to keep
my mind as open as possible as I clicked open Lil Merlin's e-mail. Despite my
skepticism about some of Narby's conclusions, something like a feedback loop
did seem to be operating between consciousness and DNA. The idea that consciousness actually carries out a crucial evolutionary task—the selection and
recombination of traits, a way to move plant alleles around and form novel
recombinations of organisms and cultures, precisely as the ayahuasca had
"taught" me—was becoming legible within the new thermodynamics, where
the very purpose of consciousness is to dissipate greater levels of energy and
information than a planet without consciousness. The strangest thing was
that this process of cultivating ever more complex descriptions and narratives
precisely in order to have them fall apart was the very description Philip K.
Dick gave of his fiction in a talk he wrote, but never delivered, called "How to
Build a Universe that Doesn't Fall Apart a Few Days Later":

It is my job to create universes, as the basis of one novel after another. And
I have to build them in such a way that they do not fall apart two days later.
Or at least that is what my editors hope. However, I will reveal a secret
to you: I like to build universes that *do* fall apart. I like to see them come
unglued, and I like to see how the characters in the novels cope with this
problem. (1995, 262)

Breathing in, I was breathing in, and breathing out, I was breathing out:
a deliberate emptying of the mind à la "The Heart Sutra's" repetition: there
is no attainment, there is no non-attainment. This focus on breath serves
to give me some tiny measure of rhetorical detachment on the meaning of

the growth and healing I was experiencing. It would be a mistake to attribute anything but a distributed causality to the ayahuasca, as Lilly's "One projects One's expectations of God onto the white noise" makes evident. These projections then serve as limits to growth, premises in which one's metaprogramming is habitually stuck. Meister Eckhart and the traditions of the perennial philosophy sometimes seek to silence the mind and its grasping after certainty such that it might undergo involution:

> Therefore God has left a little point wherein the soul turns back upon itself and finds itself, and knows itself to be a creature. (Qtd. in Inge, 39)

And finding this aspect of God in oneself means dwindling or withering the ordinary loud-mouthed self, the one that is the effect rather than the cause of the goings-on of the organism and yet takes itself to be "On the Bridge": the swaggering, worrying, anticipating ego. By shrinking this grasping aspect of the self, one can *tune to and choose* the subtler broadcasts from one's entire developmental and evolutionary itinerary. And sometimes all that is necessary is what I increasingly called, to myself and my students, dialing to zero: an interruption of the self that might enable the montage. Interruption is a tuning, and meditation was serving to interrupt the continual reprise of Me and My Story, and was cueing a new track: What We Might Become, Together.

A similar tuning takes place regarding that ego warranting phenomenon, The Truth. If The Truth is treated as a territory, a place, or "foundation" where the mollusk-like ego can attach, then Truth, despite its obvious existence and cause for wonder in human consciousness, can easily function as a fortress for the ego and its delusions—such as the delusion that it is radically distinct from the world it observes. Gene Youngblood, in his "paleocybernetic" treatment of the plant-machine gradient, argues that with ecodelics human beings rediscover their analog nature, the human organism's aspect as continuous with ecosystems and in time. We might call the awareness of this continuity the ecodelic insight:

> Man's image of himself as discretely separate from the surrounding biosphere is shaken by the discovery of the *a priori* biochemical relationship between his brain and the humble plant. (Youngblood, 144)

Youngblood writes that such a (veritably cinematic) shaken image can sometimes induce a nostalgia for the sense of control and isolation accorded by the ego, but such a discovery can also animate aspects of oneself, including those that do not rely on a radical distinction between themselves and the world. Such archetypes of self are interested less in territory than in transformation. This nomadic, journeying, tripping self begins to drop hints about routes for transformation, even points to the infinite, if only because the silencing self turns in on itself, regarding an emptiness that *is* this looping echo of a transparently empty self regarding itself, a paradoxical, infinite aspect of consciousness often marked by a phenomenology of the divine. In the Tibetan tradition articulated by fifteenth-century monk Tsong Khapa and translated by Robert Thurman, this loop of emptiness is full of marvel:

> Whatever depends on conditions,
> That is empty of intrinsic reality!
> What method of good instructions is there,
> More marvellous than this discovery?[7]

To silence the self is to begin to notice the cosmos. And yet not just any formulation can be used to compress the resulting information—clearly ego death is not merely the absence of information, nor can its story be told from anything like the perspective of an ordinary self separate from All That Is. A connection is inevitably *forged*. Any such organizing trope or refrain must have some archetypical purchase on the metaprogrammer's attention—emptiness or "letting go" is the only way through ego death, so the archetype must be *so obvious that it can be forgotten and yet relied upon*. Perhaps it is not surprising that a vine, that tendril of connection, was serving as my archetype. And my focus on DNA was now understandable as a similar rhetorical mandala for connection—it is a mandala for imagining and inhabiting the interconnection of all life.

Lil Merlin had been down visiting with Norma. Despite the fact that he had drunk ayahuasca hundreds of times, she had amazed him with her guidance and brew. She told him to contact me, her "spiritual son in North America." This testimony of Norma's was completely unexpected and it triggered a recognition that I was indeed engaged in a spiritual practice, something that had previously not even been a category in my sense of self. Lil

Merlin wrote movingly of his new understanding of death and life. I found it hard to accept this synchronicity, since as far as I knew no one knew about Norma on the online forum—I had searched—and after reading Lil Merlin's posts almost every day for over a year, the sudden experience of our interconnection carried the same charge of the miraculous as my ayahuasca journeys themselves. Early in the first most difficult night, I had suddenly wondered why exactly I had traveled all the way down to the jungle and done this to myself, and an internal voice very clearly answered:

> You don't think coming down here was your idea, do you? You're here to help Norma with her addiction clinic and work through the loss of your brother.

I had previously known nothing about Norma's work in addiction therapy. I learned about it two days later as we whistled for the pink dolphins that populate the upper Amazon. With no place to put this kind of impossible knowledge, I simply chose to accept it without asking after its "status" or mechanism. There would be plenty of time for that as I wrote up the book, and anyhow the Ayahuasca was telling me to focus now on the sound of the crickets, their noisy song a sonic handhold for me until dawn.

ON BEYOND VALIS: PART 2

With a bit of proper programming under the right conditions, with the right dose, at the right time, one can program almost anything into the noise within one's cognitive limits; the limits are only one's own conceptual limits, including limits set by one's repressed, inhibited, and forbidden areas of thought. The latter can be analyzed and freed up using the energy of the white noise in the service of the ego, i.e., a metaprogram to analyze yourself can be part of the instructions to be carried out in the LSD-25 state.
　　—John C. Lilly, *Programming and Metaprogramming in the Human Biocomputer*

It's not funny. Some force or entity melted the reality around me as if everything was a hologram! An interference with our hologram!
　　—Philip K. Dick, *VALIS* [8]

Despite the understandable bundling of silence with the epiphanies of both Eleusis and the perennial philosophy, such an opening toward enlightenment and the prohibition on its articulation often induces a desire to share. Tsong Khapa shouts "Empty!" from the rooftops, as it were, in his "Short Essence of Eloquence," a work of "praise" directed to the Buddha for the discovery that mind is itself dependent on set and setting: "dependent on conditions" and therefore "empty of reality." So, too, did John C. Lilly want to share his remarkable experiences of emptiness with other researchers—"dependent on metaprograms" and therefore "empty of reality," and with the help of Stuart Brand and the Whole Earth Catalog, a mimeographed report of his investigations of LSD-25 and the sensory-deprivation tank Lilly had pioneered made the rounds as samizdat science. It was very much as a snapshot of a moment for Lilly, written by another iteration of himself, precisely because he had continued to transform under the influence of his own metaprogramming:

> It seems as if this older work is a seminating source for other works and solidly resists revision. To me it is a thing separate from me, a record from a past space, a doorway into new spaces through which I passed and cannot return. (Lilly 1972, vii)

My own metaprogramming was now beginning to focus on the concept of information embedded in the idea that DNA and consciousness are somehow involved in a feedback loop in the distributed scheme of planetary plant intelligence suggested by the ayahuasca dialogues. In this sense, our Eleusian scene's "serpentine force" is the force of continual interconnection, the effect of consciousness on itself: the Ouroboros. As evolution has evolved ever increasing capacities to gather and then dissipate information—from the global DNA Internet of bacterial transduction to the ever increasing bandwidth of YouTube—this capacity of consciousness to affect itself in recursivity has amplified, deterritorialized, and recombined the space of all possible traits and organisms and interaction between and among organisms. Human consciousnesses, working with what Lynn Margulis and Dorion Sagan call "gradient perception," focuses individual and collective attention on the thermodynamic results of different ecosystemic interactions—when to pick fruit so that it has the highest sugar content, when to plant which strain of corn so that it will yield the greatest crop, the way a goat stomach processes milk into cheese, the best geometry for a road bicycle—garnering more and more dissipation of energy from the same system due to the discovery of more available energy (exergy) and its depletion. Information, as the "difference that makes a difference," makes a difference precisely to thermodynamic outcomes, increasing the output of entropy, and these outcomes feedback on survival as well as sexually selected fitness. Consciousness, by linking informational signs of an ecosystem to time and its syncopated order-for-free, rhythmic entrainment, processes the information of any given ecosystem such that it yields a greater dissipation of energy through the comparison with past thermodynamic outcomes of the same system, a "symmetry-breaking" of before and after that paradoxically yields yet further order.

It is in this sense that we can understand the direct apprehension of knowledge from the environment, an informational gnosis mediated not by signs but by the difference between one set of signs and another. Consciousness is *informed* by ecosystems, and the accrual of this information by consciousness and its "external symbolic media" accrues, rendering a new level of evolution, after the lithosphere and the biosphere: the noösphere.

Once formulated in this fashion, it becomes obvious that consciousness, if (I think usefully) defined in this way as the capacity to compare thermodynamic outcomes of any given ecosystem, is hardly a monopoly of human beings. Bacterial transduction and interspecies and interphyla signaling (e.g., angiosperm/insect communication) begin to map out a global layer of the planet devoted to informational transfer. Maxwell, in imagining his famous demon, offers us precisely this implicit definition of consciousness as the perception of thermodynamic differentials or gradients:

> If we conceive of a being whose faculties are so sharpened that he can follow every molecule in its course, such a being, whose attributes are as essentially finite as our own, would be able to do what is impossible to us. For we have seen that molecules in a vessel full of air at uniform temperature are moving with velocities by no means uniform, though the mean velocity of any great number of them, arbitrarily selected, is almost exactly uniform. Now let us suppose that such a vessel is divided into two portions, A and B, by a division in which there is a small hole, and that a being, who can see the indi-

vidual molecules, opens and closes this hole, so as to allow only the swifter molecules to pass from A to B, and only the slower molecules to pass from B to A. He will thus, without expenditure of work, raise the temperature of B and lower that of A, in contradiction to the second law of thermodynamics.[9]

If gnosis, the direct acquisition of information from the environment, is the post-ayahuasca symptom that itself most solidly resisted revision by me, "information" was the concept taboo that was at stake. For I appear to have been informed by ayahuasca in ways that have fed back into my life and consciousness. For even while I have been fascinated with the rhetoric of "information" and its attendant amnesias since 1977, when first I read Alvin Toffler's *The Third Wave*, I was nonetheless amazed by the demonstrative proof of ayahuasca's distributed drug action when first I began to enter into a dialogue with the plant intelligence during moments of "ordinary" consciousness. By focusing my mind through mantra and meditation, I was increasingly open to whole streams of information in my environment that had been crowded out by the incessant interior voice of the ego. Like Socrates, whose own cult of refutation focused on Knowing Nothing, I had a daemon, an uncanny and orienting internal voice and vision. During these episodes of perceiving myself deeply interconnected with my environment in ways that I had never quite anticipated, my narrative self would frequently articulate the experience in terms of an information transfer—I felt I was transcribing and being transcribed by informational networks I had not previously connected with. I knew full well that this was "nothing but" an effect of my ability to focus my attention in stillness and open my observation to other layers of complexity not usually perceived by my chattering-monkey self. Yet cognitively and affectively my mind would sometimes become alarmed that I appeared to be experiencing the cosmos as a network of living, divine information, and I wasn't by any means Philip K. Dick.

Over time, the alarm subsided and I accepted the increased interconnectivity as a real artifact of my slightly more focused attention. Instances of "synchronicity" or "coincidence" had alarmed my former self in similar fashion, sometimes awakening in me some distress at what Freud dubbed "the omnipotence of thought" in his treatment of the mathematical uncanny. But with all the programming and metaprogramming, I found that by becoming only slightly, even rhetorically, detached on the articulations my mind was offering me, I could move beyond the panic and investigate the states associated with the information transfers. Lilly writes:

> These may be personified as if entities, treated as if a network for information transfer, or realized as if self traveling in the Universe to strange lands or dimensions or spaces. If one does a further unification operation on these supraself metaprograms, one may arrive at a concept labeled God, the Creator, the Starmaker, or whatever. At times we are tempted to pull together apparently independent supraself sources as if one. I am not sure that we are quite ready to do this supraself unification operation and have the result correspond fully to an objective reality. (Ibid., x–xi)

Lilly's humor here is as instructive as Terence McKenna's sophistry: the systematic encounter with ecodelics is a process rather a state, and so it is Alfred North Whitehead's fallacy of "misplaced concreteness" that must be kept at bay, as when substituting a "result" for a process or "a pulling together." Lilly's mixture of skepticism and credulity is also pedagogical, as he professes a lack of complete certainty about, well, complete certainty. Lilly notes that it is not simply logic's other—emotion—that distorts our mediation between reality and its models. Instead, it is *the very flexibility of our capacity to experience divinity* that suggests to Lilly that he is "not sure we are quite ready" to think a concept of "labeled God, the Creator, the Starmaker, or whatever" and have "the result correspond fully to objective reality."[10]

> The quality of one's model of the universe is measured by how well it matches the real universe. There is no guarantee that one's current model does match the reality, no matter how certain one feels about the high quality of the match. Feelings of awe, reverence, sacredness and certainty are also adaptable metaprograms, attachable to any model, not just the best fitting one. Modern science knows this: we know that merely because a culture generated a cosmology of a certain kind and worshiped with it, was no guarantee of goodness of fit with the real universe. Insofar as they are testable, we now proceed to test (rather than to worship) models of the universe. Feelings such as awe and reverence are recognized as biocomputer energy sources. (Ibid., xi)

Yet neither are such feelings of "awe and reverence" to be discounted; they are instead included in the metaprogramming as energy sources for further inquiry and transformation—they *inform* the inquiry rather than resolve it. And in order to be integrated into the metaprogramming, such feelings must be inhabited, they must become part of my inquiry, they must inform it. I began to understand that in order to share these experiences I was heuristically and provisionally labeling plant intelligence, VALIS, "a concept labeled God, the Creator, the Starmaker, or whatever" and the teaching they seemed to offer, I would have to seek, in true Popperian fashion, to *falsify and maybe even dissolve* the insights I had been offered, lest I start to believe them.

ON BEYOND VALIS: PART 3, "THE EMPIRE NEVER WON!"

"What is VALIS?" Kevin said to mini. "Which deity or demi-urge is he? Shiva? Osiris? Horus? I've read The Cosmic Trigger *and Robert Anton Wilson says—"*
    —Philip K. Dick, *VALIS*

My own opinion is that belief is the death of intelligence. As soon as one believes a doctrine of any sort, or assumes certitude, one stops thinking about that aspect of existence. The more certitude one assumes, the less there is left to think about, and a person sure of everything would never have any need to think about anything and might be considered clinically dead under current medical standards, where absence of brain activity is taken to mean that life has ended.
    —Robert Anton Wilson, *The Cosmic Trigger* [11]

It turns out that *VALIS*, a fake autobiographical novel written about a man named Horselover Fat, by Philip K. Dick, became something like a guide-book to activists working to propagate and mainstream a controversial but apparently highly effective addiction treatment using Iboga, the west African root and very wedge that sundered writer Daniel P. Pinchbeck's noggin in his book, *Breaking Open the Head*.

The more Dana got into it, the more he realized having VALIS in 1980 was like being handed a roadmap to understanding gnostic substances, Ibogaine and Bwiti, back in the very beginning. But the Ibogaine story is replete with these coincidences. (De Rienzo and Beal, 112)

"Dana" is Dana Beal, and Dana Beal appears to be a pseudonym for Horselover Fat. But I anticipate myself, perhaps for good reason. *The Ibogaine Story* is perhaps best read in hypertext across a series of weeks and in no particular order. A shaggy-dog story of multiply implicating, continually interconnecting plot lines involving the Yippies, Lower East Side junkies, *High Times* magazine, the Black Panthers, Hillary Clinton, intellectual property, and the island of St. Kitts, *The Ibogaine Story* is a veritable hologram of the War against Plants; cut into it anywhere and you will find an engaging cross section of the counter culture–control culture continuum, that strange interzone where LSD, the CIA, and unmitigated internal freedom is mixed with the Internet, heroin, widespread (and raced) incarceration, and epidemics. *The Ibogaine Story* is the gnostic gospel for the post-modern-control society, an archive for an open source mystery cult devoted to plants on a fallen planet, one where, to quote Philip K. Dick, "The Empire Never Ended." And like many of his gnostic brethren, Dana or Horselover Fat, embodies a "vague sense of loss" on the part of a counter-culture and religious movement that emerged in response to the impressive effects of psychedelics as they entered the mass culture in the mid-1960s.[12]

Yet if "The Empire Never Ended," "The Mycelium Thrived through It," and *The Ibogaine Story* reminds us also of the verve and wit of a nearly invisible and underground opposition to the War on Drugs. While both major political parties in the United States have amplified and naturalized the War on Drugs and extended their range to Afghanistan, Columbia, Panama, Mexico, Canada, and through the globalizing informatic mesh, activists, psychonauts, and scientists have worked to expand the range of knowledge and information about ecodelics, often at the risk of their own freedom and livelihood. Texts like *The Ibogaine Story* help make the strategies, tactics, and experiences of this War available to the rest of us, trying to find our way toward less suffering, a Slightly Less Insane Psychedelics Policy (SLIPP). If psychedelic programming includes the rhetorical setting in which they are used, *The Ibogaine Story* provides a medium for an ecodelic programming that first and foremost respects the sheer strangeness of actuality. This "cosmic realism" is of a piece with Vernadsky's, and attempts to articulate a model of human consciousness that is thoroughly entangled with its ecological others: a plant intelligence, a strange science fiction book, and Bwiti are all aspects of human consciousness to be creatively fathomed. To the well-nigh cannibalistic rhetoric of the War on Drugs, *The Ibogaine Story* responds with puzzled fascination and a holographic model of mind: This is your brain; this is your brain on Bwiti.

I will admit that I am beginning to be enthralled by the persistence of the hologram as a trope of psychedelic discourse. In its graphomania, a telos emerges as the writing psychonaut seems to be pulled along by continual searches for some detail or story that would somehow substitute for the whole of an ecodelic experience. It is in this sense that ecodelic compositions are holographic: They asymptotically seek to (re)compose the Whole. *The Ibogaine Story* shows the hologram as the mother lode, or at least the grandmother node:

> Every fragment of a hologram contains the entire image. Your Grandmother does not reside in a "Grandmother neuron," but everywhere throughout your memory whenever you think of her. And just as many different holograms can be superimposed, so can infinite images be stacked in our brains. We have the spatial representation that maps the retinal image onto the cortex; then, in the membranes of the cells the image is transformed back into the frequency mode—the "scatter" we'd see if we saw without a lens. The brain's code for storing information resembles the interference of wave-phase relationships of a holographic plate—with the equivalent of a "reconstruction beam" zooming in on a particular coded memory when we recall something. Scattered, holographic information storage explains why stroke victims don't lose discrete parts of their memory: all memory is contained within any portion thereof. (Ibid., 136)

The hologram becomes an archetype, a way of modeling the action of Ibogaine for the participants of *The Ibogaine Story*, for whom Ibogaine is an evolutionary adjunct amplifying their connection to a unity of species across space and time. Ibogaine in the story acts as both a plant intelligence and a human genealogist linking subjects experientially (and not just diagrammatically) with what ethnobotanist Richard Evans Schultes called the "universal African ancestor." I recall along with the reader that the McKenna brothers also had recourse to the hologram when they sought a model for the biospherical action of ayahuasca and psilocybin when Dennis tuned into the totality of all living systems:

> If we imagine the harmine-DNA complex as a radio-cybernetic matrix, then we can suppose that this matrix stores information in a regressing hierarchy of interiorized reflections of itself, in a form similar to the familiar ivory balls carved one inside the other, each level free to rate independently. In response to the vibration of tryptamine-RNA charge-transfer exchanges, modulated by mind into a usable signal, information searches of any sort might be conducted through a process that we suggest might be much like the principle of retrieval of information from volume holograms. Such a process of information retrieval and image projection would never lag behind human thought. Indeed, conscious thought may be precisely this process, but occurring on a more limited scale. (McKenna and McKenna, 106)

Another way of describing holographic compression techniques is to substitute a Fourier transform of a signal for the signal itself. This means that from an informational perspective, holograms are radically distributed and, in the terms of neuroscience, "degenerate": a tiny sliver of the holographic plane replicates the whole of the plane. As an archetype for the distributed drug action of ayahuasca and my now ecodelic path, the hologram had become rhetorical-mind software for focusing attention on thinking of the increasingly hilarious interconnection that had begun to become almost commonplace in my everyday life. In particular it allowed me to imagine a unity in which truth and fiction were aspects of a whole, not opposing ontological states.

This was how, for example, I made sense of the fact that while doing some research on The Vaults of Erowid database, I found that *someone else had written up my trip report*. Trolling for the language and images of sexual selection in the online database, I came across a description of a peacock feather manifestation that was nearly indistinguishable from my own. I will admit that for a moment I wondered if I had written the report and forgotten it. I will also admit that if the report had been written up by Horselover Fat, I wouldn't have been the least bit surprised.

But the experience report was written by a different character in a different book: Gandolf. And when I re-read it, I realized this poster had had the same experience that I had on a different substance, while thinking about the substance:

> I was shown very interesting and complex imagery: Hindu gods and eyes of Horus patterns like peacock feathers swirling ad infinitum. My forehead felt like it was opening. When I realized that I was thinking about the nature of this substance, 2c-t-2, a name suddenly popped into my head with insistence: Shiva. Flames and eyes everywhere. Wow![13]

Granted, my peacock feather, Shiva, flame sequence had been induced by a different (and, at the time and place of its ingestion, legal) Shulgin compound, 2C-I.[14] But the sequence, *including the act of becoming aware that I was reflecting on the nature of the substance and the sudden, surprising emergence of Shiva and flames*, was identical. I reflected on the work of physicist Stephen Wolfram, whose cellular automata browse different attractor states even as they are composed out of otherwise random sequences of code. Over millions of iterations, unexpected and unpredictable patterns emerge. Buddha. Christ. Shiva. VALIS.

The question of whether or not my ongoing "downloads" were "true" took up less and less of my focus, as I instead attended to the remarkable interactions between the clearly fantastic teachings of these plant practices and my everyday life that was growing in response to them. This interconnection opened what I can only characterize as an ecstatically visionary space as well as an unprecedented sense of the sacred that was thoroughly surprising and yet strangely welcome to my formerly secular self. The figure of the hologram provided me, as it did physicist David Bohm, with a map for navigating the experience of being a part that contains the whole *while remembering the whole*. In the *Shiva Sutras*, I found immediate resonance with the idea that human subjectivity recapitulates the entirety of the cosmos in the sutra's enormously compressed first axiom: "1.1 Consciousness is the self."

This fundamental unity of an aspect of human experience with *all that is* itself repeats an equation written by Erwin Schrödinger, one perhaps less famous but equally profound as his equations modeling the "collapse" of the wave function. Here Schrödinger seeks to make logical sense of two aspects of living systems: (1) that they are deterministic, clearly manifesting only within the constraints of the physical laws of the Universe, and (2) that from the perspective of our consciousness we dwell in absolute freedom, capable of directing our intention in the world and thereby altering it and our experience of it. Schrödinger notes that both of these apprehensions of an aspect of living systems occur only through "immediate experience," an act of observation either of the "interior" landscape of consciousness regarding itself or the "exterior" landscape of the physical world revealed by our senses. Schrödinger notes rightly that no logical framework allows for immediate experiences that would contradict each other. Instead, to avoid contradiction, we must imagine these experiences as different aspects of a Whole:

> But immediate experiences in themselves, however various and disparate they be, are logically incapable of contradicting each other. (1992, 86)

Topologically speaking, Schrödinger is appealing here to a quite straightforward mathematical and biological characteristic of living systems, one where inside and outside do not contradict but conjoin and confuse, the domain of mobius. Schrödinger, though, continues with the language not of topology but of syllogistic reasoning:

> So let us see whether we cannot draw the correct, non-contradictory conclusion from the following two premises: (i) My body functions as a pure mechanism according to the Laws of Nature. (ii) Yet I know, by incontrovertible direct experience, that I am directing its motions, of which I foresee the effects that may be fateful and all-important, in which case I feel and take full responsibility for them. The only possible inference from these two facts is, I think, that I—I in the widest meaning of the word, that is to say, every conscious mind that has ever said or felt "I"—am the person, if any, who controls the "motion of the atoms" according to the Laws of Nature. (Ibid., 86–87)

And Schrödinger is as proleptic as Lilly, wondering at the reception his declaration of personal divinity might receive:

Within a cultural milieu (Kulturkreis) where certain conceptions (which once had or still have a wider meaning amongst other peoples) have been limited and specialized, it is daring to give to this conclusion the simple wording that it requires. In Christian terminology to say: "Hence I am God Almighty" sounds both blasphemous and lunatic. (Ibid., 87)

Of course, physics has enjoyed and suffered a proximity to divinity for much of the twentieth century, so perhaps Schrödinger's embedded code here not of Da Vinci but of the divine should not be the cause for much wonder. Yet Schrödinger is strangely confident of his reader's capacity to ignore the dominant meanings of his declaration of divinity.

But please disregard these connotations for the moment and consider whether the above inference is not the closest a biologist can get to proving also their God and immortality at one stroke. From the early great Upanishads the recognition ATHMAN = BRAHMAN upheld in (the personal self equals the omnipresent, all-comprehending eternal self) was in Indian thought considered, far from being blasphemous, to represent the quintessence of deepest insight into the happenings of the world. The striving of all the scholars of Vedanta was, after having learnt to pronounce with their lips, really to assimilate in their minds this grandest of all thoughts. (Ibid.)

And note that for Schrödinger to become mindful of this "deepest insight," one must work to "assimilate" it with mind, to grok ourselves as holographic slivers of all that is. Yet nowhere is "belief" indicated—only the understanding of Schrödinger's equation, Atman = Brahman, through a lifetime of striving.

And what does such striving of "all the scholars of Vedanta" consist of? The Upanishads are fundamentally a *genre*, more a place and a teaching than a content: "lessons at the feet of the master." Yet in these lessons at the feet of a master—and Schrödinger is indeed a master—the message is persistent and repetitive: lose the self in order to apprehend the Self. This information, if one allows oneself to be informed by it, is itself essentially transformative. Experimenting with the concept of the "hologram," I experimented with Schrödinger's real code-script, "ATMAN = BRAHMAN."

This "grandest of all thoughts," I was beginning to think, offered a way of processing the information of my plant gnosis in a way that repeats Fischer's findings: human subjects seem to process more visual information after ingesting psychedelics. It also replicates the findings of researchers studying the effects of psychedelics on engineers. These psychonautical engineers scored very well indeed on the Witkin embedded figure test, a putative measure of a human subject's ability to "distinguish a simple geometrical figure embedded in a complex colored figure."[15] Was the perception of Atman = Brahman, and its corollary observation, *"Tat Tvam Asi,"* the holographic, ecodelic insight into precisely this "simple geometric figure embedded in a complex colored figure," a whole that unifies the apparent multiplicity of the phenomenal world while recalling its existence as a whole? In the preface to his translation of the *Shiva Sutras*, engineer Subhash Kak writes that the very idea of Schrödinger's wave function itself emerges out of this Vedic insight where the part recapitulates the Whole.

> As is well known, the idea of Brahman in the Vedas being a representation of all possibilities was the inspiration behind the conception of the quantum-mechanical wavefunction defined as the sum of all possibilities. (Kak 1997)

In my first book of this trilogy, *On Beyond Living*, I noticed that Schrödinger carried out precisely this recapitulation of part and whole, only in the opposite direction, where the part (the "code-script") is made to stand for the whole (the development of a living system and the ecologies with which it is enmeshed). Now I was seeing the work Schrödinger was doing to continue the movement of this recapitulation, as if "ATMAN = BRAHMAN" was the other "code-script" left by Schrödinger, one to remind us of this remarkable capacity of consciousness to integrate the totality of all that is with the suchness of a particular moment. Here Atman, an attribute of human subjectivity, becomes a fractal or holographic attribute of Brahman, the cosmos. I now understood why I had chosen mobius as my e-mail handle. Perceiving the inside as an aspect of the outside and vice versa, I could see the way a holographic recapitulation of part and whole, pace Bohm and the McKenna brothers, enabling the algorithmic unfolding of the "Consciousness Is Self" algorithm—if we will still our minds, we can perceive the way in which our minds replicate the cosmos on another hierarchical level, just as a sliver of the hologram or a fractal can render the whole.

Was this, as the ecodelic hypothesis suggested, gradient perception in action? Or was it, perhaps "somewhat like" gradient perception, the evolution of a concept with which to process the larger scale dissipative structure that seemed to become legible in ego-death, the noösphere? To the researchers studying the effects of psychedelics on creative problem solving, this increased capacity to perceive geometric structure in the midst of a larger scale complex was most definitively an enhancement and had previously proven difficult to alter in a variety of experimental interventions:

> This shift is in the direction of enhanced ability to recognize patterns, to isolate and minimize visual distraction, and to maintain visual memory in spite of confusing color and spatial forms. Viewed as personality change, these Ss showed a shift from "field dependence" to "field independence," as defined by Witkin et al. (1962). Research has related this dimension to numerous performance variables, including autonomic stability, concept formation, resistance to suggestion in reporting illusions, and resourcefulness in ambiguous situations. . . . As measured by the embedded figures test, field-dependence-independence has been reported to be resistant to a variety of experimental interventions including stress, training, sensory isolation, hypnosis, and the influence of a variety of drugs (Witkin et al. 1962). (Tart 1972, 463)

ON BEYOND VALIS: PART 4, "I HAVE NOTHING TO DECLARE"

This recapitulation of the whole from the part has, for Schrödinger, formed a pattern, "somewhat like the particles in an ideal gas":

> Again, the mystics of many centuries, independently, yet in perfect harmony with each other (somewhat like the particles in an ideal gas) have described, each of them, the unique experience of his or her life in terms that can be condensed in the phrase: DEUS FACTUS SUM (I have become God). To Western ideology the thought has remained a stranger, in spite of Schopenhauer and others who stood for it and in spite of those true lovers who, as they look into each other's eyes, become aware that their thought and their joy are numerically one—not merely similar or identical. (1992, 87)

The Ibogaine Story narrates a similar outbreak of interconnection. If, as our sexually selecting lovers, the physics of non-locality as well as ecology teach us, everything *really is connected*, this interconnection is not necessarily tolerated for long by egoic consciousness. The hologram has to compete with the other broadcasts of the psyche, become perceived against the complex background of reality, and thus requires a focusing of attention. Hence the agony-ecstasy continuum—both extreme pain and bliss—require a letting go of the ego, an opening that, if resisted, creates a node of subjective experience where agony and ecstasy are not experientially distinct. Perceiving the One through Linkage to All That Is can suggest to the subject of perception—whoever that is!—not only that "I have become God," but that "I have become the only God!" Dana Beal notes that LSD came to be associated with real or imagined speed in street acid that became available, and "Reinforced by speed, the ego-structures wouldn't let go."

This danger of recapitulating not G*d but ego was observed by Stanislav Grof in his masterful work, *Psychedelic Therapy*. Beal, rather understandably, contextualized this sense that it was *about him* with the baroque epic of CIA research:

> I flashed that all of us in the psychedelic movement were like voluntary guinea pigs in some kind of CIA experiment that had gotten out of control. The White Light was flashing across us like searchlights on a World War I no-man's land. All around me, we were taking casualties. But *some of us would make it to the other side;* and someone would bring back something wonderful. (De Rienzo and Beal, 39)

With Terence McKenna's saucer, this surreal scene of ubiquitous experimentation and ongoing desire to prevail helps introduce more wonder and less certainty into the contextualization of and response to psychedelic experience. In Dick's *VALIS*, Beal found a character undergoing a now familiar becoming-cosmic resonant with Mitchell, Vernadsky, and now, mobius. Horselover Fat, a thinly veiled encryption of Philip K. Dick, felt the very difference between himself and the stars disappearing:

> Fat had seen an augmentation of space: yards and yards of space, extending all the way to the stars; space opened up around him as if a confining box had been removed. (Dick 1991, 49)

VALIS offered an explicit vision of this becoming-cosmic in a way that seemed to provide a template for comprehending the outbursts of interconnection apparently experienced by Beal and the Ibogaine gnostics. This was a topological re-envisioning of human being: "space opened up around him as if a confining box had been removed" precisely because the very difference between inside and outside was enveloped (implicated) by a larger-scale dissipative structure capable of linking Dick's, Fat's, mobius's only apparently interior phenomenology through a higher level topological manifold, a space of a higher logical type, itself future grist for morphogenomic browsing far from equilibrium into spaces of cognitive and haptic liberty, galactic in magnitude and only barely explored even by the shamanic tradition.

Beal found in the galactic magnitudes of the brain a giddy and perhaps vertiginous freedom. In his imaginative hacking of a technoscientific context addicted to the very paradigm of opiates, Beal was guided by the findings of nonlinear dynamics and neuroscience as much as the fictional-real fluctuations of VALIS, that Vast Active Living Intelligent System with a fake AI voice.

> If three coupled neurons make unpredictable patterns, imagine what 1,000,000,000,000 billion interacting cells could do. A working brain is more like the weather or a turbulent stream than it is like a digital computer, according to the chaos connoisseurs. In the classic lock-and-key model, one molecule of a brain chemical fits into a specific receptor on a cell membrane. But neuroscientists now know that a population of receptors can fluctuate rapidly under the influence of many microscopic conditions inside and outside the cell. (De Rienzo and Beal, 140)

Most crucial to Beal's testimony here is the verb "imagine." Here, in the galaxy of neuronal interactions where Roland Fischer suggests that paradoxes of self-referentiality can resolve in a consciousness of the next logical type, the scientific observer must himself, herself, all of us together now, experience themselves embedded in a larger scale structure of which they are an aspect manifested by the system of observation, *Tat Tvam Asi*, ecodelia.

John C. Lilly wrote about this shift in his perception of the "objects" of his study through the "feeling of weirdness":

This opening of our minds was a subtle and yet a painful process. We began to have feelings which I believe are best described by the word "weirdness." The feeling was that we were up against the edge of a vast uncharted region in which we were about to embark with a good deal of mistrust concerning the appropriateness of our own equipment. The feeling of weirdness came on us as the sounds of this small whale seemed more and more to be forming words in our own language. We felt we were in the presence of Something, or Someone, who was on the other side of a transparent barrier which up to this point we hadn't even seen. The dim outlines of a Someone began to appear.[16]

For Lilly early in his career, this feeling of weirdness was a fluctuation in the very status of the "object" of his observation, which began to manifest into existence as an ontological other worthy of reverence and from whom Lilly came to believe he could learn a great deal *and from whom Lilly could not be rigorously separated*. "The dim outlines of a Someone began to appear."

This sense of a dim outline of a Someone resonated with my own experiences, wherein I found myself in dialogue not with a cetacean but a plant intelligence, a plant planet of which I was a manifestation. During the year that followed my contact with Lil Merlin and my return trip to Peru, I was often able to manifest the experience of plant intelligence during ordinary consciousness. By focusing my attention on the surface area of all of the plants of the earth in constant respiration, for example, I could become aware of my own breathing, which allowed me, even demanded that I imagine a green planetary mesh, which would open toward the planetary totality of bacteria in continual DNA contact with itself. Emptying my mind, I sometimes found myself again in dialogue with the ayahuasca, an iridescent avatar of plant intelligence urging me to translate my interactions with it. These "translations," such as those I am narrating now, are possible only by dissolving my consciousness, going to zero in meditation and emptying my mind of everything, especially "belief." A sense of awe at my ongoing participation and awareness of the 13.7 billion year unfolding of the cosmos would often ensue. Weird.

Lilly came to cherish this feeling of weirdness in terms resonant with Schrödinger and sampling from the philosopher of Lone Pine, California, Franklin Merrell-Wolff. Wolff translated Schrödinger's equation Atman =

Brahman into the algorithm "Consciousness-Without-an-Object-Is." Here Wolff asks the reader to substitute egoic consciousness with Consciousness by asserting its ontology, sometimes inducing what the philosopher and rhetorician J. L. Austin called the "performative" aspect of an utterance, wherein to say or otherwise iterate an utterance is to begin to actualize the script or sequence of the utterance, as in "I love you!" (and I do, dear reader!). This recipe for experiencing the ordinary self as a *manifestation* is indeed a useful interpretation of the etymology of "psychedelic," which is, of course, the "manifestation" of psyche or mind. This sense of being an explication of the implicate order (Bohm) induced an "If, If, Then" algorithm of optimism for Lilly:

> If we are a manifestation of Consciousness-Without-an-Object, and if, as Wolff says, we can go back into Consciousness-Without-an-Object, then my rather pessimistic view that we are merely noisy animals is wrong. If there is some way that we can work our origins out of the basic ground of the universe, bypassing our ideas that the evolutionary process generates us by generating our brains—if there is some contact, some connection between us and Consciousness-Without-an-Object and the Void, and if we can make that contact, that connection known to ourselves individually, as Wolff claims, then there is possible far more hope and optimism than I ever believed in the past. (Lilly 1975, 175–76)

As always, though, Lilly is careful in this text, *Simulations of God*, to keep an open mind through the application of that rhetorical fluctuator, "may be":

> It may be that Wolff, like all the rest of us, is doing an over-valuation of his own abstractions. It may be that he is generating, i.e., self-metaprogramming, states of his own mind and those of others in which the ideals of the race are reified as thought objects, as programs, as realities, as states of consciousness. It may be that this is all we can do. If this is all we can do, maybe we had better do it—and see if there is anything beyond this by doing it. (Ibid., 176)

This algorithm (full of the chant, "It may be . . . It may be. . . . It may be . . . maybe") of Lilly's is a call for an additional category to add to philosopher Austin's notions of the illocutionary and the perlocutionary—the involutionary. For if our inquiries demand, as Lilly's did, that we give up the sometimes luxurious taboo on a science of subjective experience and experiment on and with the very real subjective aspect of the universe, we find, with Schrödinger, that a pattern or attractor emerges, and the dim outlines of a perennial philosophy appears where one has no choice but to explore the enormous freedom of our galaxy of possible states of consciousness. Graphomaniacal deliberation on such radical internal freedom alters the consciousness willing to experiment with it, leading to further investigations where "Consciousness-Without-an-Object" recapitulates the part for the whole, as the consciousness of self becomes recognized as holographic, a microcosm of the macrocosm, *Tat Tvam Asi*. The propensity of plant adjuncts to manifest the mindedness of ecosystems, I ask the reader to parse as "ecodelic." In the context of the global ecosystemic crisis that one most definitely does not require psychedelics to recognize, "it may be that this is all we can do. If this is all we can do, maybe we had better do it—and see if there is anything beyond this by doing it."

For many readers of Dick's "reality perturbation" compositions such as *VALIS, Ubik, Time Out of Joint,* or the *Grasshopper Lies Heavy,* it is the sometimes spooky fictional verification of real events that seem to occur in association with the novels that account for their remarkable capacity to alter consciousness, as if focusing the reader's attention on multiple orders of simulation (such as Dick's famous "fake fakes" or his "world famous imposter," who was in fact nothing but a janitor at Disneyland famous for being an imposter, or my personal favorite, a six-foot-tall hash robot that repeats, ad infinitum, "No, I don't" (have anything to declare) as it lumbers across the U.S.-Mexico border . . .) induces a kind of epistemological vertigo wherein sufficient fluctuation between fiction and reality occurs to erode belief in any particular vision. The implicate order continues in its holomovement, disallowing any fixation of knowledge into belief, manifesting the interconnection of fiction and reality and its continual fluctuation, provoking double takes:

> But Philip Dick they thought of as science fiction, even though *VALIS* is actually a book on Gnosticism and the lost plant sacrament of the early Christians, which Dick got to know all about during the sixties while married to the daughter of Episcopal Archbishop James Pike's mistress. (De Rienzo and Beal, 111–12)

Finding themselves the subject of a science fiction novel paradoxically induced a sense of humor and practicality for Beal and others in dialogue with Bwitists:

> As Fernandez points out: "For these Bwitists . . . religion was not a matter of faith. . . . It was a very pragmatic technique for understanding, predicting and controlling—in short a science or pre-science of hidden things. To believe in something despite lack of evidence or evidence to contrary, which is the Western religious condition, was foreign to their attitudes. Fang had always had good evidence for their beliefs"—via ready access to plant sacrament, like the Gnostics. (Ibid., 112)

In this context, the discovery that the Ibogaine Gnostics found a useful roadmap in a fake autobiographical novel argues less that these *Ibogainistas* had lost their reality principles and more that they were looking for useful tactics for inhabiting the always shifting ecosystem of an emerging science. Just as the very structural ambiguity of the *I Ching* initiates a rhetorical-semiotic machine capable of responding to hexagrams and their multiple, even infinite resonances, so too does the loopy self-referential Temporary Autonomous Zone of meaning called VALIS create the conditions where neither belief nor disbelief are adequate but are instead necessarily imbricated in each other as attention attractors. Lilly found in Patanjali's yoga sutras and the writings of the Sufis veritable algorithms for affirming and dissolving different premises through the focusing of attention, and the result was "yoga"—the "union with" or "yoking to God" he describes in his discovery of Wolff's "Consciousness-Without-an-Object-Is." This is perhaps *the experience* of what biologist Humberto Maturana and Francisco Varela referred to as "structural coupling," the constitutive linkage between an organism and its environment (Atman and Brahman?) even as said organism is, autopoietically speaking, *nothing but the creation of the difference between an organism and its environment*. Hence *The Ibogaine Story*'s recourse to Freud's famous zone of fluctuation, the uncanny.

In *VALIS*, in a passage on Gnostic sacraments, Beal found an Ibogaine parallel which was "just uncanny":

> ON OUR NATURE. It is proper to say: we appear to be memory coils (DNA carriers capable of experience) in a computer-like thinking system which,

although we have correctly recorded thousands of years of experiential
information, and each of us possesses somewhat different deposits from all
the other life forms, there is a malfunction—a failure—of memory retrieval.
(Ibid.)

If anything has been forgotten, though, it is belief. For Fat was quite precise in this citation from what he called his "tractate," itself a simulation of
Dick's "Exegesis."[17] Fat, in insisting on the discourse of propriety, organizes
our attention toward the precise wording of his ontological declaration: "*We
appear* to be memory coils" (emphasis mine) names, precisely an appearance
rather than an actuality. And Dick, as well as Fat, was famously skeptical of
appearance in a world subject to enormous deceit, as revealed by the fake AI
voice, through a scanner darkly.

Fat appears to seek remembrance here in the tractate, and it is the
Remembrance of Things Modeled. Like the *Diamond Sutra* (in whose Prajnaparamita lineage Fat's work should perhaps be placed), the Tractate seeks to
remind us of its status as a model even as it explores the most fundamental
attributes of consciousness and its capacity to affect itself through meditation, paradox, and repetition. This strange, phenomenologically endlessly
tuneable attribute of consciousness, what Leary called "internal freedom,"
becomes evident when we use our consciousness to focus attention on itself,
altering it, which alters the one beholding consciousness, and so on. Without the interruption of meditation, this auto-affective characteristic of consciousness spins endless imbricated narratives about itself—the chattering
monkey of ego that ironically blocks our embodied primate self and its structural coupling to an ecosystem. In this aspect too, Fat's theory is insightful,
as he offers an example of such consciousness spirals devoid of any intentional recourse to Sunyata in which instead time itself becomes a "Desart of
vast Eternity":

Over a long period of time (or "Desarts [*sic*] of vast Eternity," as he would
have put it) Fat developed a lot of unusual theories to account for his contact
with God, and the information derived therefrom. . . . This theory held
that in actuality he wasn't experiencing anything at all. Sites of his brain
were being selectively stimulated by tight energy beams emanating from far
off, perhaps millions of miles away. These selective brain-site stimulations

generated in his head the *impression*—for him—that he was in fact seeing
and hearing words, pictures, figures of people, printed pages, in short God
and God's Message, or, as Fat liked to call it, the Logos. But (this particular theory held) he really only imagined he experienced these things. *They
resembled holograms.* (Dick 1991, 23–24; emphasis mine)

In essence, Phil writes that according to Fat's theory, VALIS *caused Fat to
develop theories.* One of Phil's favorites was that Fat experienced Nothing at
All, which in its own way is the "thundering silence" of enlightenment. On
the other hand, in this case, Phil himself was nothing but a character in a
novel. He writes:

In effect, he no longer claimed that what he experienced was actually there.
Did this indicate he had begun to get better? Hardly. Now he held the view
that "they" or God or someone owned a long-range very tight information-
rich beam of energy focused on Fat's head. In this I saw no improvement,
but it did represent a change. Fat could now honestly discount his hallucinations, which meant he recognized them as such. It seemed to me a Pyrrhic
victory. (Ibid., 24)

This "Pyrrhic victory" is analogous to the practices of Tibetan dream
yoga, which exemplifies the Vedic tradition of inhabiting the "third state"
beyond true or false, waking or dreaming, where the dreamer lucidly recognizes and activates their own role within a dream and also creatively awakens to the illusory nature of ordinary reality. In short one becomes aware of
the hallucination-perception continuum and creatively acts in response to
this awareness, programming, and metaprogramming their consciousness
toward the best of all possible outcomes in compassion.

Note above, though, Fat's precise phrasing: *"They resembled holograms."*
What resembles a hologram? Any part replicating any whole. The two categories that seem to resonate with a frequency tuned by Schrödinger are
the categories of "Identity" and "Divinity." By consistently and persistently
focusing the attention on one, the other emerges in a bending or troping of
figure into ground, ground into figure, the tiniest of differences cascading
into macroscopic transformation in a repetition across time and space:

Again, the mystics of many centuries, independently, yet in perfect harmony with each other (somewhat like the particles in an ideal gas) have
described, each of them, the unique experience of his or her life in terms
that can be condensed in the phrase: DEUS FACTUS SUM (I have become
God). (Schrödinger 1992, 87)

The part, playing its part, becomes the whole. Subhash Kak's com-
mentary on the *Shiva Sutras* suggests that by observing ourselves becoming an actor in the divine play of Shiva, we can detach ourselves from an
identification with our alienated perception that extracts and abstracts our
part—"I"—from the whole—"Shiva." "Shiva" functions in Kak's commen-
tary as a larger-scale dissipative structure, a hierarchically distinct logical
framework or metaprogramming, for guiding the reader toward a recapitulation of whole in the part, the first axiom of the *Shiva Sutras*: "Self Is Consciousness." Yet this part/whole recapitulation itself relies on an intensive
labor and discipline, a repetitious dwelling in those spaces where subject
melds into object. We must become, with Edgar Mitchell, econauts, comfortable with these feelings of weirdness where all is connected to all: the
noösphere?

So Fat's recognition that in fact Nothing at All was happening to him is
yet another elaboration of the space of all possible algorithms for recapitulating the whole from the part, accessing the hologram. A cartoon shark
knocks on a door, shouting: "Hologram! Hologram for Mr. Horselover Fat!"
Phil writes that this insight into the Nothing That Happened was a "Pyrrhic
victory," noting that Fat's graphomania was in a feedback loop:

The exegesis Fat labored on month after month struck me as a Pyrrhic victory if there ever was one—in this case an attempt by a beleaguered mind to
make sense out of the inscrutable. Perhaps this is the bottom line to mental
illness: incomprehensible events occur; your life becomes a bin for hoax-like
fluctuations of what used to be reality. And not only that—as if that weren't
enough—but you, like Fat, ponder forever over these fluctuations in an
effort to order them into a coherency, when in fact the only sense they make
is the sense you impose on them, out of the necessity to restore everything
into shapes and processes you can recognize. (Dick 1991, 24)

Yet even the fictional Phil is aware of this feedback loop, and in this feedback loop further awareness might emerge, such as the awareness of *being involved in a feedback loop*. Terence McKenna, who earned the improbable title of his essay "I Understand Philip K. Dick" through both his apprehension of the totality all of that is in time (his belief in the imminent eschaton with his brother Dennis in 1974) and his struggle to make sense of the way "the world continued grinding forward in its usual less than merry way." Its "merry way" included the production of holographic models of mind and nature by the McKennas, Karl Pribram, and Philip K. Dick, resonant with physicist David Bohm's favorite analogy for the implicate order.

Terence McKenna looked not to the hologram but to his fractal understanding of Dick's VALIS to resonate with experiences of La Chorerra and after. For guidance in this transition from a glimpse of the Peacock Angel of History to that actual admixture of sudden shocking change (Nixon's departure from the White House) and the ongoing momentum of the National Security State (Ford's entry into the White House and, only two years later, George Herbert Walker Bush would become director of the CIA), McKenna experimented with the idea that VALIS's "torch had been passed," highlighting the ongoing momentum of yin as well as yang. McKenna treats Dick's VALIS experiences as quasi-verification of his own download of the world being transformed into information, what he called his "private reformist calendar":

Nixon's weary world ignored the eschatological opportunity I thought my brother's inspired fiddling with hyperspace had afforded. The world continued grinding forward in its usual less than merry way. There was only one small incident that might subsequently be construed, even within the framework of the schizoid logic that was my bread and butter then, to support my position. Unknown to me, a struggling, overweight SF writer, an idol of mine since my teens, discovered the next day that his house had been broken into, his privacy violated by the Other. How peculiar that on the first day of the new dispensation in my private reformist calendar, he had been burglarized by extraterrestrials, the CIA, or his own deranged self in an altered state. The torch had been passed, in a weird way the most intense phase of my episode of illumination/delusion ended right where Phil's began. (T. McKenna 1991)

Note here that the author of *True Hallucinations* affirms the entirety of the illumination-delusion continuum, and it is by refusing to separate these attributes of All That Is, Shiva, VALIS, that McKenna can investigate and be transformed by the bardo between them. His humor and prankster bardic way, his elfin DMT vocalizations, were eloquent calls for inquiry into ecstatic signification whose condition of possibility was the end of belief and the beginning of wonder.

McKenna begins his treatment of VALIS—an afterword to the small press publication of selections from Dick's *Exegesis, In Pursuit of VALIS*—with the recognition that "writers . . . understand narrative economy." Here his treatment of VALIS as an analogy reminds us of its role as a compression device, maybe even a hologram that recapitulates the part in the whole and in so doing squeezes the entirety of the experience down to information available to an ordinarily egoic consciousness, what mathematician Brian Rotman calls the Serial Self. Here the parallel self emerging out of ego death tells its story in terms grokable by a serial self. One experiences in a more than cognitive way the observation that a dissipative structure larger than the ego can be manifested "within" subjectivity. The condition of possibility of this transduction of the parallel into the subjective phenomenological world of any particular human is for that human's mind to become empty, interrupted, disrupted by the logos montage until it lets go of seriality and becomes aware of the simultaneity of all that is. The dim outlines of an ecosystem, Someone, begins to appear . . . McKenna treats all of history, and hence you, dear reader, as a manifestation of something higher, by which you might be lifted up:

> You are illuminated and maddened and lifted up by something great beyond all telling. It wants to be told. It's just that this idea is so damn big that it can't be told, or rather the whole of history is the telling of this idea, the stuttering rambling effort of the sons and daughters of poor old Noah to tell this blinding, reality-shattering, bowel-loosening truth. And Phil had a piece of the action, a major piece of the action. (Ibid.)

Evolution, of course, is precisely the ongoing manifestation of the implicate order, the planetary actualization of "this idea so damn big that it can't be told," the ongoing invention of theories of evolution and involution.

Hence as with Lilly's metaprograms, this involution must be actually carried out if this "interior" transformation is to syncopate with the exterior of rapid change. And the dance was tuned to a refrain: the hologram. Here in this now contagious dance, the part can perhaps only be found in the whole:

> What matters is the system that eventually emerges, not the fantasies concerning the source of the system. (Ibid.)

Ok, so my metaprogram manifests as a planetary plant intelligence interested in turning human consciousness back toward its ecodelic self. Maybe this was just an aspect of our long-neglected ecological, Darwinian, primate unconscious. If so, I am immensely grateful for the tuning and the timing!

I find myself amazed that the holographic paradigm, a kooky idea from my teen years spent reading Michael Talbot and Karl Pribram, has legs. The holographic model was of course crucial to the conceptual toolkit of Terence and Dennis McKenna in *The Invisible Landscape*, but something in my training had convinced me that "hologram" was just a useful metaphor for the process the two of them underwent along with Philip K. Dick.

My questions increasingly clustered around my apparent increased capacity to focus my attention and therefore perception. Wasn't it a bit odd that this hologram metaphor seemed to become ubiquitous, as if the fact that the universe was a hologram left a symptom anywhere you looked? Being a hologram, any slice you cut will still render a (somewhat lower resolution) whole. . . . Perhaps becoming-information—for this is the shamanic "becoming" that they, we, had all transduced—induces a need for rhetorics adequate to the experience. I guess I wasn't surprised when I learned that holograms are a very useful way to compress information and that I had indeed been compressed.

HOLOGRAM FOR DR. MOBIUS!

You get an infinite regress of folds-within-folds-within-folds like a nesting of Chinese boxes. Magnify an inch of one fold, and you'll see more folds inside, with the same rich detail repeated in miniature. It is

like a map of the English coastline each curve of which, when enlarged,
contains a smaller version of the coastline, and so on ad infinitum."
—Judith Hooper and Dick Teresi, *The Three Pound Universe* [18]

The ecodelic hypothesis continued to manifest, and I allowed it to teach me. The hologram was emerging as a refrain, a little ditty that continued to organize my thoughts about my experiences and their feedback into my life. Every now and then it seemed I could tune into this refrain, as if the implicate order were explicating itself just a bit as I focused my consciousness, and I started to notice the Ouroboros, Narby's serpentine force of consciousness on itself, in my ordinary consciousness, which more often than not took the perspective of a detached observer on mobius's life. The guy was learning a lot, but he was still on earth and subject to its constraints and the accumulated habit structures of thought and practice linked to being a joyfully embodied system. My Gaia how much healthier he was! He could only point to the infinity sign and fail to indicate how much better he feels, how ready he feels for his epic life as father and teacher, as individuated aspect of the cosmos, remembering itself, again, after 13.7 billion years.

Mobius prepared for a talk he was supposed to give on the work of Gilles Deleuze, a French thinker who had helped him think just a little bit better about his multiplicity. He focused on the refrain of the hologram as a way of narrating psychedelic experience, arguing that the refrain was perhaps the "primary process" of Deleuze and Guattari's work in *A Thousand Plateaus*, a highly distributed and even holographic work in its own right. How to exemplify the work of the refrain for the audience?

Mobius left his question alone and worked instead on preparing for a seminar he was to visit upon his return. Searching for some of his own Web-published texts to share with the seminar, he Googled himself: "Richard Doyle Aliens." Up popped the voice actor, Richard Doyle, who mobius had earlier discovered when he received some fan mail for Richard Doyle the voice actor, who apparently played the role of "mobius" for a radio show. . . . Weird. This time Richard Doyle the voice actor again manifested, via the Google oracle. I clicked and read that Doyle can now also add a voice credit for a rendition of H. G. Wells's *War of the Worlds*, this time as "Voice of the Hologram."

Terence McKenna, who is perhaps best remembered for his declarations of imminent closure, such as in his investigations of the *I Ching* and the 2012 end of time "Yeschaton," nonetheless championed "Diamond Mind":

> The message coming back at all of us is: live without closure. That's the honest position, given that you are some kind of a talking monkey, some kind of a primate, some kind of creature, on a planet, in an animal body, incarnate in a time and space. In the face of that, life without closure is the only kind of intellectual honesty there is. If you have to inoculate yourself against the various memes of closure that are around, psychedelics do that. That's why they are so politically controversial and potent because—more than any other single act that you may voluntarily undertake—they pull the plug on the myth of cultural meaning. They show that these things are provisional, and that beneath the level of culture there is lurking this erotic, time-and-space-bound, feeling-defined, pre-linguistic mode of being, which is real being. Not becoming, not caught in the various fetishistic forms of tension that commodification of culture and delayed gratification and all these other buzzwords create, but a deeper level of authentic feeling. And it was there all the time, but is denied by the culture. (T. McKenna 1998)

I guess all I would add in a remix of this compressed description both of "awareness" and "science": "ecodelics CAN do that," and deployed with the attention they call for and forth, the focusing of mind often (force majeure) induced by psychedelic experience can detach each and every one of us from our egoic consciousness, together. Attention attractors for an ecosystem, they sometimes seem to say "Hurry up please, it's time."

If Ibogaine, this distributed molecular agent of social transformation, was to do its work, it was instructing its human allies to "relax." Yet "belief" is anything in this context but relaxed; it exists on the edge of panic, a psyche disturbed and giddy with a sense of impending order and closure offered by the west African extract or the end of a book or the End of History. Yes, things were dangerously close to Making Sense. The Zen tradition would seem to be as instructive as *VALIS* on this point. When the Bodhidharma visited the Chinese Emperor, the exchange went something like this:

> "Tell me then," the emperor wanted to know, "What is the first principle of Buddhism?"

"Vast emptiness, nothing holy!" Daruma shot back.

"Who are you?" the thoroughly perplexed emperor demanded.

"I don't know!" Daruma announced, departing as suddenly as he had appeared.

Enlightenment, in this tradition, is comprehended and transmitted as a practice of "beginner's mind," an apprehension of the mind to itself as clear, and hence devoid of any "content" whatsoever. Psychedelic "programming," mantra and the philosophical and psychological literature on "performativity" are heuristic techniques for the cultivation of such open mind, which is easily described and less often achieved. This is the closure hacking of which McKenna spoke. And the Ibogainistas must know that they haven't seen anything like closure yet. In short, I have *nothing* to declare.

I was visited by a different bird deity, this one of a sparkling and iridescent blue. I saw she was a bird, but then again, no she wasn't. She wasn't because she was never arrested by anything resembling a border between her body and motion.

I do not know exactly when she arrived, but when she did so she did not cease. An animation of feather and, now, song, she sang me the story of humans acquiring speech by copying and remixing birdsong, the bird itself warbling in and out of (telepathic) speech and song. Refrains. My instructions were succinct, repeatable, and quite clear: "Listen to the feathers." And wow was I hearing them in the flutter, a stuttering buzz of pattern. When much struck with incredulity toward the shimmering, stammering, and glittering song before me, I listened only to the sounds and their patterns, then beheld whistling music go linguistic in my mind, an instantaneous translation or parallel processing of a highly unpredictable birdsong in song and speech, about speech. And its kin. Later, preparing to speak with a group of ayahuasceras and ayahuasceros, she visited again in my "ordinary consciousness," and I felt myself becoming bird: "I am the trace of the feather in language." I am here in humility to join you, introduce you to our own shamanic traditions. With words I read unto them, from page 73 of Finnegan's Wake *in honor of the late Terence McKenna and his love of the Joycean wyrd, until I came to the word "shamanah." I know nothing as I finish, a lacuna of silence.*

She said: "Listen to them. The insects."

I will admit that falling into their rhythm was the only true comfort I could feel that night, buzzing. When the information or the images or the feeling of the informa-

tion got to be too much, I emptied my mind by focusing entirely on the rhythm and song of the crickets.

In the meantime, between the beats of stridulation and song, the plant divinity again reversed the polarity of the discussion and queried me:

"What do you want to know? It's open season for questions, just ask. Anything you want."

My navigation thus far has relied on open mind, so "I" kept away, my interior filled with the clarifying light of a full moon. Empty.

"What is the Universe?"

"The Universe is a Way of Tricking Itself. Next Question."

Robert Anton Wilson says remembering the difference between the map and the territory is essential to keeping that "grandiose delusion," belief, at bay:

> Belief in the traditional sense, or certitude, or dogma, amounts to the gran-
> diose delusion, "My current model"—or grid, or map, or reality-tunnel—
> "contains the whole universe and will never need to be revised." In terms
> of the history of science and knowledge in general, this appears absurd and
> arrogant to me, and I am perpetually astonished that so many people still
> manage to live with such a medieval attitude. (ii)

In place of certainty, Philip K. Dick persistently treats this alleged *break-down* in the capacity of thought to conform to our temporal or spatial maps as the arrival of negentropy:

> When I say that they and we are waiting for spring to come, I am not
> merely using a metaphor. Spring means thermal return, the abolition of
> entropy. . . . It is spring that restores that life, restores it fully and in some
> cases, as with our species, the new life is a metamorphosis; the period of
> slumbering is a period of gestation together with our fellows that will cul-
> minate in an entirely different form of life than we have ever known before.
> (1995, 218).

Here PKD is close to Ramana Maharshi, who writes: "Reality is simply the loss of the ego." Yet perhaps here, too, as through a glass darkly, Dick has it backward, or at least mobiused: it is not the *abolition of entropy but its increased production* that is augured by the emergence and evolution of the noösphere. Through the *breakdown* of egoic thought, a new aspect of life flows through and exfoliates out of Homo sapiens, through *homo experimentalis*, and, it is to be hoped, transforming the Iron Prison: as they collectively become aware of their ecodelic aspects, humans become entangled with a developmental/evolutionary transformation that "culminates" catastrophically on a threshold of collective awareness, sending uncertainty asymptotically toward infinity, in a planetary question mark: What happened? Perhaps this is an expression of the cosmos' need to creatively destroy information to avoid what systems scientist Stanley Salthe has characterized as the "senescence" of information overload in such thermodynamic systems trolling for increased entropy production in evolution. "Evolution" here names an encounter with the often mimetic and even fictional apprehensions the cosmos periodically has of itself, so we'll see how it goes. But maybe Dick's transhuman technognosis augurs the end of all too human life and its sustainable becoming-other. Maybe!

EPILOGUE

In Darwin's Dreams

*The true student of science neglects nothing and despises nothing that may widen
and deepen his knowledge of nature, and if he is wise as well as learned he will
hesitate before he applies the term "impossible" to any facts which are widely believed
and have been repeatedly observed by men as intelligent and honest as himself.*
 —Alfred Russel Wallace, "Are the Phenomena of
 Spiritualism in Harmony with Science?"

Ayahuasca is a symbiotic ally of the human species.
 —Dennis McKenna, "Ayahuasca"

*What follows, then, represents the first collection of data on
the subject in its entirety, a first step toward the acceptance of
something that is still considered largely inadmissible.*
 —Giorgio Samorini, *Animals and Psychedelics*

*I listened. I sat. I allowed my ears to remain very, very still. I had somehow learned
to treat the very tympan of my ear as a surface for consciousness. I was the sensitive, self-aware drum upon which the Amazonian night was played. I must give way,
empty, allow myself to be calmly overtaken and overcome by the crickets' rhythmic
song and its insistent message: listen.*

*I realize that I have only begun to let go. The first night I had born witness to my
own fractalled disintegration, a screensaver in which I shattered into uncountable
chunks shattering into infinite recombinant segments which told two friends and they
told two friends and so on. . . . I had heard myself telling myself that I was letting go,
that I must give way to death if I were to survive.*

I tried to find my purge bucket somewhere in all that stridulation. There was still much to leave behind. The tub was nowhere to be found in all of the resonance, so I begin flapping around, a haptic search engine for water, and I knock over the glass with an exaggerated cartoon crash. I can hear my dad saying something about Fibber Magee's closet, and the ayahuasca drinker laughs out loud. Adonai somehow moves another vessel into my hand as the rhythms move through me in the dark. Suddenly, nothing is more obvious to me than the fact that I am inhabiting Charles Darwin's dreams, that it was his dreams that drove the notion of aesthetic selection, where it is toward beauty that evolution is tuned in the avian ecology. I see that aesthetic selection is the real principle of biotechnology. I think of bird researcher Fernando Nottebohm. A finch sang, its brain swelling with a future that would have the formality of actually occurring. Or not.

D R. DARWIN—FOR THEY KEPT CALLING HIM THAT, AND IT WAS ALL HE had to go on besides the archives and the fleeting ancestral flashes prompted by the great pyres of Shiva stuff—studied the plant, his eyes stilling, then shifting to and fro across the enormous blossoms above, his neck straining. She soared nearly three meters above the lapis pot, bending under the weight of the blooms but still urging herself skyward, toward the frosted crystal that was the floating domed roof of the jewel-encrusted Temple of Shiva. "More Bhang for Your Buck" strobed slowly in pink neon. Darwin's mind raced as he did a double take on the neon. He once again suddenly began to feel strange. He spoke aloud to his captors, looking toward the ceiling, even if it was useless.

There's no room in here, not enough. Room. "All I want is out of here."
(Burroughs 2006, 27)

Darwin's respiration became rapid and difficult. But he thought of his greenhouse back home, his plants and their love. He turned back toward the blooms and bid the panic "Adieu!," breathing from his diaphragm and playing his fingers over some of the lower blooms.

He brought a hand to his snout and, as he did each morning, inhaled. Blueberry competes with lemon for the dominant fruit today, he decided, making a note on the haptic glyph slab. The plant offered up different flavors of itself each visit, as if trying out a new greeting each time, as if it were prospecting through an infinite universe of flavinoids. Blueberry, yes, many varieties in wild type. An antioxidant, wetware prophylactic they called it here with the astringent wisdom of citrus.

There was also a difficult sound to resolve. A stridulation of stems and leaves that was more than a rustle and less than a salutation. Bells. Chimes. He felt greeted by it, solicited. "Ah, Dr. Darwin, what can I do for you today?"

Odd, he thought. I am certain that this delightful specimen of *Cannabis indica*—yet another name by Lamarck—wants a pollinator, yet in the glowing grow cavern, he had never glimpsed a single insect. And it is much taller than the Indica he had grown in that place—yes, home. England. And in the weeks after his arrival, Darwin had understood almost nothing about his hosts, but he knew this: they wanted her to reproduce. Indeed Erasmus was almost certain that he himself was a reproduction.

His vision was oddly acute since awakening—the hosts had grown him corrective lenses of soft chitin, and Darwin had at first been physically staggered by the intricacy and multiplicity of detail—a whorl of leaves and trichromes sharpened and revealed by the lenses, their iridescent arabesques spilling over into his dreams. Indeed, his hosts seemed oddly interested in the lenses and their treatment, and his weekly vision exam had become, by now, routine, if anything in a living, translucent starship can be experienced by an eighteenth-century doctor clone as routine. *They seemed to want to understand what difference the acuity would make to his interactions with the Indica.*

Yet there was much else to be puzzled by. His hosts had as yet refused to appear before him, appearing only as a series of flickering avatars on a screen, and interacting through the haptic glyphs whose differentials of pressure and response seem to transmit knowledge on an enormous scale. He had to admit that it was an effective technique, and it was certainly pleasurable, even if he did have pangs of what he recalled was guilt every now and again as he approached the bliss state, Ananda. He had slowly come to understand the glyphs as continuous units of gesture carved by the tentacled, prehensile organ he came to imagine as the anandamide-meshed mind of his host, somewhere beyond the screen.

As he did each day, Darwin would leave the Temple with more questions than he brought in. By now, he wrestled and caressed the glyphs more or less without thought, and they helped him navigate The Way, as they called it. The Way, as the continually responding avatars had expressed it, was not simply Dr. Darwin's way toward the freedom to leave the pod. It was The Way by which evolution would become aware of itself, *if only for an eternal instant.*

When it came to The Way, Dr. Darwin felt something like fervor vibrationally transmitted through the multiple media putatively connecting his hosts and himself. The glyphs, wrapping themselves around his waist while he struggled to maneuver them with his hand toward his favorite spot, had become charged with a difference that was as unmistakable as it was difficult to articulate. After the first few sessions of struggle and despair, Dr. Darwin gave up and quickly settled into the best strategy for becoming effectively and deliriously entangled with his host's wet, slippery interface, and before long he had began to work his hips, arms, and hands in the Hawaiian style, and hula.

CUT TO:
ERAMSUS DARWIN AT A WRITING DESK. (He dips his peacock plume into an ink pot and begins to write.)

VOICEOVER:
Dear Charles,
I now, improbably, know of your existence, and of the good work you did to forward The Idea, to weave the long filament of life back into ourselves, and to move our heads only slightly away from our own while focusing our gaze on the work of "that entangled bank," as you so eloquently put it. For when we contemplate the ways in which your bowerbird, even while he builds his temples to reproduction, is equally dependent upon the birds flitting and the insects singing upon the bank near which the fish feed, we remember a single green, shimmering unity in constant communication with itself.

(Erasmus pauses to slowly re-dip the plum and sighs with pleasure)

VOICEOVER:
Yes, we are both men who sing the praises of the plants. You wondered, even, if insectivorous plants might use "psychic tissue" for their uncanny action. To find out, you exposed them to other plants, or at least their essences, many of them found in my pharmacy. *Datura.* Do plants *mind*? I

am in contact with one here, and I feel a tug toward destiny from it. Perhaps I better understand my illustrator—rather, the Plant's illustrator—William Blake. Perhaps plants are our common way, as if the slow unfolding of leaves out of themselves could remind us of our evolution out of the love for the plants. You were always closest to the birds. "On the whole, birds appear to be the most aesthetic of all animals, excepting, of course, man, and they have nearly the same taste for the beautiful as we have." Quite right my boy, and you were always one for wondering about what *use* evolution might have for all of this prettiness. Dream of such wonder!

(Erasmus pauses to re-dip the peacock plume)

Well, I've got to go. The starship seems to exfoliate around this cycle and I don't want to miss the Festival of Creative Destruction. The non-locals heap up the terrabytes of recombinant Shiva stuff to make way for the To Come, and each year we get deliciously and deciduously closer to Omega and the pyres burn brighter, in a Clear Light that suffuses All That Is. I am not sure how we used to language it, but the information pyres are something to see, even in this galaxy. "Tiger, Tiger, Burning Bright!"

Your Gramps (or someone very much like him),
Erasmus

CUT TO:
MIDDLE DISTANCE SHOT. THE PAGE IS COVERED IN SWIRLING HIERO-GLYPHS FROM WHICH AN ETHEREAL SMOKE CURLS, TALKING IN ELFIN TONGUES (many channels of noise and communication share the soundtrack here)

CUT TO:
TIMOTHY LEARY, BEAMING, HOLDS A 1980s VIDEOGAME JOYSTICK WITH A LIGHT GRIP. A SPECTACULAR GALACTIC HORIZON SPILLS OUT BEFORE HIM.

LEARY (cackling): Mr. Alpert, I need full power if we are to overcome the
    neurogenetic threshold.

ALPERT (smiling serenely, from a wheelchair): We're letting go with everything we have, captain.

LEARY (beaming): Then let go of what we don't have! We are approaching the life-mind chiasmus for this quadrant. It is time to hold on to Nothing! Govinda, how much time have we got?

LAMA GOVINDA (in robes, dressed as Flavor Flave, wears an enormous alarm clock around his neck): "The light is clear. Tara is in the building. Going to full chant. Yeah, you know what time it is." He begins to chant OM MANI PADME HUM.

WHIRLING, WHORLING ARCHETYPICAL FORMS OVERTAKE THE SCREEN. A GOAT, SNORTING, CHEWS ON A MUSHROOM MORPHING INTO PAN. A FLUTE PLAYS AN ICARO, TENDERLY.

LEARY (in bardic garb of golden threads, stands over a pot, stirring. The joystick has become a shillelagh swizzle stick stirring a jeweled cauldron covered with arabesques. He plops in a variety of jewels)

CLOSE ON:
A HEXAGONAL BERNARD CELL BEGINS TO FORM OUT OF THE BUBBLING BREW

LEARY: In the name of St. Albert, let everyone within reach of these molecular and plant allies open to the absurd, galactic depths of your own noggin.

LEARY CROSSES HIMSELF, BOWS WITH HANDS TOGETHER, AND DESCENDS INTO A LOTUS POSITION.
THE PAN FIGURE MORPHS INTO A BEJEWELED LEARY, FILLING THE SCREEN.

LEARY (now filling the pixelated screen, speaking to himself): "Om Mani Padme Hum. Om Mani Padme Hum. This is the Timeship Castelia, peercasting all channels. Om Mani Padme Hum. This is the Timeship Castelia, peercasting on all channels."

A PLANT FILLS THE SCREEN, FIRST WITH SHAMROCK CLUSTERS,

THEN LOTUSED VINES GROWING OUT FROM LEARY'S NAVEL AND
CONNECTING TO THE SHIP.

CUT TO:
TERENCE MCKENNA (in an eddy of smoke, leaning into a microphone at a
TechnoCosmic Mass)

TERENCE: The saucer motif guides us to a lens through which the transcendental object at the end of time can take a good look at itself and
grok the futuring of its DNA, a glimpse of the sustainable, refrainable,
Peacock Angel.

CLOSE ON:
A SILVER SAUCER MORPHS INTO THE OCELLUS OF AN OWL
BUTTERFLY.

CLOSE ON:
COVER OF CASTANEDA'S *TALES OF POWER* FEATURING OWL
BUTTERFLY.

CUT TO:
AERIAL SHOT OF THE BIG BLUE MARBLE, THROBBING WITH SOUND.
IT IS THE VOICE OF DENNIS MCKENNA. WORDS SPILL FROM HIS
MOUTH IN ANCIENT FONTS THAT BECOME GLYPHS MORPHING
OGHAM SCRIPT INTO ENGLISH

DENNIS MCKENNA: Holographic compression techniques are ideal for the
flood of information that ensues when you sprout a star antenna tuning
into the totality of all living things.

CUT TO:
DAVID BOHM (sitting with KRISHNAMURTI and laughing)

BOHM: Yes, sample from the implicate order. That's one way to evolve.
Awareness.
KRISHNAMURTI (laughing): "A sample. . . of Nothing!"

CUT TO:

MIDDLE DISTANCE SHOT. PIXELATED "SOLAR FLARES" LEAP OUT
FROM THE RAIN FOREST CANOPY, SPARKLING IN INFINITELY DIF-
FERENTIATING HUES. PERIODICALLY, ONE FLARE CONNECTS IN A
FLASH OF LIGHTNING WITH ANOTHER, PIXELS EXPLODING AND
GATHERING INTO THE CONTOURS OF ERASMUS DARWIN'S SPECTA-
CLED FACE. THE FACE BEGINS TO SPEAK:

VOICEOVER:
Dear Charles,
The information pyres are part of the Grand Arts of Informatic Ananda. My
hosts (or The One Host!) seem to enjoy something like collective mind, but
like our ordinary waking selves, this ecstatic collective needs to shut down
periodically, interrupt itself, restart. We must all become ever more individuated, together. The festivals are like sleep, *en vacances*, with what later scholars called "lucid dreaming." If you hated each wave of the sea on *The Beagle*,
you would come to love the undulations of the pixels in their incessant differentiation. And every now and then, you would want the love to stop.

A FLARE EXPLODES, CONGEALING INTO A GREEN TRANSLUCENT
PUDDING THAT BEGINS TO THROB SUGGESTIVELY BEFORE AGAIN
COHERING INTO THE FACE OF ERASMUS.

VOICEOVER:
And you saw it all, didn't you? What face winked back when you drank the
Kava Kava in the South Pacific? Was it mine? Yes, the peacock gave you
nausea. A symptom, that nausea, an interruption of the inside and outside,
before and after. Are you starting to wake up?

CUT TO:

A JUNGLE SCENE IN THE UPPER AMAZON. LORD ALFRED WALLACE
SITS CROSS-LEGGED NEXT TO A CRACKLING, PIXELATED FIRE. WITH
HIM IS RICHARD EVANS SCHULTES, A HARVARD PROFESSOR, WIL-
LIAM S. BURROUGHS, A WRITER, AND NORMA PANDURO, AYAHUAS-
CERA OF THE UPPER AMAZON. WALLACE FLICKERS IN AND OUT OF
PIXELATION. BURROUGHS WORKS SOME SCISSORS OVER OLD COPIES

OF *EXPOSURE,* A 1940S PULP MAGAZINE WITH A STORY ABOUT YAGE,
WHILE SCHULTES MOUNTS SOME HERBARIUM SHEETS WITH THE
LEAVES OF *PSYCHOTRIA VIRIDIS.* BURROUGHS BEGINS TO CROON
SLOWLY AND A CLAY POT BUBBLES AN ORANGE LIQUID OVER
WHICH NORMA WHISTLES.

WALLACE: It's not that I mind being forgotten, actually.
BURROUGHS (interrupts his crooning and his cutting)
BURROUGHS: It's certainly better than being remembered. Osteopaths of
   the spirit, cutting in and out like an old radio.

BURROUGHS RETURNS TO HIS CUTTING AND CROONING. SCHULTES
ROLLS AND LIGHTS A THICK JOINT. HE IS JOINED BY RABBI ZALMAN
SCHACTER-SHALOMI WHO ARRIVES WITH HIS PALMS TOGETHER,
BOWING.

WALLACE: It's what they forget along with me: the evolution of Mind. You
   know I thought of it in the peak of a malarial reverie. What does that
   say about Mind and evolution? Perhaps the mosquito evolved such that I
   might have the idea of evolution.
SCHULTES (holding in his smoke): I always wondered about Spruce's ship-
   ment of *B. caapi.*
WALLACE: All I mean is: the order is there to be found. It's all a teaching
   for the evolution of mind. It's not just for the past. Ever present, ubiquitous, immanent. Forever and ever.
RABBI ZALMAN SCHACTER-SHALOMI (rising from his bow): Amen.
   Glory be to Gaia, on high and down low, the cthonic, the chronic.

THE RABBI ACCEPTS THE JOINT FROM SCHULTES. BURROUGHS
PAUSES IN HIS CROONING:

BURROUGHS (to Schultes): Wondered?
SCHULTES: What if Bentham had drunk it? Would our prisons be any
   different?
BURROUGHS (laughing): Panopticon woulda had a mandala in the center
   to focus the attention.

WALLACE: And the inmates would sing hallelujah!

RABBI ZALMAN SCHACTER-SHALOMI (smiling): With an emphasis on the "Jah."

NORMA: Yage suena como *Jehová*.

THE RABBI PICKS UP A GUITAR, KICKS DOWN ON A DELAY PEDAL, AND BEGINS TO PLUCK DUB RHYTHMS IN SYNCOPATION WITH THE CLAY POT BUBBLE. HE BEGINS TO CHANT.

> Kaaaaaah-Vah
>
> Kaaaaaaa-Vah
>
> OY
>
> Kava Kava
>
> Kava Kava
>
> OY
>
> It gives you funny faces
>
> Charles

THE REST OF THE GROUP ECHOES INTO LAUGHTER WHILE THE PIXELATED FIRE EXPLODES INTO A FLARE.

CUT TO:

AN AERIAL SHOT OF ANTARCTICA ZOOMS IN FROM SPACE TO REVEAL THE OUTLINES OF ERASMUS DARWIN'S FACE. HE BEGINS SPEAKING. THE GROWING CRACK IN THE ICE SHEET SERVES AS DARWIN'S MOUTH.

VOICEOVER:

You saw. You noticed that in all its myriad forms, she is One. Remember the filament? In my old man talk I was talking about evolution, well, you know.

OPENING GUITAR SAMPLE FROM THE BEATLES' "REVOLUTION."

CLOSE ON: THE BIG BLUE MARBLE SHOT FROM SPACE

VOICEOVER (in documentary voice):

From this account of reproduction it appears, that all animals have a similar origin, viz. from a single living filament; and that the difference of their forms and qualities has arisen only from the different irritabilities and sensibilities, or voluntarities, or associabilities, of this original living filament; and perhaps in some degree from the different forms of the particles of the fluids, by which it has been first stimulated into activity. And that from hence, as Linnæus has conjectured in respect to the vegetable world, it is not impossible, but the great variety of species of animals, which now tenant the earth, may have had their origin from the mixture of a few natural orders.

CUT TO:

THE TRANSLUCENT STARSHIP WITH ITS INTERIOR NOW CRAWLING WITH *B. CAAPI* VINES AND GOLDEN JEWEL HEADED ANACONDAS. GORDON WASSON, IN HIS SIXTIES, INSPECTS AN *AMANITA MUSCARIA* MUSHROOM WITH A HAND LENS. LYNN MARGULIS's FACE FILLS THE STARSHIP SCREEN.

MARGULIS: Mixture, that's my cue. You must begin to fold the muted bacterial chatter into your mix.

LEARY: I am beginning to feel strange . . .

MARGULIS: In the name of the one hundred plus phyla on the blue liquid orb of Our Ancestors, I request your presence at a teaching.

NORMA'S FACE NOW FILLS THE SCREEN.

LEARY: Yes Maestra! We'll meet you in Iquitos. (to Alpert) Where's Lilly?

ALPERT: He's floatin', captain.

LEARY: Well, tell him to make room for two more. We are engaging the transtemporal seminar, attempting nonlinear resonance now . . .

THE SHIP BEGINS TO VIBRATE WITH THE SOUND OF "AUM."

WASSON (to the camera): Behold a hologram of God.

CUT TO:

CLOSE ON: THE EARTH FROM SPACE, AGLOW WITH A CLEAR LIGHT. PULL BACK TO REVEAL ASTRONAUT EDGAR MITCHELL, BEHOLDING THE COSMOS IN AWE.

NOTES

INTRODUCTION: GLIMPSING THE PEACOCK ANGEL

Chapter epigraph: Jünger, 34.

1   "Big whorls have little whorls that feed on their velocity, and little whorls have smaller whorls and so on to viscosity." (Richardson, 66)

2   In 2003, extreme heat waves caused the deaths of more than 20,000 people in Europe, and more than 1,500 in India.

3   If so, what might Schwartz and Randall's "An Abrupt Climate Change Scenario and Its Implications for National Security" (2003) have in common, *a priori*, with that notorious memo of August 6, 2001? While National Security Advisor Condoleezza Rice has characterized the PDB document detailing Al Qaeda intentions as "historical" in nature, in fact it was part of a different rhetorical genre altogether: scenario planning (U.S. Department of State 2004). This is a genre whose very medium is uncertainty, and as such must be read less for "information" about the future than "inclinations" toward it. This capacity to evaluate the difference of the future relies on a detachment from the present—what we "know," now—and a connection to the strangeness of the future—what we might become, then (U.S. Global Change Research Information Office 2002).

4   "5) Rehearse adaptive responses. Adaptive response teams should be established to address and prepare for inevitable climate driven events such as massive migration, disease and epidemics, and food and water supply shortages." (Schwartz and Randall)

5   This recognition is isomorphic to the recognition of Gaia itself, emerging as it did from James Lovelock and Lynn Margulis's meditation on the earth as seen from space (Lovelock, vii).

6   This ecosystemic aspect of human subjectivity was named by Ted Nelson (1974) with his neologism "intertwingled."

7   An interview with a founder of the Deep Ecology movement, Arne Naess, was published as this book was in production, and offers intriguing insights on and possible confirmation of this "ecodelic" hypothesis (Cf. Schroll and Rothenberg).

8   Qtd. from Smokeloc on The Vaults of Erowid, posted January 18, 2002. Available online at http://www.erowid.org/experiences/exp.php?ID=5684 (accessed March 28, 2008).

9   "The purpose of the following is to show that bureaucratic self-interest and predatory public finance in the form of explicit or implicit sin taxes have been and continue to be the primary determinants of public policy in the area of illicit drug control." (Benson and Rasmussen, 164)

10   So too should we be mindful of etymology, whose alluvial and labyrinthine accretions of tropes suggest that mind is first and foremost a verb, an awareness of borders at play in the London Underground imperative: "mind the gap." A spatial practice of the body—and not simply or primarily a representational one—consciousness or mind is thus an essentially interactive and embodied practice subject to both habit, volatility, and becoming.

11   Section epigraph: Deleuze, 3.

12   So, for example, in the Controlled Substance Analog Enforcement Act of 1986, the United States seeks to prohibit even the *intention* to experience psychedelic states, regardless of the materials used: "with respect to a particular person, which such person represents or intends to have a stimulant, depressant, or hallucinogenic effect on the central nervous system that is substantially similar to or greater than the stimulant, depressant, or hallucinogenic [*sic*] effect on the central nervous system of a controlled substance in schedule I or II." A literal reading of this statute—a reading thinkable within the present encroachment on civil and cognitive liberties—suggests that even the reading of this book is itself in violation of this statute, as I fully intend to induce "hallucinogenic" effects. For more on rhetoric as a psychedelic tool, see chapter 2.

13   Cf. D. McKenna 2006; and Shulgin and Shulgin 1997.

14   As potent treatments for addiction, psychedelics such as LSD, mescaline, and Iboga have demonstrated great efficacy even in the context of the prohibition, with heroic treatments taking place under difficult conditions. The remarkable work of ethnobotanist and animal ethnologist Giorgio Samorini suggests that psychedelics function as "depatterning factors" for the diverse set of creatures that eat them (85).

15   Ayahuasca was itself the object of "biopiracy," so not all ecodelics have always been open source, but the plants and molecules treated as sacraments in these traditions have proved difficult to control or otherwise "schedule." Cf. The Center for International Environmental Law and Nwabueze.

16   See the discussion of Leary's contact with poet and yage drinker Allen Ginsberg in Leary, 1983.

17   Cf. Vernadsky 2005.

18   Leary, a political prisoner of the prohibition hounded across the planet, went in and out of thirty-six jails and came out smiling. Leary has become a scapegoat for "What Went Wrong with Psychedelics" in a manner not supported by a scholarly analysis of the historical record.

19   Section epigraph: Sagan, 354.

20   See MAPS.org for a model of a nonprofit open-source pharmaceutical company:

"In essence, MAPS is a non-profit pharmaceutical company. MAPS' top priority projects involve researching the potential of 3,4 methylenedioxymethamphetamine (MDMA, Ecstasy) and marijuana to become FDA-approved prescription medicines. MAPS has estimated that it would take 5 million dollars and 5 years to conduct the research necessary to prove to the satisfaction of the FDA the safety and efficacy of either MDMA or marijuana for one clinical indication." (MAPS.org. n.d.)

21 The ethical response to the massive opportunity cost of the drug war, for example, is itself incalculable.

22 Dawkins has shown himself, though, to be a good empiricist, willing to don the "god helmet" to test hypotheses about the ontology and nature of religion. Would he perhaps also be willing to assay the now frequent claim that ayahuasca is a symbiotic ally of the human species?

23 Leary et al. write in *The Psychedelic Experience*: "Of course, the drug dose does not produce the transcendent experience. It merely acts as a chemical key—it opens the mind, frees the nervous system of its ordinary patterns and structures. The nature of the experience depends almost entirely on set and setting. Set denotes the preparation of the individual, including his personality structure and his mood at the time. Setting is physical—the weather, the room's atmosphere; social-feelings of persons present towards one another; and cultural-prevailing views as to what is real. It is for this reason that manuals or guide-books are necessary. Their purpose is to enable a person to understand the new realities of the expanded consciousness, to serve as road maps for new interior territories which modern science has made accessible." (Leary et al., 11)

24 For the ayahuascera, such singing and whistling no doubt also serves to regulate the breath and allow for traversal as well as survival of these domains of multiplicity. See also rhythm entrainment and symmetry breaking throughout (Collier and Burch).

25 I sample this distinction between awareness and consciousness from Amit Goswami's *The Self Aware Universe*. Goswami argues that an awareness of form resolves a form of the Prisoner's Dilemma through the emergence of immanent communication strategies (1993, 98).

26 "Infinite" is itself, to paraphrase Robert Anton Wilson, "a word," and perhaps it labels an experience of radical choice even as no ego is in any position to deliberate (Ott, 19).

27 Section epigraph: Leary 1966, 89.

28 "It is a very great safeguard to learn by heart instead of writing. It is impossible for what is written not to be disclosed. That is the reason why I have never written anything about these things, and why there is not and will not be any written work of Plato's own." (Hamilton and Cairns, 1567)

29 Cf. Beja-Pereira.

30 For the evolution of the hand and its role in the evolution of mind, see F. Wilson, 34.

31 Section epigraph: Fischer 1972, 189.

32 Section epigraph: Metzner and Leary, 1.

33 For more on the nature of rhetorical "software," see Doyle 1997, 6; Doyle 2003.

1. THE FLOWERS OF PERCEPTION

Chapter epigraphs: Jünger, 36; Schultes, Hofmann, and Rätsch, 27.

1 This distinction will be actively abandoned as soon as it has done its job, but for now: the Web is alive with arguments concerning the distinction between dissociatives and psychedelics, in which case over the counter medications such as Robotussin, taken in the proper dosages, would qualify. "Rihkal: Robodrinkers I have Known and Loved" remains to be written.

2 Qtd. from a post by Nanobrain, December 8, 2003, on The Vaults of Erowid. Available online at http://www.erowid.org/experiences/exp.php?ID=28969 (accessed November 20, 2009).

3 Hence perhaps it is no surprise that while "eyeballing" compounds of such potency continues to be reported in trip reports, it is almost always accompanied by a disclaimer indicating that the writer knows full well that she or he is an idiot for doing so. The putative skill of visualizing the approximate mass of research compounds may be a dangerous meme indeed—this would make an excellent focus for a harm-reduction study.

4 In a footnote, Fischer defines ergotropic arousal as denoting "behavior patterns preparatory for positive action, and is characterized by increased sympathetic activity and an activated psychic state."

5 The enormously prolific and insightful output of mathematician Paul Erdös, coupled with the altered states of consciousness he seemed to inhabit, would also be suggestive here. Cf. Hoffman 1998, 256–58.

6 Cf. Hacking.

7 Claude and Ey state that a "hallucinogenic substance," such as mescaline, produces an "osmose du reel et de l'imaginaire" (Klüver, 62).

8 See chapter 2 of McCloud. Summary available online at http://en.wikibooks.org/wiki/Creative_Writing/Comics (accessed November 12, 2009).

9 See, for example, the integration of this connectivity into "transhumanism" by biologist Julian Huxley (1957, 13).

10 Note that unlike the famous tongue twister involving Peter Piper, "perpetual perishing" appears profoundly repeatable.

11 *Gonzalez v. O Centro Espirita Beneficente Uniao do Vegetal,* 546 U.S. 114, U.S. Sup. Ct., 2006.

12 Cf. Harner on the familiar.

13 Here we might sample Sun Ra, who echoes Gorgias in "Nothing Is."

14 In humans, this sensitivity to the outside is necessary to survival, as its systematic withdrawal in long-term solitary confinement indicates. In turn, Elizabeth Wilson has written persuasively that this outside must be inhabited by a psyche, suggesting a divergence between the will to truth proper to epistemology or ontology and the topology of "interiority."

15 Qtd. at http://www.mindcontrolforums.com/hambone/lillyint.html (accessed November 20, 2009).

2. RHETORICAL MYCELIUM

Chapter epigraphs: Maria Sabina qtd. at The Vaults of Erowid, http://www.erowid.org/plants/mushrooms/mushrooms_quote1.shtml (accessed November 20, 2009); Michaux, 9; Pynchon 1995, 411.

1 Cf. DeLillo, 416.

2 "Eloquence," *OED Online*.

3 Ecstasy: "The state of being 'beside oneself,' thrown into a frenzy or a stupor, with anxiety, astonishment, fear, or passion." It is striking that this definition itself has recourse to a figurative description, marked by the scare quotes. *OED Online*.

4 Cf. Walker, 181.

5 "De Plant. i.; 815a 16. Anaxagoras and Empedokles said that plants are moved by desire, and assert that they have perception and feel pleasure and pain. . . . Empedokles thought that sex had been mixed in them. (Note 817a 1, 10, and 36.) i.; 815 b 12. Empedokles et al. said that plants have intelligence and knowledge." (Fairbanks 1898, 220)

6 Qtd. at The Vaults of Erowid, http://www.erowid.org/plants/mushrooms/mushrooms_quote1.shtml (accessed November 20, 2009).

7 First collected in 1923 by W. A. Murrill in a sugar cane field near Montgomery, Alabama, this mushroom has never been reported in the United States since. In Mexico this entheogenic psilocybe is common in the states of Veracruz, Puebla, and Oaxaca, where it can be found fruiting in clusters from bagasse and also in geologically disturbed areas throughout spring and summer. Its common name is the "Landslide Mushroom."

8 Psilocin (designated 4-hydroxy-N,N-dimethyltryptamine) and serotonin, the human neurotransmitter (designated 5-hydroxytryptamine), differ by only one hydroxy molecule. In this respect, the mushrooms are mirror images of the human brain (Stain Blue Press).

9 "*Naked Lunch* is a blueprint, a How-To Book. . . . How-to extend levels of experience by opening the door at the end of a long hall. . . . Doors that only open in *Silence*. . . . *Naked Lunch* demands Silence from The Reader. Otherwise he is taking his own pulse." (Burroughs 1959, 203)

10 Lucretius called this smallest conceptual difference the "clinamen."

11   Michaux was, he wrote, "at odds" (7) with mescaline, and also reports a capacity for listening that is instructive. After a catastrophic error in dosage, Michaux underwent what he called an "Experiment in Schizophrenia" wherein, he writes, "I could talk to myself as if I were someone else" (139).

12   Consider, for example, the use of the term "intercourse" in this translation of *The German Ideology*: "Communism differs from all previous movements in that it overturns the basis of all earlier relations of production and intercourse, and for the first time consciously treats all natural premises as the creatures of hitherto existing men, strips them of their natural character and subjugates them to the power of the united individuals." (Marx and Engels 1970, 86) "Bourgeois marriage is, in reality, a system of wives in common and thus, at the most, what the Communists might possibly be reproached with is that they desire to introduce, in substitution for a hypocritically concealed, an openly legalized system of free love. For the rest, it is self-evident that the abolition of the present system of production must bring with it the abolition of free love springing from that system, i.e., of prostitution both public and private." (Marx and Engels 2004, 18)

13   Section epigraphs: Ginsberg qtd. in Burroughs 1975, 56; quote from a post by Gus on The Vaults of Erowid, September 23, 2001, available online at http://www.erowid.org/experiences/exp.php?ID=7395 (accessed November 20, 2009); quote from a post by Navi on The Vaults of Erowid, May 17, 2001, available online at http://www.erowid.org/experiences/exp.php?ID=6909 (accessed November 20, 2009).

14   "Louis Althusser," *Wikipedia*, 2009. Available online at http://en.wikipedia.org/wiki/Louis_Althusser (accessed November 21, 2009).

15   Of ninety-two posted trip reports for psilocybin mushrooms on The Vaults of Erowid, twenty-four mention "heal." Available online at http://www.erowid.org/experiences/exp.cgi?S1=8&S2=-1&S3=-1&C1=-1&Str= (accessed April 1, 2008).

16   "Heal, v.¹," *OED Online*.

3. RHETORICAL ADJUNCTS AND THE EVOLUTION OF RHETORIC

Chapter epigraphs: Maria Sabina qtd. in Rothenberg 2003, 36; Leary 1983, 131; Jünger, 36; Emerson, 4–5.

1   Cf. Kennedy 1998, 1.

2   Section epigraphs: Rostand 2008, 28; Roger Nelson qtd. at http://noosphere.princeton.edu/homepage.html (accessed November 21, 2009); Darwin 1909a, 287.

3   This direction to "picture to ourselves" resonates with the persistent rhetorical strategy of what is perhaps the most famous "psychedelic" song, "Lucy in the Sky with Diamonds" by the Beatles: "Picture yourself in a boat on a river. . . . Picture yourself in a train at a station."

4   "Perception of music is not 'mere perception' but perception allied to the presence of a different, more fundamental system." (Josephson and Carpenter, 281)

5   U-Spray Inc., n.d., "Grasshopper Control." Available online at http://www. bugspray.com/catalog/products/page753.html (accessed November 21, 2009).

6   "It is suggested that intimidating eyespots act by mimicking the eyes of the large avian predators preying on the small insectivorous passerines which are among their natural enemies." (Blest, 253; see also J. Huxley 1938.)

7   Cf Prusinkiewicz and Lindenmayer 1990, 55.

8   Such a need for varying the theme is itself a theme in *Cyrano*:

> CYRANO:
> Ah no! young blade! That was a trifle short!
> You might have said at least a hundred things
> By varying the tone—like this, suppose,
> Aggressive: "Sir, if I had such a nose
> I'd amputate it!" Friendly: "When you sup
> It must annoy you, dipping in your cup;
> You need a drinking-bowl of special shape!"
> Descriptive: "Tis a rock! a peak! a cape!
> —A cape, forsooth! Tis a peninsular!"
> Curious: "How serves that oblong capsular?
> For scissor-sheath? Or pot to hold your ink?" (Rostand 2008, 19)

9   Cf. Miller, 78.

10   CYRANO:
> 'Tis enormous!
> Old Flathead, empty-headed meddler, know
> That I am proud possessing such appendice.
> 'Tis well known, a big nose is indicative
> Of a soul affable, and kind, and courteous,
> Liberal, brave, just like myself, and such
> As you can never dare to dream yourself,
> Rascal contemptible! For that witless face
> That my hand soon will come to cuff—is all
> As empty. (Rostand 2008, 18)

11   Section epigraph: A. Huxley 2002, 343.

12   From the noöspheric perspective I have been seeking to articulate, there is at once little and much to be amazed by in the GCP's results. The sequence goes like this:

(1) Using extraordinary amounts of order, produce noise.

(2) Make predictions about order correlating with major media events which command the individual researcher's attention.

(3) Apply attention and determine the quantity of the deviation from randomness, if any. In other words, the effect of the application of *researcher attention*

must be incorporated into the GCP model. The order researchers find is actual, and is perhaps nothing less than the discovery of an implicate order through the application of attention. Hence while there may indeed be a collective effect of focused attention, there may also be an effect whereby the production of noise operates as an attention attractor for then locating order. This would add an interesting extension to the Heisenberg uncertainty principle: a project such as the GCP, in studying attention, may be able to measure the effect of researcher attention or global attention, but not both. Research in the GCP must then toggle back and forth between a position that studies its own output as an artifact of the application of attention, the researcher's, and a system studied for artifacts of global attention. In one case, the changes from expected randomness are the artifact of researcher consciousness, and on the other level, changes from expected randomness are artifacts of the application of global attention as measured by media exposure. In other words, the GCP must control for the application of researcher attention and yet it cannot do so without applying such attention. It is crucial to note that the GCP results are no less remarkable if they are found to be the effect of researcher attention than if they were an effect of the collective media. The results in either case strongly suggest a recursive relationship between human consciousness, the application of attention, and the ontology of any order that might emerge from any observable system.

13 Section epigraph: Samorini, 42.

14 And the use of plants among primates is not limited to courtship adjuncts. Chimpanzees in Tanzania gather leaves from Aspila. These leaves contain thiarubine-A, a powerful antibacterial and antifungal agent, and are traditionally in indigenous medicines of Nigeria, Cameroon, and Ghana." (Cf. Rodriguez et al. 1985, 419)

15 Section epigraphs: Quote from a post by Mr. Nice Guy, on The Vaults of Erowid, May 14, 2003, available online at http://www.erowid.org/experiences/exp. php?ID=16807 (accessed November 22, 2009); A. Huxley 2004, 130; Klüver qtd. in T. McKenna 1992, 232; Quote from a post by Souldier on The Vaults of Erowid, October 5, 2000, available online at http://www.erowid.org/experiences/exp. php?ID=2845 (accessed November 22, 2009); W. A. Stoll qtd. in Hofmann 1980, 42; Crowley available online at http://www.erowid.org/chemicals/absinthe/ absinthe_writings2.shtml (accessed November 22, 2009); Quote from a post by thechubbygoblin on The Vaults of Erowid, December 19, 2003, available online at http://www.erowid.org/experiences/exp.php?ID=29343 (accessed November 22, 2009); quote from a post by Gandolf on The Vaults of Erowid, April 4, 2001, available online at http://www.erowid.org/experiences/exp.php?ID=6023 (accessed November 22, 2009); "Trip Toys" available online at http://www.erowid.org/psychoactives/basics/basics_trip_toys.shtml (accessed November 22, 2009); Sopa, 11.

4. LSDNA

Chapter epigraphs: Hofmann, 16; Mullis, 5; Crick 1975, 23.

1   "The fact that micro-organisms are alive is without legal significance for purposes of the patent law." (*Diamond v. Chakrabarty*, 447 U.S. 303, U.S. Sup. Ct., 1980)

2   In this context, the distinction between living and nonliving systems disappears, even as it was the very impetus of Watson and Crick's quest. Why else be interested in the secret of life if life is nothing in particular?

3   Of course, the "code script"—and its subsequent rhetorical transformations of code, program, and network—was an element of a system to DO something: uncover the mysteries of life or, as François Jacob put it, "to triumph over death." In this sense the emphasis on the replication activity of DNA rather than its communicative capacities is the return of the repressed of Schrödinger's machinic and ultimately cryptographic, rendering of life. I have discussed this topic in greater length in Doyle 1997.

4   It is not the case, for example, that Hofmann had no options: as his own account details, many more ways of fathoming LSD-25's effects were available, including the animal testing alluded to earlier.

5   The question of caution is an interesting one here, seemingly oxymoronic to any notion of self-experiment, as the very status of the self is at stake. Hofmann writes that during the terrifying peaks of his hallucinogenic panic, he worried over the possibility of ever communicating the fact of LSD-25's unforeseeable effects to his family. This retroactive recognition highlights the true experimental conclusion of his work: that there is no preparation, decided or otherwise, for the encounter with the qualitative difference of LSD-25: "Would they ever understand that I had not experimented thoughtlessly, irresponsibly, but rather with the utmost caution, and that such a result was in no way foreseeable? My fear and despair intensified, not only because a young family should lose its father, but also because I dreaded leaving my chemical research work, which meant so much to me, unfinished in the midst of fruitful, promising development. Another reflection took shape, an idea full of bitter irony: if I was now forced to leave this world prematurely, it was because of this lysergic acid diethylamide that I myself had brought forth into the world" (18).

6   Section epigraph: Qtd. in the documentary "Psychedelic Science," by Bill Eagles, which aired February 27, 1997. Available online at http://ibogaine.mindvox.com/index.html?Media/BBC-Ibogaine.html~mainFrame (accessed November 22, 2009); Crick, 1975, 23.

7   Ibid.

8   A version of this chapter was first published in 2002 before the journalistic disclosures provided uncanny verification for the article's observation that "Timothy Leary and Francis Crick were speaking the same language, the language

of information where the organic and the machinic enfold each other helically
and, sometimes, the capacity for replication goes through the ceiling." See Doyle
2002, 153; see also Crick 1975.

9   See "2009 Peruvian Political Crisis," Wikipedia, 2009. Available online at http://
en.wikipedia.org/wiki/2009_Peruvian_political_crisis (accessed November 22,
2009).

5. HYPERBOLIC

1   As researchers Alexander Shulgin and Dennis McKenna have both pointed
out, DMT, the visionary molecule contained in all ayahuasca brews, abounds in
nature. Cf. "DMT Is Everywhere" (Shulgin and Shulgin 1997, 247–84).

2   "If one were specifically interested only in experiencing ayahuasca, it would
be more cost-effective to home brew a batch with ingredients ordered from
an ethnobotanical supplier. With the help of an experienced friend as a sitter,
one could have an intense entheogenic experience in the safety and comfort of
home or in an isolated natural setting. This do-it-yourself approach could poten-
tially be far more enlightening than what one might experience after traveling
all the way to South America." Cf. Stuart.

3   Section epigraph: Fischer 1971, 901.

4   Cf. Landauer.

5   "From within the system an observer will never be able to know very much
about the true microscopic state of that system. Every part of space is computing
its future as fast as possible, while information pours in from every direction."
(Fredkin)

6   Hyperbole, *OED Online*.

7   Cf. Tiefert 1994.

8   The next chapter will intensify the focus on self-experiment as a mode of scientific communication.

9   Given that living systems go through a constant cycle of birth, development,
regeneration, and death, preserving information about what works, and what
does not, is crucial for the continuation of life. This is the role of the gene and,
at a larger scale, biodiversity, to act as information databases about self-organization strategies that work (Schneider and Kay).

10 Erich Jantsch, *The Self Organizing Universe*. Available online at http://www.
music-mind.com/pmwiki/index.php/CommunityAndFeedback/Book6 (accessed
November 22, 2009).

11 Section epigraph: Evans-Wentz, 104.

12 "The Heart Sutra," trans. by the Nalanda Translation Committee. Available
online at http://www.dharmanet.org/HeartSutra.htm (accessed December 1,
2009).

13 Section epigraph: The Reluctant Messenger, "Here Begins the 'Root Verses of

the Bardo.'" Available online at http://reluctant-messenger.com/tibetan-bardo-verses.htm (accessed December 1, 2009).

14   True, the Merry Pranksters composed a forty-hour film devoted to their own psychedelic collective, but what the Pranksters offer in terms of the programming of psychedelic experience through aesthetic practice, the McKenna brothers fashion in terms of a group scientific experiment.

15   Section epigraphs: Guenther, 80; DMT volunteer qtd. in Strassman, 206; cf. also Narby 1998.

16   Elfdancer 2007.

17   Jim DeKorne's excellent work, *Psychedelic Shamanism: The Cultivation, Preparation and Shamanic Use of Psychotropic Plants*, offers a lengthy and instructive discussion of the connection between shamanic practice and abduction. DeKorne was a volunteer in Strassman's study.

18   This refusal of the metaphorical nature of the DMT vision is one of the most consistent symptoms of its ingestion, both in Strassman's reports and the more anonymous media of the Internet trip report: "It wasn't a planet that 'kind of looked like a lion,' it absolutely WAS a lion's head so immense as to be a planet."

19   Cf. Kaplan.

20   Quote from a post by Twitchin, June 1, 2002, on the discussion board at alt.drugs.psychedelics.

21   Cf. Wolfram, 739.

22   For all intents and purposes, Dick was Gumby, damn it: "This brings me to my frightening premise. I seem to be living in my own novels more and more" (Sutin, 2).

23   Fischer notes that these "as-if" experiences are kin to both hallucinated states and dreams in the difficulty they pose for interpretation: "The symbolic metaphorical language and logic is labeled by reference to the very context within which it is expressed which makes the interpretations as difficult as, for instance . . . learning English in English" (Fischer 1972, 166). This is, of course, precisely the situation of the human infant.

24   Cf. Lyman and Varian.

6. THE TRANSGENIC INVOLUTION

Chapter epigraphs: Kac; Lamarck, 109–10; Johnston, 341.

1   See the work of Richard Glen Boire and Wrye Sententia. Their work, both conceptual and pragmatic, is crucial to any future of human consciousness as a freely accessible platform of introspection—*homo experimentalis*.

2   Charles Darwin writes of the effects of "spangles" in the courtship of birds, discussed further below: "In this attitude the ocelli over the whole body are exposed at the same time before the eyes of the admiring female in one grand bespangled expanse" (1909a, 405).

3   Cannabis is a dioecious plant—it has at least two sexes. If isolated from pollen, the female buds continue to grow and ripen. The sticky THC-laden trichromes grow larger and larger in the solicitation and attempted attraction of pollen. If allowed to seed, females more or less cease THC production and the potency of the crop is much diminished. Males produce early pollen-laden flowers, so the trick in the cultivation of sinsemilla is to "rogue out the males" as early as possible if one is growing from seed. Cloning methods help the grower avoid this sometimes tricky and urgent process of sexual selection. Cuttings are taken from a select female and grown repeatedly, giving the plant an almost Raëlian quality of immortality. For more on the importance of sexual selection, see below.

4   In this regard—her need for a kind of "activation" by an audience and light—Alba is akin to the vain peacock as described by Charles Darwin in *The Descent of Man*: "this latter bird, however, evidently wishes for a spectator of some kind, and, as I have often seen, will shew off his finery before poultry, or even pigs" (1909a, 401).

5   Section epigraph: Darwin 1909b, 528.

6   DJ Short's formulation here seems to highlight the uncertainty of such a proposition and the scarcity of windows for such an assay. Given cannabis's notorious, albeit contestable, effects on short-term memory, one looks in vain for a heuristic whereby one could indeed "make certain" that this was the first ingestion.

7   DJ Short's locution of "psyche" here is instructive, as it harkens to a more expansive understanding of self than that of the ego whose death so famously occurs under the influence of LSD.

8   It might be worth noticing that in the opening paragraph of *On the Origin of Species*, Darwin describes himself as similarly "struck": "WHEN on board HMS *Beagle*, as naturalist, I was much struck with certain facts in the distribution of the inhabitants of South America, and in the geological relations of the present to the past inhabitants of that continent." (1909b, 21)

9   Indeed, Darwin noted that charm was sometimes a more crucial evolutionary ally than the much vaunted "battle": "We shall further see, and it could never have been anticipated, that the power to charm the female has sometimes been more important than the power to conquer other males in battle." (1909a, 231)

10 "The impassioned orator, bard, or musician, when with his varied tones and cadences he excites the strongest emotions in his hearers, little suspects that he uses the same means by which his half-human ancestors long ago aroused each other's ardent passions, during their courtship and rivalry." (Ibid., 586)

11 "These feathers have been shown to several artists, and all have expressed their admiration at the perfect shading. It may well be asked, could such artistically shaded ornaments have been formed by means of sexual selection?" (Ibid., 407)

12 See Gong 2003. For a remarkable resource on bioluminescence (and the difficulties of imaging it), see Mills 1999.

13 Here I follow the coinage of "psychedelic" (Osmond) as a "mind manifesting." This marks less the fantastic character of the effects of these compounds than the dissolution of the boundary between mind and nature. Both Alba and cannabis thus are psychedelic in the sense that they are indeed manifestations or exfoliations of human consciousness into, out of, the ecosystem. Following Julian Jaynes's analysis, Alba is also "psychedelic" to the extent that she manifests "psyche" as "life." Jaynes writes of the ancient Greek sense of the word *psyche*: "its primary use is always for life. . . . After the Homeric poems, Trytaues, for example, uses *psyche* in that sense . . . as does Alcaeus. And even as late as the fifth century B.C., Euripides uses the phrase 'to be fond of one's *psyche*' in the sense of clinging to life. . . . Some of the Aristotelian writings also use *psyche* as life, and this usage even extends into much of the New Testament. 'I am the good shepherd: the good shepherd giveth his psyche for his sheep' (John 10:11). Jesus did not mean his mind or soul." (288–89)

7. FROM ZERO TO ONE

Chapter epigraphs: Arberry, 357; Schrödinger 1992, 89–90.

1 A combination of cocaine and heroin.

2 Section epigraphs: Dick 1991, 147; Lilly 1972, vi.

3 Cf. Wallace.

4 "There is an additional caution in the use of these substances; the self programmer must be strong enough to experience these phenomena and not make difficult to reverse mistakes in reprogramming or difficult to correct errors in new commitments in the external world. This is an area of human activity for the most experienced and strongest personalities, with the right training. *I do not recommend the use of these methods except under very controlled and studied conditions with as near ideal as possible physical and social environment and as near ideal as possible help from thoroughly trained empathic matching persons. The subject's short term and long term welfare must control all actions, all speech, and all transactions between each pair of persons present, unconsciously and consciously.*" (Lilly 1972, 68)

5 "The Lac Operon," *MIT Biology Hypertextbook*. Available online at http://www.sciencegateway.org/resources/biologytext/pge/lac.html (accessed November 22, 2009).

6 John Lash, "The Gnostic Gallery." Available online at http://www.metahistory.org/GnosticGallery2.php (accessed December 1, 2009).

7 Available online at http://www.bodhicitta.net/Essence%20of%20Eloquence.htm (accessed November 22, 2009).

8 Section epigraphs: Lilly 1972, 77; Dick 1991, 160.

9 James Clark Maxwell qtd. in "Maxwell's Demon," *Wikipedia*, 2009. Available online at http://en.wikipedia.org/wiki/Maxwell's_demon (accessed December 2, 2009).

10 When a flotation tank seemed to drop out of the sky for me and my collabora-

tor, I took it in stride, laughing at the strangeness of the interconnection, but not interpreting it. Not wondering at the amazing interconnection for too long helped me to assemble a small crew to quickly extract the tank from the basement where it had gone unused for some years.

11 Section epigraphs: Dick 1991, 185; R. Wilson, ii.

12 By now there is something like an implicit consensus that psychedelics "failed" and that this was essentially Timothy Leary's fault. Yet a careful reader who looks to Leary's writings during the period or gives any attention to the historical and technological context of the time will not support this simple scapegoating of Leary for the loss of mainstream psychedelic science. This scapegoating ignores the remarkable work of many scientists and psychonauts in continuing the investigation into consciousness with the help of these technologies.

13 Quote from a post by Gandolf on The Vaults of Erowid, April 4, 2001, available online at http://www.erowid.org/experiences/exp.php?ID=6023 (accessed November 22, 2009).

14 "Smart Shops" in Amsterdam offered a diverse array of phenethylamines in the late twentieth and early twenty-first century.

15 "Every S but one male improved on this measure. Mean time for locating the 12 figures was 404 sec. for the pre-session testing and 234 sec. for the testing during the session. Performance enhancement was significant ($\times$2 = 8.64, df = 13, p < 0.01). These are extremely high and significant changes (individual improvements in speed up to 200 percent) compared with changes observed in previous research with this instrument." (Witkin et al. 1962; qtd. in Tart 1972, 463)

16 John C. Lilly, "A Feeling of Weirdness." Available online at http://deoxy.org/lil_weirdness.htm (accessed December 3, 2009).

17 Cf. Doyle 2003.

18 Section epigraph: Qtd. in De Rienzo and Beal, 143.

REFERENCES

Alexander, Christopher. 2002–2005. *The Nature of Order.* 4 vols. Berkeley, CA: Center for Environmental Structure.

Allott, Robin. 1989. "The Motor Theory of Language." In *Studies in Language Origins*, ed. Walburga von Raffler-Engel, Jan Wind, and Abraham Jonker, vol. 2, 123–57. Philadelphia, PA: John Benjamins. Available online at http://cogprints.org/3269/01/motor-ii.htm (accessed April 16, 2008).

Arberry, A. J., trans. 2000. *The Discourses of Rumi.* Ames, IA: Omphaloskepsis. Available online at http://www.omphaloskepsis.com/ebooks/pdf/discour.pdf (accessed December 1, 2009).

Aurobindo, Sri. 2006. *The Life Divine.* Pondicherry, India: Sri Aurobindo Ashram.

Austin, James H. 1998. *Zen and the Brain: Toward an Understanding of Meditation and Consciousness.* Cambridge, MA: MIT Press.

Ayer, A. J. 1952. *Language, Truth and Logic.* New York: Dover.

Barad, Karen. 2007. *Meeting the Universe Halfway: Quantum Physics and the Entanglement of Matter and Meaning.* Durham, NC: Duke University Press.

Barnard, Mary. 1963. "The God in the Flower Pot." *The Psychedelic Review* 1, no. 2. Available online at http://www.maps.org/psychedelicreview/v1n2/012244bar.pdf (accessed April 16, 2008).

Bateson, Gregory. 2000. *Steps to an Ecology of Mind: Collected Essays in Anthropology, Psychiatry, Evolution, and Epistemology.* Chicago, IL: University of Chicago Press.

Baudelaire, Charles. 2004. "The Poem of Hashish." Trans. Aleister Crowley. Whitefish, MT: Kessinger. Available online at http://users.lycaeum.org/~sputnik/Ludlow/Texts/Rats/poem.html (accessed April 13, 2008).

Beach, Horace. 1996–1997. "Listening for the Logos: A Study of Reports of Audible Voices at High Doses of Psilocybin." *Newsletter of the Multidisciplinary Association for Psychedelic Studies MAPS* 7, no. 1: 12–17. Available online at http://www.maps.org/news-letters/v07n1/07112bea.html (accessed April 5, 2008).

Beja-Periera, Albano, et al. 2003. "Gene-culture Coevolution between Cattle Milk Protein Genes and Human Lactase Genes." *Nature Genetics* 35, no. 4: 311–13. Available online at http://www.nature.com/ng/journal/v35/n4/pdf/ng1263.pdf (accessed November 12, 2009).

Benjamin, Walter. 2006. *On Hashish.* Ed. Howard Eiland. Cambridge, MA: Harvard University Press.

Benson, Bruce L., and David W. Rasmussen. 1996. "Part 1. Predatory Public Finance and the Origins of the War on Drugs 1984–1989." *The Independent Review* 1, no. 2: 163–89. Available online at http://www.drugpolicy.org/library/predatory_public_p1.cfm (accessed March 25, 2008).

Blest, A. D. 1957. "The Function of Eyespot Patterns in Lepidoptera." *Behaviour* 11, nos. 2–3: 209–58.

Bohm, David. 1990. "A New Theory of the Relationship of Mind and Matter." In *Philosophical Psychology* 3, no. 2: 271–86. Available online at http://evans-experientialism.freewebspace.com/bohmphysics.htm (accessed November 11, 2009).

———. 2003. *The Essential David Bohm*. New York: Routledge.

Boire, Richard Glen. "On Cognitive Liberty, Part IV." In *Journal of Cognitive Liberties* 4, no. 1: 15–25. Available online at http://www.cognitiveliberty.org/jcl/jcl_online.html (accessed June 14, 2010).

Borges, Jorge Luis. 1964. "The Analytical Language of John Wilkins." In *Other Inquisitions, 1937–1952*. Austin: University of Texas Press. Available online at http://www.alamut.com/subj/artifice/language/johnWilkins.html (accessed November 13, 2009).

Burroughs, William S. 1959. *Naked Lunch*. New York: Grove Press.

———. 1975. *Yage Letters*. San Francisco: City Lights Publishers.

Burroughs, William S., and Allen Ginsberg. 2006. *The Yage Letters Redux*. Ed. Oliver C. G. Harris. San Francisco: City Lights Publishers.

Carrington, Sean C. M. 1998. "Introduction to the Flowering Plants," Available online at http://www.cavehill.uwi.edu/FPAS/bcs/bl14apl/flow1.htm (accessed November 12, 2009).

The Center for International Environmental Law. 2003. "The Ayahuasca Patent Case." Available online at http://www.ciel.org/Biodiversity/ayahuascapatentcase.html (accessed November 11, 2009).

Collier, John, and Mark Burch. 1998. "Order from Rhythmic Entrainment and the Origin of Levels Through Dissipation." *Symmetry: Culture and Science Order/Disorder*, Proceedings of the Haifa Congress 9, nos. 2–4: 165–78.

Crick, F. H. C. 1966. "The Genetic Code—Yesterday, Today, and Tomorrow." *Cold Spring Harbor Symposia on Quantitative Biology* 31: 3–9.

———. 1975. "The Poetry of Michael McClure: A Scientist's View" *Margins* 18: 23–24.

Darwin, Charles. 1889. *The Various Contrivances by Which Orchids Are Fertilized by Insects*. London: J. Murray.

———. 1896. *The Variation of Animals and Plants under Domestication*. New York: D. Appleton. Available online at http://charles-darwin.classic-literature.co.uk/variation-of-animals-and-plants-under-domestication-v2/ebook-page-25.asp (accessed April 1, 2008).

———. 1909a. *The Descent of Man*. London: John Murray.

———. 1909b. *On the Origin of Species*. New York: P. F. Collier & Sons.

———. 1916. *The Expression of the Emotions in Man and Animals.* New York: D. Appleton.

Dawkins, Richard. 2004. *The Ancestor's Tale: A Pilgrimage to the Dawn of Evolution.* Boston: Houghton Mifflin.

DeKorne, Jim. 1994. *Psychedelic Shamanism: The Cultivation, Preparation and Shamanic Use of Psychotropic Plants.* Port Townsend, WA: Loompanics.

Deleuze, Gilles. 1993. *The Fold: Leibniz and the Baroque.* Minneapolis: University of Minnesota Press.

DeLillo, Don. 1988. *Libra.* New York: Viking.

De Rienzo, Paul, and Dana Beal. 1997. *The Ibogaine Story: Report on the Staten Island Project.* New York: Autonomedia.

Desmond, Adrian, and James Moore. 1994. *Darwin: The Life of a Tormented Evolutionist.* New York: W. W. Norton & Co.

Dick, Philip K. 1991. *Valis.* New York: Vintage.

———. 1996a. *Selected Letters of Philip K. Dick, 1938–1971.* Grass Valley, CA: Underwood Books.

———. 1996b. *The Shifting Realities of Philip K. Dick: Selected Literary and Philosophical Writings.* Ed. Lawrence Sutin. New York: Pantheon.

DJ Short. 2003. "The Art of Selection and Breeding Fine Quality Cannabis." *Cannabis Culture* 41: 94–98. Available online at http://www.cannabisculture.com/v2/articles/2788.html (accessed June 14, 2010).

———. 2004. *Cultivating Exceptional Cannabis.* Oakland, CA: Quick American Archives.

Dobkin De Rios, Marlene. 1996. *Hallucinogens: Cross-Cultural Perspectives.* Long Grove, IL: Waveland Press.

Doctorow, Cory. 2004. "Solving and Creating Captchas with Free Porn." Available online at http://boingboing.net/2004/01/27/solving-and-creating.htm (accessed November 11, 2009).

Donald, Merlin. 1991. *Origins of the Modern Mind.* Cambridge, MA: Harvard University Press.

———. 2000. "The Central Role of Culture in Cognitive Evolution: A Reflection on the Myth of the 'Isolated Mind'." Available online at http://psycserver.psyc.queensu.ca/donaldm/reprints/TheCentralRole8.PDF (accessed November 10, 2009).

Doyle, Richard. 1997. *On Beyond Living: Rhetorical Transformations of the Life Sciences.* Stanford, CA: Stanford University Press.

———. 2002. "LSDNA: Consciousness Expansion and the Emergence of Biotechnology." *Philosophy and Rhetoric* 35, no. 2: 153–74.

———. 2003. *Wetwares: Experiments in Postvital Living.* Minneapolis: University of Minnesota Press.

Easwaran, Eknath. 1987. *The Upanishads.* Tomales, CA: Nilgiri Press.

Eckhart, Meister. 2007. *Meister Eckhart's Sermons.* Trans. Claud Field. New York: Cosimo, 2007.

Elfdancer. 2007. "Psychedelic Rollercoaster." Available online at http://www.erowid. org/experiences/exp.php?ID=61113 (accessed June 15, 2010).

Elfstone. 1989. "Comments on the Psilocybin Mushroom." Available online at http://www.erowid.org/plants/mushrooms/mushrooms_article3.shtml (accessed November 20, 2009).

Ellis, Havelock. 1898. "Mescal: A New Artificial Paradise." *The Contemporary Review,* January. Available online at http://nepenthes.lycaeum.org/Ludlow/Texts/mescal.html (accessed March 31, 2008).

Emerson, Ralph Waldo. 1848. *Essays, Orations, and Lectures.* London: W. Tegg.

Evans-Wentz, W. Y. 2000. *The Tibetan Book of the Dead.* Oxford, UK: Oxford University Press. Available online at http://reluctant-messenger.com/Tibetan-Book-Dead_Evans-Wentz.htm (accessed November 22, 2009).

Fairbanks, Arthur. 1898. *The First Philosophers of Greece.* New York: Scribner.

The Federal Analog Act, 21 U.S.C. § 813. 1986.

Feyerabend, Paul. 1988. *Against Method: Outline of an Anarchistic Theory of Knowledge.* Atlantic Highlands, NJ: Humanities Press.

Fischer, Roland. 1971. "A Cartography of Ecstatic and Meditative States: The Experimental and Experiential Features of a Perception-Hallucination Continuum Are Considered." *Science,* November 26, 897–904.

———. 1972. "On Separateness and Oneness."*Confinia Psychiatricia* 15, nos. 3–4: 165–94.

———. 1978. "On Dissipative Structures in Both Physical and Information Space." *Current Anthropology* 19, no. 1: 66.

———. 1994. "On the Remembrance of Things Present: A Neuro-Epistemological Approach." In *Repetition,* ed. Andreas Fischer, 255–58. Türbigen: Gunter Narr Verlag.

Fischer, R., and G. M. Landon. 1972. "On the Arousal State-Dependent Recall of 'Subconscious' Experience: State Boundness." *British Journal of Psychiatry* 120: 159–72.

Fredkin, Edward. 1982. "Finite Nature." Available online at http://www.ai.mit.edu/projects/im/ftp/poc/fredkin/Finite-Nature (accessed August 27, 2010).

Furst, Peter. 1972. *Flesh of the Gods: The Ritual Use of Hallucinogens.* New York: Praeger.

———. 1976. *Hallucinogens and Culture.* Novato, CA: Chandler & Sharp Publishers. Available online at http://www.erowid.org/library/books/hallucinogens_and_culture.shtml (accessed June 14, 2010).

Gans, Deborah, and Zehra Kuz. 2003. *The Organic Approach to Architecture.* Great Britain: Wiley Academy Press.

Global Consciousness Project. 2005. "Registering Coherence and Resonance in the World." Available online at http://noosphere.princeton.edu/index.html (accessed April 13, 2008).

Goldman, Jennifer. 2003. Interview by Zalman Schachter-Shalomi. The Beyond Intractability Project. Available online at http://www.beyondintractability.org/audio/10864/ (accessed June 14, 2010).

Gong, Z., et al. 2003. "Development of Transgenic Fish for Ornamental and Bioreactor by Strong Expression of Fluorescent Proteins in the Skeletal Muscle." *Biochemical and Biophysical Research Communications* 308, no. 1: 58–63.

Goody, Jack. 1993. *The Culture of Flowers.* New York: Cambridge University Press.

Goswami, Amit. 1993. *The Self-Aware Universe: How Consciousness Creates the Material World.* New York: Putnam's Sons.

Govinda, Lama Anagarika. 1969. *Foundations of Tibetan Mysticism.* San Francisco, CA: Weiser Books.

Gracie and Zarkov Productions. 1985. "DMT: How and Why to Get Off: A Note from Underground." Available online at http://www.lycaeum.org/drugs.old/other/gandz/gandz.dmt.html (accessed April 5, 2008).

Grof, Stanislav. 2005. *Higher Wisdom: Eminent Elders Explore the Continuing Impact of Psychedelics.* SUNY Series in Transpersonal and Humanistic Psychology. Albany: SUNY Press.

Guenther, Herbert, trans. 1971. *The Life and Teaching of Naropa.* Oxford, UK: Oxford University Press.

Hacking, Ian. 1983. *Representing and Intervening: Introductory Topics in the Philosophy of Natural Science.* New York: Cambridge University Press.

Hall, David, Mark Kirkpatrick, and Brian West. 2000. "Runaway Sexual Selection When Female Preferences Are Directly Selected. *Evolution* 54, no. 6: 1862–869.

Hamilton, Edith, and Huntington Cairns, eds. 1961. *The Collected Dialogues of Plato.* Princeton, NJ: Princeton University Press.

Haraway, Donna. 1997. *Modest_Witness@Second_Millennium.FemaleMan©Meets_Onco-Mouse™: Feminism and Technoscience.* New York: Routledge.

Harner, Michael. 1968. "The Sound of Rushing Water." *Natural History*, July. Available online at http://deoxy.org/rushingw.htm (accessed June 14, 2010).

———. 1973. *Hallucinogens and Shamanism.* New York: Oxford University Press.

———. 1990. *The Way of the Shaman.* San Francisco, CA: Harper & Row.

Hebbard, F. W., and R. Fischer. 1966. "Effect of Psilocybin, LSD, and Mescaline on Small, Involuntary Eye Movements." *Psychopharmacologia (Berl)* 9: 146–56.

Heidegger, Martin. 1977. *The Question Concerning Technology and Other Essays.* New York: Harper Perennial.

Hoffman, Paul. 1998. *The Man Who Loved Only Numbers: The Story of Paul Erdös and the Search for Mathematical Truth.* New York: Hyperion.

Hofmann, Albert. 1980. *LSD: My Problem Child.* Trans. Jonathan Ott. New York: McGraw-Hill. Available online at http://www.erowid.org/library/books_online/lsd_my_problem_child/ (accessed November 22, 2009).

Hofstadter, Douglas R. 2008. *I Am a Strange Loop.* New York: Basic Books.

Huxley, Aldous. 2002. *Island.* New York: HarperCollins.

———. 2004. *The Doors of Perception and Heaven and Hell.* New York: HarperCollins.

Huxley, Julian. 1938. "Darwin's Theory of Sexual Selection and the Data Subsumed by it, in the Light of Recent Research." *The American Naturalist* 72, no. 742: 416–33.

———. 1957. "Transhumanism." In *New Bottles for New Wine*. London: Chatto & Windus. Available online at http://www.transhumanism.org/index.php/WTA/more/huxley (accessed March 30, 2008).

Inge, William Ralph. 1976. *Mysticism in Religion*. Westport, CT: Greenwood.

Jarvik, M. E., et al. "Lysergic Acid Diethylamide (LSD-25): Effect on Attention and Concentration." *Journal of Psychology* 39: 373. Available online at http://www.erowid.org/references/refs_view.php?ID=1538&C=Hof&S=jarvik&SField=All (accessed June 14, 2010).

Jaynes, Julian. 1990. *The Origin of Consciousness in the Breakdown of the Bicameral Mind*. Boston, MA: Houghton Mifflin.

Johnston, James. 1891. *The Chemistry of Common Life*. New York: D. Appleton. Available online at http://nepenthes.lycaeum.org/Ludlow/Texts/johnston.html#R12 (accessed December 3, 2009).

Josephson, Brian D., and Tethys Carpenter. 1994. "Music and Mind: A Theory of Aesthetic Dynamics." In *On Self-Organization,* ed. R. K. Mishra, Dieter Maass, and Eduard Zwierlein. New York: Springer. Available online at http://www.tcm.phy.cam.ac.uk/~bdj10/mm/articles/kaiserslautern.txt (accessed April 12, 2008).

Josiah Macy Jr. Foundation. 1951. *Cybernetics: Circular, Causal, and Feedback Mechanisms in Biological and Social Systems, Volume 6*. New York: Josiah Macy Jr. Foundation.

Judson, Horace. 1979. *The Eighth Day of Creation*. New York: Simon and Schuster.

Judson, Olivia. 2002. *Dr. Tatiana's Sex Advice to All Creation*. New York: Metropolitan Books.

Jünger, Ernst. 2000. "The Plant as Autonomous Power." *The Entheogen Review* 9, no. 1: 34–36.

Kac, Eduardo. n.d. "GFP Bunny." Available online at http://www.ekac.org/gfpbunny.html (accessed November 10, 2009).

Kak, Subhash. 1997. "Consciousness and Freedom According to the Shiva Sutras." Available online at http://www.beezone.com/SivaSutras/sivasutras.htm (accessed June 11, 2010).

———. 2001. "The Siva Sutras." Available online at http://www.saivism.net/etexts/sivasutras.asp (accessed November 23, 2009).

Kaplan, Louis. 1986. *Gumby: The Authorized Biography of the World's Favorite Clayboy*. New York: Harmony Books.

Kay, J. J. 1984. "Self-Organization in Living Systems." PhD thesis, University of Waterloo.

Kennedy, George. 1998. *Comparative Rhetoric: An Historical and Cross-Cultural Introduction*. New York: Oxford University Press.

Kent, James. 2004. "The Case against DMT Elves." Available online at http://www.tripzine.com/listing.php?id=dmt_pickover (accessed December 14, 2009).

Khorana, H. G., et. al. 1966. "Polynucleotide Synthesis and the Genetic Code." *Cold Spring Harbor Symposia on Quantitative Biology* 31: 39.

Klüver, Heinrich. 1928. *Mescal: The "Divine" Plant and its Psychological Effects*. London: Kegan Paul, Trench, Trubner, and Company.

———. 1966. *Mescal, and Mechanisms of Hallucination*. Chicago: University of Chicago Press.

Krippner, Stanley. 1970. "The Effects of Psychedelic Experience on Language Function." In *Psychedelics: The Uses and Implications of Hallucinogenic Drugs*, ed. Bernard Aaronson and Humphry Osmond. New York: Doubleday & Co. Available online at http://www.druglibrary.org/schaffer/lsd/krippner.htm (accessed April 6, 2008).

Kuhl, Patricia. 2003. "Human Speech and Birdsong: Communication and the Social Brain." *Proceedings of the National Academy of Sciences* 100, no. 17: 9645–46. Available online at http://www.ncbi.nlm.nih.gov/pmc/articles/PMC187796/ (accessed November 8, 2009).

Lalvani, Haresh. 2005. "Genomic Architecture." *Journal of Architecture and Computation*. Available online at http://www.comparch.org/articles/genomic_architecture (accessed April 15, 2008).

Lamarck, Jean Baptiste. 1914. *The Zoological Philosophy: An Exposition with Regard to the Natural History of Animals*. New York: MacMillan.

Landauer, R. 1961. "Irreversibility and Heat Generation in the Computing Process." *IBM Journal of Research and Development* 5: 183–91.

Lash, John. 2007. "Entheogenic Revelation." Unpublished manuscript. Available online at http://www.metahistory.org/psychonautics/Eadwine/EadwinePsalter.php (accessed June 11, 2010).

Leary, Timothy. 1966a. "Programmed Communication During Experiences with DMT." *Psychedelic Review* 8: 83–95. Available online at http://www.erowid.org/chemicals/dmt/dmt_journal5.shtml (accessed March 31, 2008).

———. 1966b. *Psychedelic Prayers after the Tao Te Ching*. Kerhonksen, NY: Poets Press.

———. 1968. *High Priest*. New York: World Publishing.

———. 1983. *Flashbacks: An Autobiography*. Boston, MA: Houghton-Mifflin.

———. 1994. *Chaos and Cyberculture*. Berkeley, CA: Ronin Publishing.

———. 1995. *The Psychedelic Experience: A Manual Based on the Tibetan Book of the Dead*. New York: Citadel Press. Available online at http://www.erowid.org/library/books_online/psychedelic_experience/psychedelic_experience.shtml (accessed June 14, 2010).

Leary, Timothy, and Joanna Leary. 1973. *Neurologic*. Self-published.

Leary, Timothy, et al. 1964. *The Psychedelic Experience: A Manual Based on the Tibetan Book of the Dead*. New York: University Books. Available online at http://www.lycaeum.org/books/books/psychedelic_experience/tib1.html (accessed March 29, 2008).

Le Roy, Edouard. "The Origins of Humanity and the Evolution of Mind." In *The Biosphere and Noösphere Reader*. Ed. Paul R. Samson and David Pitt. New York: Routledge, 60–70.

Lewin, Louis. 1964. *Phantastica: Narcotic and Stimulating Drugs: Their Use and Abuse.* Trans. P. H. A. Wirth. New York: Dutton.

Lilly, John. 1972. *Programming and Metaprogramming in the Human Biocomputer.* New York: Julian.

———. 1974. "A Feeling of Weirdness." In *Mind in the Waters: A Book to Celebrate the Consciousness of Whales and Dolphins.* New York: Scribner, 71–75. Available online at http://deoxy.org/lil_weirdness.htm (accessed June 14, 2010).

———. 1975. *Simulations of God: The Science of Belief.* New York: Simon & Schuster.

Lovelock, James. 2000. *Gaia: A New Look at Life on Earth.* Oxford, UK: Oxford University Press.

Luna, Luis Eduardo, and Stephen F. White. 2000. *Ayahuasca Reader: Encounters with the Sacred Vine.* Santa Fe, NM: Synergetic Press.

Lyman, Peter, and Hal R. Varian. 2003. "How Much Information." Available online at http://www.sims.berkeley.edu/how-much-info-2003 (accessed December 6, 2009).

Maharshi, Ramana. 1988. *The Spiritual Teachings of Ramana Maharshi.* New York: Random House.

MAPS.org. n.d. "Mission Statement and Purpose." Available online at http://www.maps.org/mission.html (accessed March 29, 2008).

Margulis, Lynn, and Dorion Sagan. 1990. *The Origins of Sex: Three Billion Years of Genetic Recombination.* New Haven, CT: Yale University Press.

———. 1997. *What Is Sex?* New York: Simon & Schuster.

———. 2002. *Acquiring Genomes: The Theory of the Origins of the Species.* New York: Basic Books.

Martindale, Colin, and Roland Fischer. 1977. "The Effects of Psilocybin on Primary Process Content in Language." *Confinia Psychiatrica* 20: 195–202.

Marx, Karl, and Friedrich Engels. 1970. *The German Ideology.* New York: International Publishers. Available online at http://www.marxists.org/archive/marx/works/1845/german-ideology/cho1d.htm#d2 (accessed April 6, 2008).

———. 2004. *The Communist Manifesto.* Whitefish, MT: Kessinger.

May, Rollo. 1994. *The Courage to Create.* New York: W. W. Norton.

McCloud, Scott. 1993. *Understanding Comics: The Invisible Art.* Northampton, MA: Tundra.

McFarlane, Thomas J. n.d. "Franklin Merrell-Wolff's Aphorisms." Available online at http://www.integralscience.org/gsc/#aphorisms (accessed June 14, 2010).

McKenna, Dennis. 1998. "Ayahuasca: An Ethnopharmacologic History." Available online at http://www.neurosoup.com/entheogens/ayahuasca_history.pdf (accessed June 15, 2010).

———. 2003. *Ayahuasca Reader: Encounters with the Amazon's Sacred Vine.* Ed. Luis Eduardo Luna and Steven F. White, 154–60. Santa Fe, NM: Synergetic Press.

———. 2006. Conference presentation on Shamanism, Iquitos, Peru.

McKenna, Dennis, and Terence McKenna. 1975. *The Invisible Landscape: Mind, Hallucinogens, and the I Ching.* New York: Seabury Press.

McKenna, Terence. 1991. "I Understand Philip K. Dick." Available online at http://www.sirbacon.org/dick.htm (accessed December 2, 2009).

———. 1992. *Food of the Gods: The Search for the Original Tree of Knowledge: A Radical History of Plants, Drugs, and Human Evolution*. New York: Bantam Books.

———. 1993. *True Hallucinations: Being an Account of the Author's Extraordinary Adventures in the Devil's Paradise*. San Francisco: HarperSanFrancisco.

———. 1998. "Live at the Wetlands Preserve, NYC." Available online at http://www.abrupt.org/LOGOS/tm980728.html (accessed December 2, 2009).

Metzner, Ralph. 1999. *Ayahuasca: Human Consciousness and the Spirits of Nature*. Philadelphia: Running Press.

———. 2005. *Sacred Vine of the Spirits: Ayahuasca*. Rochester, VT: Park Street Press.

Metzner, Ralph, and Timothy Leary. 1967. "On Programming Psychedelic Experiences." *Psychedelic Review* 9: 4–19. Available online at http://www.maps.org/psychedelicreview/n09/n09005met.pdf (accessed November 9, 2009).

Michaux, Henri. 2001. *Miserable Miracle*. Trans. Louise Varèse. New York: New York Review of Books.

Microscopy Resource Center. n.d. Butterfly Wing Scale Digital Image Gallery: Tawny Owl Butterfly. Available online at http://www.olympusmicro.com/micd/galleries/butterfly/tawnyowl010.html (accessed April 16, 2008).

Miller, Geoffrey. 2000. *The Mating Mind: How Sexual Choice Shaped the Evolution of Human Nature*. New York: Doubleday.

Mills, Claudia. 1999–present. "Bioluminescence of Aequorea, a hydromedusa." Available online at http://faculty.washington.edu/cemills/Aequorea.html (accessed December 2, 2009).

Mitchell, Edgar. 2008. *The Way of the Explorer: An Apollo Astronaut's Journey through the Material and Mystical Worlds*. Franklin Lakes, NJ: New Page Books. Available online at http://www.edmitchellapollo14.com/bookexcerpt.htm (accessed March 25, 2008).

———. n.d. "Nature's Mind: The Quantum Hologram." Available online at http://www.edmitchellapollo14.com/naturearticle.htm (accessed March 25, 2008).

Mitchell, S. Weir. 1896. "Remarks on the Effects of Anhelonium Lewinii (The Mescal Button)." *British Medical Journal* 2: 1625–629.

Mullis, Kary. 1998. *Dancing Naked in the Mind Field*. New York: Pantheon.

Munn, Henry. 1973. "The Mushrooms of Language." In *Hallucinogens and Shamanism*, ed. Michael J. Harner. Oxford, UK: Oxford University Press. Available online at http://www.druglibrary.org/schaffer/LSD/munn.htm (accessed April 08, 2008).

Narby, Jeremy. 1998. *The Cosmic Serpent: DNA and the Origins of Knowledge*. New York: Jeremy P. Tarcher/Putnam.

———. 2001. *Shamans through Time: Five Hundred Years on the Path to Knowledge*. New York: Jeremy P. Tarcher/Penquin.

Nelson, Theodor. 1974. *Computer Lib/Dream Machines*. Rev. ed., 1987. Redmond, WA: Tempus Books of Microsoft Press.

Noman. 2007. "DMT for the Masses." Revised version of an article that appeared in *The Entheogen Review* 15, no. 3. Available online at http://www.erowid.org/plants/mimosa/mimosa_chemistry1.shtml (accessed November 10, 2009).

Nottebohm, F. 2005. "The Neural Basis of Birdsong." *PLoS Biol* 3, no. 5: e164.

Nwabueze, Remigius. 2007. *Biotechnology and the Challenge of Property.* Farnham, UK: Ashgate.

Ott, Jonathan. 1993. *Pharmacotheon: Entheogenic Drugs, Their Plant Sources and History.* Kennewick, WA: Natural Products Co.

Pilkington, Mark. 2002. "Animals and Psychedelics." Available online at http://www.forteantimes.com/reviews/books/422/animals_and_psychedelics.html (accessed November 22, 2009).

Pinchbeck, Daniel. *Breaking Open the Head: A Psychedelic Journey into the Heart of Contemporary Shamanism.* 2004. New York: Broadway Books.

Pollan, Michael. 2001. *The Botany of Desire: A Plant's-Eye View of the World.* New York: Random House.

Popper, Karl. 2002. *The Logic of Scientific Discovery.* New York: Routledge.

Price, Peter W. 2002. "Species interactions and the evolution of biodiversity." In *Plant-Animal Interactions: An Evolutionary Approach,* 3–25. Oxford, UK: Blackwell Science Limited.

Prusinkiewicz, Przemyslaw, and Aristid Lindenmayer. 1990. *The Algorithmic Beauty of Plants.* New York: Springer-Verlag.

Pynchon, Thomas. 1966. *The Crying of Lot 49.* Philadelphia, PA: Lippincott.

———. 1995. *Gravity's Rainbow.* New York: Penguin.

Rabinow, Paul. 1996. *Making PCR: A Story of Biotechnology.* Chicago: University of Chicago Press.

Ramachandran, V. S. n.d. "Mirror Neurons." *The Reality Club.* Available online at http://www.edge.org/3rd_culture/ramachandran/ramachandran_p1.html (accessed June 15, 2010).

Rees, Alun. 2004. "Nobel Prize Genius Crick Was High on LSD When He Discovered the Secret of Life." *The Sunday Mail,* August 8. Available online at http://www.serendipity.li/dmt/crick_lsd.htm (accessed April 12, 2008).

Reynolds, Craig W. 1994a. "An Evolved, Vision-Based Model of Obstacle Avoidance Behavior." In *Artificial Life III,* ed. C. Langton. Santa Fe Institute Studies in the Sciences of Complexity Proceedings, vol. 16, 327–46. Redwood City, CA: Addison-Wesley.

———. 1994b. "Competition, Coevolution and the Game of Tag." In *Artificial Life IV,* eds. Rodney Allen Brooks and Pattie Maes. Cambridge, MA: MIT Press.

———. 1994c. "Evolution of Corridor Following Behavior in a Noisy World." In *From Animals to Animats 3: Proceedings of the Third International Conference on Simulation of Adaptive Behavior,* eds. Dave Cliff, Philip Husbands, Jean-Arcady Meyer, and Stewart W. Wilson. Cambridge, MA: MIT Press.

———. 1994d. "Evolution of Obstacle Avoidance Behavior: Using Noise to Promote

Robust Solutions." In *Advances in Genetic Programming*, ed. K. E. Kinnear Jr. Cambridge, MA: MIT Press.

Richardson, Lewis Fry. 1922. *Weather Prediction by Numerical Process*. Cambridge, UK: Cambridge University Press.

Ridley, Matt. 2003. *The Red Queen: Sex and the Evolution of Human Nature*. New York: Perennial.

Rodriguez, E., et al. 1985. "Thiarubrine A, a Bioactive Constituent of Aspilia (Asteraceae) Consumed by Wild Chimpanzees." *Cellular and Molecular Life Sciences* 41, no. 3: 419–20.

Roseman, Bernard. 1966. *LSD: The Age of Mind*. North Hollywood, CA: Wilshire Books.

Rosen, Robert. 1999. *Essays on Life Itself*. New York: Columbia University Press.

Rosenthal, Ed. 2001. *The Big Book of Buds: Marijuana Varieties from the World's Great Seed Breeders*. Oakland, CA: Quick American Archives.

Rostand, Edmund. 2008. *Cyrano De Bergerac*. Trans. Gladys Thomas and Mary F. Guillemard. Lawrence, KS: Digireads.com Publishing. Available online at http://www.gutenberg.org/dirs/etext98/cdben1oh.htm (accessed April 12, 2008).

Roszak, Theodore, Mary E. Gomes, Allen D. Kanner, eds. 1995. *Ecopsychology: Restoring the Earth, Healing the Mind*. San Francisco, CA: Sierra Club Books.

Rothenberg, Jerome. 2003. *Maria Sabina: Selections*. Berkeley, CA: University of California Press.

Rotman, Brian. 1993. *Ad Infinitum*. Palo Alto, CA: Stanford University Press.

———. 2008. *Becoming Beside Ourselves: The Alphabet, Ghosts and Distributed Human Being*. Durham, NC: Duke University Press.

Sagan, Carl. 1980. "Who Speaks for the Earth?" In *Cosmos: A Personal Voyage*. Arlington, VA: Public Broadcasting Service. Available online at http://www.cooperativeindividualism.org/sagan_cosmos_who_speaks_for_earth.html (accessed June 15, 2010).

Sagan, Lynn. 1964. "Communications: An Open Letter to Mr. Joe K. Adams." In *Psychedelic Review* 1, no. 3: 354–56.

Salthe, Stanley, and Gary Fuhrman. 2006. "The Cosmic Bellows: The Big Bang and the Second Law." *Cosmos and History* 1, no. 2: 295–318.

Samorini, Giorgio. 2002. *Animals and Psychedelics: The Natural World and the Instinct to Alter Consciousness*. Rochester, VT: Park Street Press.

Samson, Paul R., and David Pitt. 1999. *The Biosphere and Noosphere Reader*. New York: Routledge.

Schneider, E. D., and J. J. Kay. 1994. "Life as a Manifestation of the Second Law of Thermodynamics." *Mathematical and Computer Modeling* 19, no. 6–8: 25–48. Available online at http://www.nesh.ca/jameskay/www.fes.uwaterloo.ca/u/jjkay/pubs/Life_as/text.html (accessed November 22, 2009).

———. 1995. "Order from Disorder: The Thermodynamics of Complexity in Biology." In *What Is Life: The Next Fifty Years. Reflections on the Future of Biology*, eds.

Michael P. Murphy and Luke A. J. O'Neill, 161–72. Cambridge, UK: Cambridge University Press. Available online at http://www.redfish.com/research/SchneiderKay1995_OrderFromDisorder.htm (accessed November 10, 2009).

Schrödinger, Erwin. 1980. "The Present Situation in Quantum Mechanics." Trans. John D. Trimmer. *Proceedings of the American Philosophical Society* 124: 323–38; *Quantum Theory and Measurement* 124: 323–38. Originally published as "Die gegenwartige Situation in der Quantenmechanik." *Naturwissenschaftern* 23: 807–12, 823, 844–49.

———. 1992. *What Is Life?* Cambridge, UK: Cambridge University Press.

Schroll, Mark, and Donald Rothenberg. 2009. "Psychedelics and the Deep Ecology Movement: A Conversation with Arne Naess." *MAPS Bulletin* 19, no. 1: 41–43. Available online at http://www.maps.org/news-letters/v19n1/ (accessed June 15, 2010).

Schultes, Richard Evans, Albert Hofmann, and Christian Rätsch. 1979. *Plants of the Gods.* Rochester, VT: Inner Traditions. Available online at http://www.erowid.org/library/books/plants_of_the_gods.shtml (accessed November 10, 2009).

Schultes, Richard Evans, and Siri Von Reis. 1995. *Ethnobotany: Evolution of a Discipline.* Portland, OR: Discoridies Press.

Schwartz, Peter, and Doug Randall. 2003. "An Abrupt Climate Change Scenario and Its Implications for National Security." Report for Global Business Network. Available online at http://www.gbn.com/ArticleDisplayServlet.srv?aid=26231 (accessed March 25, 2008).

Sententia, Wrye. 2004. "Neuroethical Considerations: Cognitive Liberty and Converging Technologies for Improving Human Cognition" *Annals of the New York Academy of Sciences* 1013: 221–28.

Shannon, Claude Elwood. 1949. *The Mathematical Theory of Communication.* Champaign, IL: University of Illinois Press.

Shapin, Steven, and Simon Schaffer. 1985. *Leviathan and the Air-Pump.* Princeton, NJ: Princeton University Press.

Shulgin, Alexander, and Ann Shulgin. 1991. *Pihkal: A Chemical Love Story.* Berkeley, CA: Transform Press. Part II available online at http://www.erowid.org/library/books_online/pihkal/pihkal.shtml#index (accessed June 15, 2010).

———. 1997. *Tihkal: The Continuation.* Berkeley, CA: Transform Press. Part II available online at http://www.erowid.org/library/books_online/tihkal/tihkal.shtml (accessed June 15, 2010).

Siegel, Ronald. 1989. *Intoxication: Life In Pursuit of Artificial Paradise.* New York: Dutton Adult.

———. 2005. *Intoxication: The Universal Drive for Mind-Altering Substances.* Rochester, VT: Park Street Press.

Silveira, Linda A. 2007. "Experimenting with Spirituality: Analyzing *The God Gene* in a Nonmajors Laboratory Course." *CBE Life Sci Education* 7, no. 1: 132–45. Available online at http://www.ncbi.nlm.nih.gov/pmc/issues/162045/ (accessed December 11, 2009).

Simon, H. A. 1971. "Designing Organizations for an Information-Rich World." In *Computers, Communication, and the Public Interest*, ed. Martin Greenberger, 37–73. Baltimore: Johns Hopkins University Press.

Sopa, Lhundup. 2001. *Peacock in the Poison Grove*. Somerville, MA: Wisdom Publications.

Stafford, Peter, and Bonnie Golightly. 1967. *LSD: The Problem Solving Psychedelic*. London: Award Books. Available online at http://www.erowid.org/library/books/lsd_probtem-solving.shtml (accessed June 15, 2010).

Stafford, Tom, and Matt Webb. 2004. *Mind Hacks: Tips and Tools for Using Your Brain*. Sebastopol, CA: O'Reilly Media.

Stain Blue Press. 1996–2000. "Psilocybe cubensis/Psilocybe subcubensis." Available online at http://www.stainblue.com/cubensis.html (accessed November 20, 2009).

Stamets, Paul. 1999. "Earth's Natural Internet." Available online at http://www.wholeearth.com/issue/2098/article/86/earth's.natural.internet (accessed November 20, 2009).

Stapp, Henry. 2007. *The Mindful Universe: Quantum Mechanics and the Participating Observer*. New York: Springer.

Strassman, Richard. 2001. *DMT: The Spirit Molecule: A Doctor's Revolutionary Research into the Biology of Near-Death and Mystical Experiences*. Rochester, VT: Park Street Press.

Stuart, R. 2002. "Ayahuasca Tourism: A Cautionary Tale." *MAPS Bulletin* 12, no. 2: 36–38. Available online at http://www.maps.org/news-letters/v12n2/12236stu.pdf (accessed November 22, 2009).

Suchman, Lucy. *Human-Machine Reconfigurations: Plans and Situated Actions*. Cambridge, UK: Cambridge University Press, 2007.

Sutin, Lawrence. 1991. *In Pursuit of VALIS: Selections from the Exegesis*. San Francisco, CA: Underwood-Miller.

Swenson, Rod. "The Law of Maximum Entropy Production." Available online at http://www.lawofmaximumentropyproduction.com/ (accessed June 15, 2010).

Tart, Charles T. 1972. *Altered States of Consciousness*. New York: Doubleday.

———. n.d. "On Being Stoned." The Psychedelic Library. Available online at http://www.psychedelic-library.org/tart6.htm (accessed April 16, 2008).

Teilhard de Chardin, Pierre. 1959. *The Phenomenon of Man*. New York: Harper. Available online at http://arthursbookshelf.com/other-stuff/phenom10.html (accessed June 15, 2010).

Tiefert, Marjorie. 1994. Notes to "Rime of the Ancient Mariner." Available online at http://etext.lib.virginia.edu/stc/Coleridge/poems/notes.html#Kubla3 (accessed November 22, 2009).

Tolle, Eckhart. 2005. *A New Earth: Awakening to Your Life's Purpose*. New York: Penguin.

Uhl, Christopher. 2003. *Developing Ecological Consciousness: Path to a Sustainable World*. Lanham, MD: Rowman & Littlefield.

U.S. Department of State. 2004. Declassified Intelligence Briefing: Presidential Daily Briefing, August 6, 2001. Available online at http://www.gwu.edu/~nsarchiv/NSAEBB/NSAEBB116/pdb8-6-2001.pdf (accessed June 15, 2010).

U.S. Global Change Research Information Office. 2002. *U.S. Climate Action Report 2002*. Third National Communication of the United States of America under the United Nations Framework Convention on Climate Change. Available online at http://www.gcrio.org/CAR2002/ (accessed March 25, 2008).

The Vaults of Erowid. 2007. "Oracle at Delphi May Have Been Inhaling Ethylene Gas Fumes." Available online at http://www.erowid.org/chemicals/ethylene/ethylene_history1.shtml (accessed April 5, 2007).

Vernadsky, V. I. 1998. *Biosphere*. Trans. D. B. Langmuir. New York: Copernicus.

Vernadsky, Vladimir I. 2005. "The Biosphere and the Noösphere." Executive Intelligence Review, February 18, 2005. Available online at http://www.larouchepub.com/other/2005/site_packages/vernadsky/3207bios_and_noos.html (accessed March 25, 2008).

Wade, Nicholas. 2006. "A Peek into the Remarkable Mind Behind the Genetic Code." *The New York Times*, July 11. Available online at http://www.nytimes.com/2006/07/11/science/11book.html?_r=1 (accessed November 11, 2009).

Walker, Jeffrey. 2000. *Rhetoric and Poetics in Antiquity*. Oxford, UK: Oxford University Press.

Wallace, Alfred Russel. 1885. "Are the Phenomena of Spiritualism in Harmony with Science?" *The Sunday Herald* (Boston), April 26. Available online at http://people.wku.edu/charles.smith/wallace/S379.htm (accessed June 15, 2010).

Wallace, B. Alan. 2006. *Contemplative Science: Where Buddhism and Neuroscience Converge*. New York: Columbia University Press.

Wasson, Gordon. 1957. "Seeking the Magic Mushroom." *Life*, June 10. Available online at http://www.imaginaria.org/wasson/life.htm (accessed November 21, 2009).

———. 1968. *Soma: Divine Mushroom of Immortality*. New York: Harcourt, Brace and World.

Watts, Alan W. 1962. *The Joyous Cosmology: Adventures in the Chemistry of Consciousness*. New York: Pantheon. Available online at http://www.druglibrary.org/schaffer/lsd/jccontnt.htm (accessed February 11, 2010).

Wenzel B., N. Elsner, and R. Heinrich. 2002. "mAChRs in the Grasshopper Brain Mediate Excitation by Activation of the AC/PKA and the PLC Second-Messenger Pathways." *Journal of Neurophysiology* 87. Available online at http://jn.physiology.org/cgi/content/abstract/87/2/876 (accessed April 12, 2008).

Wertheim, Margaret. 2004. "Crocheting the Hyperbolic Plane: An Interview with David Henderson and Daina Taimina." The Institute for Figuring. Available online at http://www.theiff.org/lectures/05a.html (accessed April 15, 2008).

Wigner, Eugene. 1967. *Symmetries and Reflections: Scientific Essays of Eugene P. Wigner*. Bloomington: Indiana University Press.

———. 1970. "Physics and the Explanation of Life." *Foundations of Physics* 1, no. 1: 35–45.

Wilson, Elizabeth. 2004. *Psychosomatic: Feminism and the Neurological Body*. Durham, NC: Duke University Press.

Wilson, Frank R. 1998. *The Hand: How Its Use Shapes the Brain, Language, and Human Culture*. New York: Pantheon.

Wilson, Robert Anton. 1977. *Cosmic Trigger: Final Secret of the Illuminati*. Berkeley, CA: And/Or Press.

Witkin, H. A. 1950. "Individual Differences in Ease of Perception of Embedded Figures." *The Journal of Personality* 19, no. 1: 1–15.

Witkin, H. A., et al. 1962. *Psychological Differentiation*. New York: Wiley.

Wolfram, Stephen. 2002. *A New Kind of Science*. Champaign, IL: Wolfram Media.

Youngblood, Gene. 1970. *Expanded Cinema*. New York: Dutton.

INDEX

and Philip K. Dick, 49, 298; and trip reports, 50

consciousness, 10, 30, 105, 122, 124, 146–47, 186, 214, 231, 250, 284, 289, 294–96; and Aldous Huxley, 69–71, 77, 81–84, 95–96; altered, 31, 24, 159, 165, 217, 223, 259, 269, 192, 320n5; and Amit Goswami, 76, 79–80, 82, 319n25; and the author, 271, 275, 277, 283, 286, 290–91, 300, 302, 305; and ayahuasca, 73, 134, 199, 208, 277; and Buddhism, 89, 217–18; cellular, 190; and Christopher Uhl, 8–9; cosmic, 9, 269; definition of, 276; and effect on itself, 191, 273, 275, 300, 318n10; egoic, 12, 72, 152, 288, 291, 298, 301; global, 136, 157–59; human, 12, 30, 35, 44–45, 78, 148, 157, 160, 175, 177, 189, 208, 213, 222–23, 232, 243, 252, 254, 272, 281, 299, 323–24n12, 327n1, 329n13; and Kary Mullis, 192, 194–96, 269; and John C. Lilly, 40, 263, 266, 270, 291, 293; and the McKennas, 12, 97, 110–11, 213, 255, 286; ordinary, 52, 63, 88, 132, 160, 231, 266, 268, 277; and psychedelics, 22, 25, 37, 43, 50, 52, 113, 133, 167, 169, 176–77, 189, 216, 244, 246, 268, 271, 318n10; self-, 184, 292; and sexual selection, 145, 148, 170, 235, 252, 276; states of, 14, 21, 50, 52, 133, 292; and Timothy Leary, 113, 129, 177, 184, 190–91, 260, 294, 319n23, 330n12; and the War on Drugs, 14, 19, 21, 281

contact high, 43, 80, 230, 231, 239, 240, 261

contagion, 44. 53–54, 299

cosmic realism, 83, 280

cosmos, 7, 21, 76, 192, 220, 269, 316; and Aldous Huxley, 71, 80, 92, 123, 161; and the author, 41, 125, 257, 277, 283, 286, 290, 300; and Carl Sagan, 194; and consciousness, 145, 165, 266; distinction between self and,

12, 222, 273; and Galileo, 74–75, 113; Hindu concept of, 79, 222, 283, 286; and information, 80, 252, 304; and interconnection, 9, 10, 41, 63, 77, 80, 125; and psychedelics, 40, 72, 114, 165, 222–23, 257

Crick, Francis, 38, 174, 177, 192, 195, 325, 326; LSD and, 194

crickets, 136, 274, 303, 305

cryptographic paradigm, 194, 325n3

The Culture of Flowers (Goody), 59

Cyrano de Bergerac (Rostand), 150–57, 323n8, 323n10

D

Darwin, Charles, 40, 68–70, 94, 127, 152, 158, 164, 208–9, 211, 244, 252, 257, 299, 305, 306–8, 312, 314; and *The Descent of Man*, 37, 136, 137, 141–42, 155, 327n2, 328n4, 328n9; and Neo-Darwinism, 20, 76; and *On the Origin of Species*, 83, 130–31, 137, 207, 245, 328n8; and sexual selection, 31, 32, 37, 75, 103, 136–48, 162, 247–49

Darwin, Erasmus, 40, 306–9, 312, 314

Dawkins, Richard, 20, 27, 319n22

Dekorne, Jim, 49, 327n17

Deleuze, Gilles, 14, 300

Delphi, 87–88, 107

De Rios, Marlene Dobkin, 25, 33

Derrida, Jacques, 106

The Descent of Man (Darwin), 37, 137, 141, 142, 155, 328n4

"Diamond Mind," 41, 51, 301

Diamond v. Chakrabarty, 176, 325n1

Dick, Philip K., 81, 214, 256, 259, 271, 277, 303–4, 327n22; and psychedelics, 49, 50, 230–31; and *VALIS*, 39, 230–31, 274, 279–80, 288–89, 292, 294–99

Dionysus, 88

divine, 13. 136, 150, 167, 196, 211, 267, 269, 273, 296; and Aldous Huxley, 62; and

Slightly Less Insane Psychedelics Policy
(SLIPP), 280
Smokeloc, 12, 20, 62
Socrates, 27, 107, 154, 277
Sonea, Sorin, 253
Spruce, Richard, 201–8, 313
Stamets, Paul, 122–23
Stapp, Henry, 76, 79, 80, 82
stigmergy, vii, 43, 85, 95, 205, 256
Stolaroff, Myron, 35
Strassman, Richard, 227–28, 230, 327n18
St. Theresa, 178
subjectivity, 211, 212, 216; and Edgar
Mitchell, 10–11, 52; human experi-
ence of, 162, 283, 298, 317n6; and real-
ity, 21, 62, 71, 75, 110
Sunyata, 268, 294
Swimme, Brian, 9
synecdoche, 18, 41
synesthesia, 45, 61, 63, 209, 270

T
Tao Te Ching, 188
Tart, Charles, 243
Tat Tvam Asi, 7, 24, 52, 66, 71, 79, 209, 286,
289, 292
Taussig, Michael, 133–35
telepathy, 85, 98, 206
telos, 17, 40, 163, 216, 222, 230, 236, 252,
281
THC, 237–38, 246, 328n3. *See also* cannabis; hashish
thermodynamics, 82, 164, 210, 212, 271,
277
Tihkal (Shulgin and Shulgin), 45
The Transcendental Other, 12, 110
transhuman, 18, 212; and Julian Huxley,
35, 320n9; and psychdelics/ecodelics,
13, 19–22, 24, 43, 56, 83, 104, 111, 113–15,
123–25, 128, 132, 169, 230, 232, 263, 304
Travelling Stoner Problem, 240–42
trip reports, 17–18, 50, 97, 227, 322n15;

and Aldous Huxley, 53, 54, 91, 95; as
data, 40, 45, 47, 51, 91, 169; and DMT,
16, 327n18; and *the* mistake, 47, 320n3;
as rhetorical software, 36–37, 52. *See
also* trope
trope, 3, 21, 27, 30, 40, 46, 48, 67, 77, 97,
169, 238, 273, 281, 318n10; and Aldous
Huxley, 58, 60, 63, 77, 85; and DMT,
20; and rhetoric, 40, 77, 150; and synecdoche, 41. *See also* trip reports
True Hallucinations (T. McKenna), 213,
219–20, 222, 225–26, 232, 298
Twain, Mark, 3

U
UDV, 257–58
Uhl, Christopher, 8, 9
universe, as way of tricking itself, 223
Upanishads, v, 7, 26, 66, 260, 285

V
Varela, Francisco, 293
The Vaults of Erowid, ix, 11, 88, 103; and
trip reports, 46–47, 51, 122, 169, 282
Vedanta, 285
vegetalista, 3
Vermeer, 48
Vernadsky, V. I., 7, 82–84, 160, 280; and
The Biosphere, 82; and the noösphere,
11, 14, 83–85. 96, 124, 131, 141, 143, 175
vibration, 54, 108, 137, 145, 217, 282
Villavincencio, Manuel, 205
VMAT2, 223
vomit, 42, 167, 199, 202–4
Von Frisch, Karl, 67, 70
Von Neumann, John, 76

W
Wallace, Alfred, 40, 138, 196, 207, 305,
312–14
War on Drugs, 13, 18, 72, 227, 280–81